Eisenhower Public Library
4613 N. Oketo Avenue
Harwood Heights, Il. 60706
708-867-7828

DEMCO

# Laboratory Disease

# Laboratory Disease

*Robert Koch's Medical Bacteriology*

Christoph Gradmann
*Translated by* Elborg Forster

The Johns Hopkins University Press
*Baltimore*

Originally published as *Krankheit im Labor: Robert Koch und die medizinische Bakteriologie* © Wallstein Verlag, Göttingen

© 2009 The Johns Hopkins University Press
All rights reserved. Published 2009
Printed in the United States of America on acid-free paper
9 8 7 6 5 4 3 2 1

The Johns Hopkins University Press
2715 North Charles Street
Baltimore, Maryland 21218-4363
www.press.jhu.edu

Library of Congress Cataloging-in-Publication Data
Gradmann, Christoph, 1960–
  [Krankheit im Labor. English]
  Laboratory disease : Robert Koch's medical bacteriology / Christoph Gradmann ; translated by Elborg Forster.
    p. ; cm.
  Translation of: Krankheit im Labor : Robert Koch und die medizinische Bakteriologie / Christoph Gradmann. c2005.
  Includes bibliographical references and index.
  ISBN-13: 978-0-8018-9313-1 (hardcover : alk. paper)
  ISBN-10: 0-8018-9313-5 (hardcover : alk. paper)
  1. Koch, Robert, 1843–1910. 2. Medical bacteriology—History I. Forster, Elborg, 1931– II. Title.
  [DNLM: 1. Koch, Robert, 1843–1910. 2. Bacteriology—history—Germany. 3. History, 19th Century—Germany. 4. History, 20th Century—Germany. QW 11 GG4 G732k 2009a]
  QR46.G6713 2009
  616.9'2010092—dc22      2008050488

A catalog record for this book is available from the British Library.

*Special discounts are available for bulk purchases of this book. For more information, please contact Special Sales at 410-516-6936 or specialsales@press.jhu.edu.*

The Johns Hopkins University Press uses environmentally friendly book materials, including recycled text paper that is composed of at least 30 percent post-consumer waste, whenever possible. All of our book papers are acid-free, and our jackets and covers are printed on paper with recycled content.

*For Friederike*

# Contents

CHAPTER I Introduction   1

CHAPTER II Lower Fungi and Diseases: Infectious Diseases between Botany and Pathological Anatomy, 1840–1878   19

CHAPTER III Tuberculosis and Tuberculin: History of a Research Program   69

CHAPTER IV Of Men and Mice: Medical Bacteriology and Experimental Therapy, 1890–1908   115

CHAPTER V Traveling: Robert Koch's Research Expeditions as Private and Scientific Undertakings   171

A Perspective   230

*Acknowledgments*   235
*Notes*   237
*Bibliography*   289
*Index*   313

# Laboratory Disease

CHAPTER I

# Introduction

## I.1. Some Questions about the History of Robert Koch's Medical Bacteriology

"'Ah! now I see,' said the visitor. 'Not so very much to see after all. Little streaks and shreds of pink. And yet those little particles, those mere atomies, might multiply and devastate a city! Wonderful!'" In H. G. Wells's short story "The Stolen Bacillus," readers find themselves in a bacteriological laboratory.[1] The view through the microscope reveals a reality with which humankind has lived since the late nineteenth century: a world filled with minuscule, ubiquitous living organisms that cause diseases but are invisible to the naked human eye. Technical aids are needed to see these bacteria, but this does not make them any less real. On the contrary, the protagonist of Wells's story comes to feel that the microscope allows him to see an otherwise hidden but elementary force. He addresses an observation to the scientist in the laboratory: "'Scarcely visible,' he said, scrutinizing the preparation. He hesitated. 'Are these—alive? Are they dangerous now?'"

The answer to both questions is "yes," and this certainty, which is shared by the visitor and the scientist, illustrates one of the most important cultural facts of modern times: that there are minuscule organisms such as bacteria and viruses, and that these organisms are dangerous to humans because they cause diseases.

Both scientists and lay people accept this idea, which takes on different connotations in different spheres of life. To most physicians the idea of a contagious disease without a biological pathogen simply does not make sense.[2] To lay people the existence of bacteria is a reason to use household cleansers that promise a germ-free environment. With the help of these cleansers we try to protect ourselves from any contact with bacteria—even though there is something absurd in the attempt to avoid contact with organisms that are both invisible and ubiquitous.[3]

The history of the science dealing with pathogenic bacteria is the subject of this book. This study is essentially that of a discipline, medical bacteriology. One of the new, experimental fields of medicine in the second half of the nineteenth century, it produced fundamental research concerning the causation of disease by microorganisms.[4] This history is inextricably bound up with the biography of the German physician Robert Koch (1843–1910). Starting after 1875, Koch himself or members of his circle of associates and students introduced important new methods of studying bacteria, such as staining, solid culture media, and microphotography. He also formulated criteria for proving a bacterial etiology—which came to be known as Koch's postulates—and developed animal models for the study of such diseases. The importance of these studies for the history of medicine can hardly be overstated: though modified over time, bacteria staining, solid culture media, and the Petrie dish remain fundamental techniques in the microbiology lab to this day. Likewise, recent debates about the criteria for proving the involvement of pathogens still refer to Koch's postulates, and the bacteria identified by Koch and his associates as causing anthrax, tuberculosis, diphtheria, and so forth are still considered the pathogens of these diseases. The dominance of the Koch school in medical science in the years around 1900 is well known; it is depicted in Ludwik Fleck's classic study of the subject, for instance, as the perfect example of a particularly successful and rigid thought style.[5] The dominant influence of bacteriological hygiene in the public health of that era is also undisputed.[6] But we know much less about the origin and the development of medical bacteriology as an experimental field within medicine. And this is precisely what this book sets out to investigate. On the basis of an examination of Robert Koch's scientific work, I intend to elucidate the process by which the methods and subjects of the medical-bacteriological laboratory were formulated and developed.

There is no dearth of literature about the history of bacteriology and about the life and work of its founder. Some twenty biographies are available, and the secondary literature is well-nigh immeasurable.[7] However, most of the biographical studies are somewhat dated. Methodological innovations in the more recent history of science, in connection with studies of specific laboratories or "practical turn" studies based on such sources as laboratory notes are rarely found in the historiography about Koch.[8] Thus we have remarkable studies concerning the history of theories and ideas about the bacteriological concept of disease, which address the concept of Koch's postulates and the importance of bacterial specificity;[9] but the experimental practices of Koch and his associates that gave rise to these concepts have until now hardly been examined in their own right.[10] This book therefore attempts to understand Robert Koch's bacteriology, its genesis, its content, and its development from the perspective of his experimental practice. Its source materials are Koch's and other physicians' published works, plus such materials as diaries, letters, and (laboratory) notes, which permit a detailed reconstruction of scientific work.[11]

The approach I have chosen attempts to establish a relationship between the details of scientific work in the bacteriological laboratory and Koch's biography. This perspective is associated with the hypothesis that looking at the experimental practice brings to light different foundations of the scientific field and its development than a history-of-ideas approach that starts with the production of theories and postulates a unified cognitive structure of the scientific field. This insight was inspired by the thesis formulated some time ago by Georges Canguilhem to the effect that medical bacteriology should be understood more as a successful practice than as a theoretical innovation, and that this is what distinguishes it from older nineteenth-century theories of disease rooted in the tenets of natural history.[12] For Koch's work this reasoning is particularly enlightening. A comparison of the German bacteriologist with the French microbiologist Louis Pasteur, or with his own students Emil von Behring and Paul Ehrlich, makes Koch look positively adverse to theory. Theoretical statements of major importance are rare in his writings, and when they appear, they vary considerably depending on the circumstances, as in the case of the criteria for determining the involvement of a pathogen that have become known as Koch's postulates.[13] Rather, his innovations were the development of investigative methods, such as microphotography and the pure culture, as well as the formulation of models showing the bacterial etiology for several infectious diseases. Taken as a whole, these concepts, techniques, and methods added up to a suggestive and far-reaching solution to the problem of infectious disease itself, a bacteriological style of thinking that molded the perceptions of a whole generation of laboratory physicians, created certain realities, and blocked the view onto others.[14]

In approaching the history of medical bacteriology from the perspective of Koch's laboratory, one also encounters biographical questions. The personal element—the experience of his own work and that of others, dealing with the working conditions existing in the laboratory—and the scientist's underlying personal motivation are essential to the choice of topics investigated and the strategies used.[15] Yet this account is less intended as a biography than as a study in the history of science concerning Koch's medical bacteriology, its presuppositions, its content, and its effects. No claim to completeness, as is usually associated with the story of a life, is made here. Instead, the framework of this account is provided by the history of medical bacteriology. A way to combine issues in the history of science with biographical questions is suggested by Gerald Geison's *The Private Science of Louis Pasteur* (1995). That book and the present one share the premise that diaries, laboratory notes, and similar materials are particularly apt for shedding light on a scientist's creativity and personal style. Unlike published studies, which tend to present a finished and idealized picture of scientific work, such materials openly show the process of the research. They are valuable even without the implied assumption that they will provide unmediated access to the scientist's work. The fact is—as Geison emphasizes, following Frederick L. Holmes—that this is simply a genre of texts in which a researcher's individual style is most clearly revealed.[16] Occasionally such notes starkly contradict the published findings, indicating that the research had taken a different path than was claimed in the publication. This is the case, for instance, with Koch's studies of tuberculin. But just as Geison was not interested in unmasking or even condemning Pasteur,[17] I only want to illuminate the biographical dynamic between laboratory work and published science in Koch's case. There is a difference between Geison's approach and my own, however, and it is essentially of a heuristic nature. The present book always refers to laboratory notes, rather than laboratory logs. Whereas Geison could consult a voluminous collection of laboratory logs, Koch documented his work on innumerable pieces of paper that, moreover, have not been preserved in their entirety.[18] These provide interesting glimpses into his work in specific instances but rarely permit a precise reconstruction of how the laboratory work was conducted.

In addition to studying Koch's works in the narrow sense, I will pay special attention to their place in the scientific and social contexts of their time. Two aspects are so important that it is necessary to broach them right here in the introduction. The first is that the character of the rapidly developing field of medical bacteriology was decisively molded by its relations with other fields and by the areas of its practical use. Second, the public staging for a field that in the course of Koch's career had been transformed from an exotic body of knowledge to one of the leading disciplines in experimental medicine was a matter of major importance. Both aspects call for a brief elucidation.

Unquestionably one of the characteristics of developing medical bacteriology was that it borrowed from other scientific disciplines such as botany and experimental pathology and in this way built up a new area of medical knowledge. Moreover, it had a wide spectrum of applications in such fields as clinical medicine, public health, preventive medicine, and therapy. The first question to be asked, then, is in what ways botany or experimental pathology were involved in the founding of this discipline. The next question is how Koch acquired these different bodies of knowledge and transformed or expanded them, and what kinds of problems arose from combining, for instance, biological and medical issues. For all the advantages that this constellation implied for the field's initial stage in the late 1870s, a certain tension between medicine and biology, the clinic and the laboratory, was also part of the new discipline's baggage.[19] Moreover, the dangers of so prominent a position in the new field are obvious: the very factors that facilitated this transfer of knowledge also entailed the danger of exaggerating the validity of stated claims and thereby enlarging the circle of potential critics. In this sense we shall have to ask what use Koch's medical bacteriology made of these heterogeneous foundations and how other disciplines, such as the more biologically oriented protozoology, or internal medicine, reacted to the claims of validity put forward by competing medical bacteriologists.

Remarkably, toward the end of his career, Koch repeatedly indicated that he felt his position within bacteriology to lie in an uncertain and dangerous terrain. What he had liked to idealize as virgin research territory in the 1870s had changed into a fiercely contested battlefield a generation later. In 1904, the year of his retirement, he summarized his experiences by saying that "in my work my lot was particularly hard and I encountered more criticism, unjustified criticism, than anyone else."[20] This autobiographical experience gives rise to an important question: what accounts for this failure to make his scientific insights totally convincing, a failure that Koch repeatedly lamented, and that starkly contradicts the image of a successful new discipline? To what extent can it be explained by the development of medical bacteriology itself—by the particular scientific problems with which it dealt—and to what extent can it be attributed to new (and old) competition from other persons or disciplines? And finally one must ask whether Koch's complaints about this failure did not reflect the way he personally experienced the development of the discipline rather than the facts themselves. In that case, his self-perception would be an important indication that over time certain tensions arose between the development of the discipline and Koch's biography.

A hypothetical answer to this question might be formulated in accordance with Georges Canguilhem's reflections on the special, more practical than theoretical, objectives of bacteriology. This means that Koch's complaints about the lack of acceptance for his science can be seen as inherent in the pragmatic objectives of his discipline and its heterogeneous foundations. Following this thesis, I shall look at the history of bacteriology as one of internal contradictions and external competition, a stance that was caused precisely by its technical rather than conceptual objectives and by its very success.

This brings us to the second aspect, the public staging of research. Particularly in the 1880s, the decade of the microbe hunters, the public image of bacteriology as a science that yielded exciting discoveries came into being. Whereas the epoch-making importance of other innovations in the history of medicine is frequently revealed only through the efforts of later historians, the innovative, even spectacular character of medical bacteriology was apparent from the very beginning.[21] The discovery of hitherto unknown and terribly dangerous creatures such as pathogenic bacteria and the promise of finding ways to combat them successfully conferred a special character on the science of bacteriology and influenced contemporaries' imaginations much beyond the science involved.[22] The excitement caused by Lister's treatment of wounds, Koch's search for pathogens, or Pasteur's serum therapies created an image of medical bacteriology as a science filled with discoveries and sensations, which Paul de Kruif has celebrated in his highly popular book *Microbe Hunters*. To his contemporaries this discipline was, for good or for evil, the very essence of laboratory medicine. When Bernhard Naunyn stated in 1905 "medicine will be a science, or it will not be,"[23] he was referring to medical bacteriology, as did Ottomar Rosenbach, who in his 1893 book *Arzt contra Bakteriologe* [Physician vs. bacteriologist] had drawn the picture of a soulless laboratory medicine that had lost all touch with the physician's art.

In the final analysis, the reputation of medical bacteriology was founded on the expectation of its practical uses. Above everything else, contemporaries expected that the incomparable tangibility of disease in the bacterium would make it possible to combat simply the pathogen rather than the disease, thus liberating humankind from the great epidemics by means of medical bacteriology. In the popular consciousness of the time, this conviction is expressed in an identification of the pathogen with the disease. In cartoons, poems, and works of popular science, for instance, "the tubercle bacillus" often stands in for the disease.[24] For the scientific concept of the subject, the definition of the pathogen as the necessary cause of the disease had to be complemented by the control of the pathogen as the logical point of intervention in the ensuing disease.[25] This promise of combating diseases through their pathogens must be seen as the context for the development of bacteriological diagnostics, chemotherapy for infectious diseases, disinfection, purification of drinking water, personal hygiene, and so forth. Such ideas indicate that one of the reasons for the popularity of medical-bacteriological knowledge was the fact that it seemed to be applicable to the everyday practices of public health institutions, the clinic, and even one's bathroom.

The prestige of the bacteriologists was thus closely related to the subject matter of their research: the more dangerous the microorganism studied, the more prestigious the researcher. It was not the existence of bacteria, but the dangerous qualities ascribed to them and the promise that these could be brought under control that brought fame to Koch and his school. It stands to reason that the character of bacteriology as a science filled with discoveries and sensations had an impact on its development in general and on the biography of its founding father in particular.[26] From the middle of the 1880s on, Robert Koch's choice of research projects and the strategies needed to carry them out were to some extent motivated by his desire to live up to the expectations aroused by his earlier successes. We will see that after 1885 his research on tuberculin and his work in tropical medicine were decisively shaped by these considerations.

To understand medical bacteriology historically, it is important to see that to a considerable extent, the bacteriologists' effectiveness in the fight against infectious diseases was not a reality but an aspiration, and that it remained so for decades. New research has shown that the definitions of diseases put forth by bacteriologists initially brought little change to day-to-day clinical practice, and that it could take a long time before newly available bacteriological diagnoses were acted on.[27] Hardly anyone, of course, doubts that the advent of medical bacteriology wrought a fundamental change, but historians have now replaced the year 1880 as the turning point with a more extensive transitional period that includes the subsequent decades. Throughout this period, medical bacteriology was caught up in complicated and tense relations with other medical disciplines and with the nascent microbiology. Very gradual as well was the diffusion of bacteriological knowledge in everyday clinical practice in such areas as clinical diagnostics and serum therapy.[28] From the perspective of the history of science, the rise of a bacteriological explanation for infectious diseases thus presents itself as a phenomenon that, as Nancy Tomes and John Harley Warner put it, is easy to see as a whole but difficult to grasp in detail.[29] This finding will be taken into account in the present study, which extends over the long period between about 1840 and 1910. The task at hand is to use the example of Koch's laboratory to explore and elaborate this view of a process characterized by internal contradictions, external resistances, and time lags.

The issue outlined so far will be treated in four independent studies. The first of these deals with the research on the genesis of pathogenic bacteria. The time frame envisaged lies between Jacob Henle's work *Von den Miasmen und Contagien* (1840) and Koch's studies on traumatic infections (1878). Here the aim is to investigate the relation between botany and pathological anatomy that surfaced in the discussions about the issue of living pathogens. The question is whether some rudiments of a biology of infectious diseases came to be formulated in analogy to one or the other of these fields. The highly complex discussion that filled these four decades will further our understanding of the prehistory of Koch's work on traumatic infections. By this I do not mean to imply that there was a direct or even inevitable development from Henle to Koch. What I do intend to show is in what way Koch's work on traumatic infections represented a turning point in the contemporary research on the pathology of infectious diseases. What was the importance of this subject in relation to the experimental pathology of the 1870s, how original were Koch's early studies, from what premises did they start, and what were the competing approaches of that time?

The second study deals with the dynamic of a research program about tuberculosis in the years after 1881. Koch's work on this subject is particularly well suited for an investigation of the impact and the internal development of medical bacteriology. To begin with, and from the epidemiological and clinical point of view, tuberculosis was the most important infectious disease of its time. More important for the present book, however, is the fact that Koch owed much of his fame as a scientist to the identification of its bacterial pathogen, the tubercle bacillus, which he presented to the public in March 1882. His studies on the etiology of tuberculosis used a plethora of methods that would mold the traditional perceptions of a generation of physicians in fundamental ways. Yet the focus of the chapter is not so much Koch's effect on others as his own continued work on the subject, which led to the development of tuberculin between 1888 and 1890, and his effort to present a kind of bacteriological therapy. Here it will be important to find out what effects can be ascribed to the transfer of an etiological understanding of disease to the therapeutic arena. In this case the source material allows for a precise reconstruction of many years of experimentation that eventually led to tuberculin as a medication. It also elucidates the special dynamic of a mature research program. Koch's problem was that he wanted to combine a large amount of established knowledge about the disease with new questions about its therapy while attempting to live up to his own expectations as the leading medical bacteriologist of his era.

The framework of the third study is the transfer of bacteriological knowledge to clinical medicine beginning in the 1890s. In its initial stages medical bacteriology had succeeded in shifting research on infectious diseases from the patient's bedside to the laboratory and in replacing the human patient with an animal model. This reductionist strategy was brought up short when scientists attempted to bring bacteriological therapy to the patients themselves. What was needed now was a kind of bedside experimental medicine. Examining the clinical trials for tuberculin in 1890–92 and a therapeutic experiment connected with sleeping sickness in 1906–8 will illuminate whether it was possible to make clinico-therapeutic use of knowledge gained in the laboratory. Two questions above all will be raised in this chapter: one is about the relationship between medical bacteriology and the clinical medicine of the time, with which the new field, with its definitions and diagnoses of disease, began competing as soon as medical bacteriologists attempted to develop therapies for sick patients. The second question concerns changes in the doctor-patient relationship, which resulted from the application of—presumably—highly effective and specific medications. In what way were the first experiences with this bacteriological medicine particularly suited to raise questions about the paternalism that typically informed the physician's self-understanding at that time? What did the experience with "bacteriological" medicine have to do with the formation of a medical ethics discourse, calling for patient consent and patients' rights, including the right to information, that characterized the years around 1900?[30]

The theme of the fourth and final study concerns the significance of travel in the history of bacteriology and in Koch's biography. Bacteriological hygiene was an experimental, but not necessarily a laboratory, science in the narrow sense. The results it produced in the laboratory could be reproduced outside that sphere, and bacteriologists made considerable efforts in that direction; one has to think only of disinfection, purification of drinking water, or bacteriological diagnostics. The objective was to turn the world into a laboratory, and bacteriological hygiene was a science of traveling experts. In this sense, Robert Koch's passion for travel was a central aspect of his research activity. Choosing the examples of the cholera expedition of 1883–84 and the sleeping sickness expedition of 1906–7 makes it possible, given the great distance in time between them, to assess the intellectual and scientific importance of this theme over the entire span of Koch's career. Here one will have to ask whether traveling was a form of scientific work specific to bacteriological hygiene and what the nature was of Koch's private passion for this way of working and living. The comparison between two expeditions that took place twenty years apart shows how the attitude of the founding father Koch to "his" discipline changed over time. Not least—this being the century of Alexander von Humboldt and Henry Morton Stanley—travel to remote places created special public attention to science at home. The analysis of Koch's two expeditions therefore also illustrates the changing relations of bacteriology to politics and public opinion. This chapter focuses on a dynamic mutual relation between the development of a discipline and the individual conduct of a scientist's life over this entire period. The example of travel is germane to the central issue of this book, which is the relationship between the development of a science and the biography of an individual.

To facilitate the reader's orientation in the text, I will supply a brief biographical sketch as part of this introduction. It is not meant to compete with the existing biographies of Koch, and certainly I do not intend to preempt potential new work.[31] It simply reports the present state of our knowledge, broaches the issues in the field, and supplies a framework for the four case studies.

## I.2. Robert Koch—A Brief Biography

Heinrich Hermann Robert Koch was born December 11, 1843, in the mining town of Clausthal in the Harz Mountains. His parents, Hermann and Mathilde Koch, might be said to belong in the upper strata of the petty bourgeoisie, given the father's rise from mining foreman to leading official in the Upper Harz mining industry.[32] Aside from an interest in collecting and cataloging plants, which an uncle living in Hamburg fostered as much as he could, the schoolboy did not show any inclinations or talents that might have pointed to his future career.[33]

More suggestive is the fact that Koch as a young man harbored plans and dreams of emigration (usually to America). Considering that no fewer than six of his eight brothers had taken this step, such longings were perfectly understandable.[34] For Koch's parents the plans of this son, the only one to prepare for university studies, was a constant source of worry. These plans seem to have ended only with Koch's 1867 marriage to Emmy Fraatz, who was also opposed to emigration.

After graduating from the Clausthal gymnasium in the spring of 1862, Koch enrolled at nearby Göttingen University, where he initially studied botany, physics, and mathematics. He soon realized, however, that without the requisite financial background, his professional goal of becoming a world-traveling naturalist was unrealistic, and so he decided in 1863 to study medicine as a way to make a living. In the nineteenth century this was not an unusual consideration, one that had, for example, also caused Hermann von Helmholtz to study medicine despite his love of physics.[35] Koch soon turned out to be an outstanding student; for instance, he won a prize offered by the Faculty of Medicine for the best paper on a topic in anatomic microscopy. His doctoral thesis treated a theme in physiology: the origin of the so-called butric acid in the human organism. The study involved a heroic self-experiment: the ingestion of several kilograms of butter. Its findings were published in Henle's *Zeitschrift für rationelle Medizin*.[36] Along with the physiologist Georg Meissner and the pathologist Karl Hasse, the anatomist Henle was one of Koch's teachers in Göttingen. Thanks to a paper Henle had published in 1840 on living agents of contagion and the criteria needed to prove their involvement, he is often assumed to have taught bacteriology to the prospective researcher. However, Koch himself expressly denied this.[37] And indeed, when he left Göttingen as a newly minted doctor in 1866, essentially all he took away was a thorough training in anatomy and physiology.

For the time being the young physician decided against an academic career and instead sought to set up a reasonably lucrative practice after his marriage in 1867. But the different positions he held over the following years were always so financially unsatisfactory that his youthful plans for emigration revived. Meanwhile, two of these temporary posts, one in 1866 during a cholera epidemic in Hamburg, and the other in 1870–71 as a field hospital physician in the Franco-Prussian War, at least gave him the opportunity to learn in depth about such infectious diseases as cholera and sepsis.

It was not until 1872 that Koch succeeded in obtaining a materially satisfactory position as county public health officer (Kreisphysicus) in Wollstein, Prussia (today Wolstyn, Poland).[38] Shortly after his arrival there, he began his bacteriological studies, choosing as his subject matter anthrax, an epizootic that was frequent in his district and counted among the most thoroughly studied infectious diseases at the time. Its transmission by rod-shaped structures that could be identified in the blood of affected animals was considered likely at least since Casimir Joseph Davaine had published his studies in 1863.[39] Koch was able to describe the entire life cycle of the *Bacillus anthracis*, basing his description on the bacteriology of Ferdinand Julius Cohn.[40] By showing that the bacterium had a stage where it took the form of spores, Koch not only explained such phenomena as the temporary "disappearance" of the bacteria, but also proposed the robustness of the spores as an explanation for certain epidemiological peculiarities, among them the continuing contagiousness of certain places—pastures, for example—that seemed to contradict the extraordinary fragility of the anthrax bacillus.[41] Koch thus had not discovered the bacillus but rather completed the description of its life cycle, conducted infection experiments to show its etiological role, and suggested appropriate control measures, such as improved methods of disposing of animal cadavers.

In the anthrax studies, Koch made use of Cohn's (micro)biology to treat a medical problem. It was only logical that in order to publish his study in 1876, he should contact Cohn, whom he had only known through his publications. Cohn received Koch's work enthusiastically, helped him publish it in the *Beiträge zur Biologie der Pflanze*, which he edited, and also liberated him intellectually from the isolated existence of a country doctor. He introduced him to a number of his Breslau colleagues. The pathologist Julius Cohnheim helped Koch expand his knowledge of experimental pathology, and the latter's prosector, Carl Weigert, taught him histology and staining techniques.[42]

Still working in Wollstein during the years 1877 to 1880, Koch used the specific case of anthrax etiology to produce general studies on the methodology and technique of examining pathogenic bacteria. In 1877 he published his "Verfahren zur Untersuchung, zum Konservieren und Photographieren der Bakterien," which, most importantly, introduced microphotography. This method, which he had developed thanks to improved microscopes and special staining techniques, enabled him to publish his findings through the medium of photography, which was considered by his contemporaries to reflect reality.[43] Closely related to this area were his studies of traumatic infections, which are particularly important in two respects. It was here that Koch developed his method for animal experimentation, and this was the first time that he systematically reflected on the criteria for experimentally demonstrating the presence of pathogens, criteria that later became known as Koch's postulates.[44] Actually, the term "postulates" as coined later by Friedrich Loeffler is somewhat misleading, for Koch's criteria varied considerably over the years, depending on the stage his methods had reached and the subject under study. In 1877 they did not yet have their "classical form"—(1) isolating the pathogen in the affected tissue, (2) growing the pathogen as a pure culture, and (3) reproducing the disease in animals by means of the cultured pathogen—for the simple reason that the bacteriologist developed his method for obtaining the so-called "pure culture" only years later.[45]

In 1879, Koch finally took steps to leave Wollstein and to embark on an academic career. Unfortunately, the efforts of his Breslau colleagues to have him appointed to an associate professorship were unsuccessful, and his position as municipal physician in Breslau turned out—by contrast to his rather lucrative practice in Wollstein—to be an economic disaster. Finally his appointment to the newly founded Imperial Health Office (IHO)[46] in 1880, in which Cohnheim was instrumental, provided Koch with a research position, the title of Geheimer Regierungsrat, and the directorship of a small research team. The first associates to join Koch were Georg Gaffky and Friedrich Loeffler. This nucleus soon grew to include other associates, among them Ferdinand Hueppe and the chemist Bernhard Proskauer.

The five years between 1880 and 1885 that Koch spent at the IHO—in the end as its deputy director—were his scientifically most productive period and laid the foundation for his fame. He and his associates began by continuing to develop new methods of investigation. The study "Zur Untersuchung von pathogenen Mikroorganismen," published in 1881 in the first volume of *Mittheilungen aus dem Kaiserlichen Gesundheitsamt*, described a whole repertoire of methods, which by then had come to include the production of solid culture media that permitted the growing of pure cultures.[47] Added to this was work on the various methods of disinfection and a heated controversy with Pasteur, in which Koch allowed himself to become embroiled from 1882 on. The controversy was ostensibly about questions of priority concerning the etiology of anthrax and about Koch's critical assessment of Pasteur's anthrax vaccine. According to more recent research,[48] it seems fairly certain that Koch used this controversy to sharpen and disseminate the methodology of his school. In doing so, however, he not only introduced a nationalist tone into scientific debates but also embarked on what turned out to be a greatly exaggerated and in the end untenable criticism of Pasteur's work. Koch's overemphasis on bacterial specificity, the pure culture, and similar matters thus misled him into viewing Pasteur's investigation of the phenomenon of bacterial virulence as a result of inadequate technique and to lump the French microbiologist together with certain German microbiologists who firmly believed in bacterial transformation.[49]

Koch' scientific fame originated in Berlin on March 24, 1882, when he presented a description of the etiology of the epidemiologically most significant infectious disease of the nineteenth century, tuberculosis.[50] To establish a bacterial etiology, Koch continued to do the kind of animal testing that had been conducted in the 1860s by Jean Antoine Villemin, who had demonstrated the transmissibility and identity of such different diseases as lupus and phthisis as forms of an underlying disease, namely tuberculosis. Koch complemented Villemin's concept not only by adding a bacterium as a cause of disease, but also by establishing the demonstration of a pathogen (rather than clinical findings) as a valid criterion for the definition of infectious diseases. The use of experimental pathology in the laboratory thus became a valid method of examination. The highly complex theories generally accepted in German medicine, in which tubercular processes were considered to be transformations of other diseases, were swept away overnight.[51] The attention this work had aroused can also be gauged by the fact that it brought Koch an appointment as Geheimer Regierungsrat.[52]

The etiology of tuberculosis was a scientific sensation. By contrast, Koch's success in identifying the "comma bacillus" of cholera in India in 1884 bore all the marks of a solid public undertaking under the auspices of the imperial authorities.[53] Not only did Koch succeeded in showing the public a bacillus whose existence had been known for almost thirty years, this discovery also unfolded in the form of a spectacular race with French microbiologists: two teams precipitously left for the Near East in the summer of 1883, following the outbreak of an epidemic there, but neither of them was able to identify the pathogen. Koch continued from there to India, where he succeeded in identifying the *Vibrio cholerae*. His findings were presented in the form of pure cultures of the pathogen and led to an improved understanding of the pathological anatomy of the disease. However, they did not satisfy his self-imposed criteria as long as he was unable to reproduce the disease in animals. Nonetheless, Koch postulated a bacterial etiology for cholera, thereby triggering a bitter debate with the hygienist Max von Pettenkofer that continued over many years. The cholera studies of the Berlin bacteriologists were definitely meant as a challenge to the Pettenkofer school in Munich.[54]

The discovery of the cholera pathogen was to be Koch's last scientific success for some time. In the years after 1885 the scientist seems to have experienced a period of personal and professional crisis. Koch finally tried to leave the Imperial Health Office to become the director of a yet-to-be-established independent imperial institute for bacteriology.[55] These plans, however, did not materialize, and the bacteriologist eventually found himself as the holder of a chair of hygiene, which had been created at Berlin University over the objections of the faculty.[56] However important this step was for the field of bacteriological hygiene—which thereby changed from a "secret science" to an academic discipline—the situation was most unfortunate for Koch himself, a scientist who had not yet achieved his Habilitation (a postdoctoral examination establishing eligibility to a professorship in Germany), who lacked significant teaching experience, and who occupied a position in a faculty that mostly viewed him critically. Added to this were health and personal problems, which culminated in 1890 in the separation from his wife, Emmy.

Beyond all this, Koch faced a strategic problem in the conduct of his research: his success so far had rested mainly on the search for microbes. It seemed unlikely that this would remain a promising strategy much longer. Moreover, the practical application of this knowledge had so far been limited to an unspecific preventive medicine, to such measures as disinfection or the purification of water. Therapeutic perspectives, which would seem to follow from the incomparable tangibility of the pathogen, had not come to light so far. The Paris microbiologists, by contrast, could register some accomplishments in the field of serum therapy, which Koch had decidedly neglected. The greatest of these accomplishments, Pasteur's rabies serum of 1885, led to the establishment of the Institut Pasteur and placed Pasteur in the outstanding position to which Koch had unsuccessfully aspired.

The increasing stress created by his difficulties at work and at home account for much of Koch's conduct in the tuberculin affair, which seems to have been an attempt to present a "bacteriological" therapy as a means of living up to his self-imposed expectations.[57] Announced in August 1890 and commercially available by early November, this medication immediately triggered a widespread euphoria that matched the reaction to Koch's earlier achievements. But then the first doubts were voiced only a few weeks later. Increasingly, and at times dramatically, some of the patients died. Engaged opponents contradicted Koch's opinion of the effect of tuberculin on the basis of clinical and pathological evidence. He had initially presented tuberculin, an extract of tuberculosis bacteria suspended in glycerin, as a secret remedy without divulging its composition or the technique of its production. As the incidence of successful treatments declined and Koch was obliged to publish a description of the medication in early 1891, two reasons for keeping tuberculin's composition secret became clear: to serve his own financial interest and to conceal certain weaknesses in the research. The effective ingredients of his medication were not known to Koch himself, who had unsuccessfully attempted to analyze and isolate them; and no one ever saw the animals he claimed to have cured with tuberculin. The theory of its curative effect was obviously based on pure guesswork, and as a result the failure of tuberculin, which became clear by the spring of 1891, caused considerable damage to the scientist's prestige. Nor did the separation from his first wife in 1890 and the affair with the seventeen-year-old art student Hedwig Freiberg, whom Koch married in 1893, help to restore his reputation.[58]

In the spring of 1891 Koch, who the previous autumn had taken a leave of absence from his professorship in order to devote all his time to tuberculin, faced the prospect of an embarrassing return to his unloved university position. But then he was spared this humiliation. The Prussian Ministry of Education and Cultural Affairs had used the short-lived euphoria about the medication to persuade the Prussian parliament to grant the funds for an Institute for Infectious Diseases, predecessor to today's Robert Koch Institute. In the summer of 1891 Koch became the first director of this institute, with what at the time was a very high yearly salary of 20,000 marks, although in exchange he had to agree to several hard conditions, especially giving up the right to a private practice.

The Hamburg cholera epidemic of 1892–93, in which the methods of combating epidemics propagated by Koch's school of hygiene yielded outstanding results, restored Koch's prestige but at the same time demonstrated the limits of what was politically and scientifically feasible.[59] Attempts to quickly pass bold legislation for a nationwide bacteriology-based method of dealing with epidemics had to embark on a ten-year odyssey through the Reichstag and its commissions. Nor did Koch achieve an unambiguous victory over Pettenkofer, who had made a symbolic statement with his notorious self-experiment with cholera cultures, for the Munich hygienicist became only slightly ill. While Koch was working on the Hamburg epidemic, it became clear to him that the epidemiology and the clinical implications of the infectious diseases he had studied were far more complicated than the simple invasive models he had been using would suggest. Unlike his prominent students Behring, Ehrlich, and Wassermann, Koch allowed the rise of immunology in the 1890s to pass him by. Instead, his work on cholera in the years 1892–93 marked a turn toward epidemiology in his research. Koch now took note of such phenomena as atypical or subclinical infections and in subsequent years recognized the significance of the healthy carrier for the epidemiology of infectious diseases. The simple invasion models for these diseases, which among other shortcomings paid scant attention to the gap between infection and the outbreak of the disease, now became more differentiated. This new understanding made it possible, for example, to mount an extremely effective antityphus campaign early in the century, one that focused not on the search for victims but on the identification and isolation of healthy carriers and persons who constantly excreted the pathogen.[60]

With this in mind, Koch increasingly turned to tropical medicine in the 1890s,[61] for here the research on epizootics allowed him to combine his scientific interest in epidemiology and the control and prevention of epidemics with his passion for travel. After 1895 Koch clearly spent more time traveling and residing abroad than in his institute, whose running he delegated to Richard Pfeiffer and Ludwig Brieger, the directors of the departments of research and of clinical studies, respectively. No doubt the new field of vector-dependent parasitical infections, which allowed Koch to continue his favorite work on the etiology of disease, enhanced the appeal of tropical medicine for him. He usually traveled at the behest of the imperial government, but on occasion also worked on the control of epizootics for the British government: thus he was involved in a successful study of rinderpest in South Africa in 1896, whereas his expedition to study and combat the so-called east coast fever in Rhodesia did not produce results.[62]

Two of Koch's studies of tropical diseases deserve special attention. Having repeatedly worked on malaria, he was able, following Ronald Ross's clarification of the etiology of this disease, to contribute to the knowledge of the life cycle of its pathogen, *Plasmodium falciparum*. To be sure, there were voices that accused Koch of exploiting the findings of others.[63] After the turn of the century Koch also became deeply involved in the study of African sleeping sickness. Following his retirement in 1904 at the age of sixty—unusually early for that time—he began to prepare for a major sleeping-sickness expedition (1906–7) designed to contribute to the fight against the epidemic that was just then raging in East Africa.[64] The decision to treat the victims with atoxyl, however, turned out to be unfortunate. Of doubtful therapeutic value, this treatment produced severe side effects, so that Koch once again failed to achieve a clear-cut success in the area of therapeutics.

After the sleeping-sickness expedition, Koch's travels were mostly private undertakings. While traveling in Japan and in the United States, he was showered with honors.[65] In Germany as well he was decorated with the Pour le mérite, Prussia's highest medal, in 1906. As early as 1902, Koch had become a member of the French Academy of Science—two years before he was appointed to the Prussian Academy of Science as successor to the recently deceased Rudolf Virchow. In 1905 he received the Nobel Prize for the most important work of his life, his studies of tuberculosis. However, he was distressed that his former student Emil von Behring, with whom he had had a falling out, received this honor four years before him.

Koch died on May 27, 1910, while visiting Baden-Baden.[66] His body was cremated and kept in a mausoleum specially added to the Institute for Infectious Diseases—one more remarkable parallel to his French rival Pasteur.[67] The creation in 1907 of a Robert Koch Foundation for Combating Tuberculosis and the designation of his institute as the Robert Koch Institute[68] are marks of the continuing fame bestowed on the physician who, like few others, embodied "German science" in the late nineteenth century.[69]

CHAPTER II

# Lower Fungi and Diseases

Infectious Diseases between Botany and
Pathological Anatomy, 1840–1878

## II.1. Overview of the Prehistory of Medical Bacteriology

"Turning now . . . to my scientific career and in particular to my work in bacteriology, I should like to begin by mentioning that I did not receive any direct stimulus for my subsequent scientific preoccupations, for the simple reason that bacteriology did not exist at the time."[1] If we accept Robert Koch's personal assessment, this chapter, which examines the preconditions of medical bacteriology, is irrelevant, since it looks for something that did not exist. It was only with Koch's anthrax studies of 1876 that the new discipline entered obviously uncharted territory.[2] That is why Koch, in that 1909 lecture, depicted the research of the preceding years as a vanished and almost incomprehensible world that had little importance for his work. In the mid-1870s there simply was not much to build on. What did characterize the situation were erroneous theories, like those of a certain Naegeli, who had "declared that he had examined thousands of forms of schizomycetes [as bacteria were called at the time] without finding a reason to divide them into specific forms . . . His student Buchner, who later became one of our most able researchers, claimed to have artificially changed anthrax bacteria into bacillus subtilis and vice versa."[3] The only thing worth mentioning from the perspective of 1909 was a "tiny" segment of biology, which was linked with the name of Ferdinand Julius Cohn. Cohn was "the only botanist who . . . attempted to classify the bacteria systematically,"[4] thereby creating important preconditions for medical bacteriology.

Does this mean that Koch's early work on anthrax, wound infections, and the methods of bacteriological investigation were a breakthrough from nothing? His own memories seem to have gone in that direction. Other contemporaries agreed, such as the pathologist Julius Cohnheim, who in 1876 witnessed the demonstration of Koch's work on anthrax in Breslau. He immediately recognized Koch's experiments as milestones: "Just forget all of this and go to see Koch; the man has made a fabulous discovery."[5] One might look for a personal connection, but none is to be found. The microbiological studies of the botanist Cohn were extremely important for Koch's work,[6] but Cohn denied any personal influence. "When the press reports that Koch was my student and that he worked in my laboratory, this is incorrect. When I met Koch, he was already the great scientist whom the whole world now knows and admires."[7]

This much uniqueness is curious, and taking a closer look at the historical period when Koch began his work is challenging. Was there really no such thing as medical bacteriology until the mid-1870s? Were contemporary biology and medicine really so unimportant for Koch's early work? The very fact that two years after his anthrax studies he published a polemical attack on the above-mentioned Naegeli[8] raises some questions: could it be that Carl von Naegeli—and Hans Buchner, who was included in the same attack—were protagonists of what was a very minor field at the time, and that the attack merely served to enhance the reputation of the attacker at the expense of a couple of hapless straw men? Or was the contemporary work on bacteria actually more sustained, and possibly more fruitful, than Koch's attitude would lead one to believe? Under this assumption, this chapter asks, how did the biologists and physicians of the 1860s and 1870s discuss the problem of living pathogens, how intense was this discussion, and what were its themes? What were the existing concepts of the biology of infectious diseases and what, in particular, was the importance of microbiology for the understanding of infectious disease in general? And finally, what did Koch contribute to this discussion?

This chapter, then, aims to approach the early history of medical bacteriology by describing continuities with the medico-biological knowledge of the time. And yet Koch's statement that there was no bacteriology at all at the time must also be taken seriously and historically verified. It is plausible that the assertion of such a break was more than an easy means of creating a heroic contrast with the past, considering that the historiography of the discipline generally shares this view. The assertion of such a break can be tested with an easy thought experiment: many concepts and objects in Koch and Cohn's bacteriology—such as Cohn's descriptions of the basic shapes of bacteria as globules, rods, threads, or spirals—seem to have continuity with modern knowledge. With only slight deviations, these basic morphological characteristics of bacteria are familiar to every student of medicine or bacteriology to this day. Thus the bacillus that Koch identified as the pathogen for anthrax seems to be identical with the *Bacillus anthracis* we know today. Yet other concepts and objects that were in use in the same period now strike us as exotic. Does anyone still believe that there are miasmatic-contagious diseases, and who would know what is meant by the pathogenic morph of a fungal entity, how to define disease monads, or what group of microorganisms are designated as *Coccobacteria septica*? These concepts are unknown to today's professionals, who would consider the phenomena to which they refer difficult to understand or nonexistent. Yet around 1870 they were generally accepted and linked to influential scientific views. In his work on pathogenic microorganisms, Robert Koch also had to deal with such positions. This being the case, we will have to inquire into Koch's attitudes toward such concepts and particularly toward the theories to which they were linked. We must also find out to what extent exactly Koch's bacteriology brought about the turning point we now discern in that period.

This question addresses not only the relation between Koch's bacteriology and its time, but also its own development. In retrospect it appears as a fully developed system of methods, techniques, and theories that featured such aspects as animal experiments, solid culture media, the staining of bacteria, and the bacterial etiology of infectious diseases. However, it is unlikely that all of these elements were in place from the beginning and that, once introduced, their content, scope, and use remained unchanged. Looking for the origins of Robert Koch's medical bacteriology thus leads inevitably to an investigation of the gradual differentiation of this body of knowledge itself. What were Koch's initial methods and techniques, which subjects did he use to develop them, and how were they subsequently modified?

This chapter endeavors to draw a complex picture of the historical period that in the course of one generation of medico-biological research led up to the works Koch published in the late 1870s. The paths this endeavor takes are not of course necessarily those taken by Koch himself. They include those on which he went astray, but then, even these have their own importance. The point is to elucidate how these "errors" became points of entry that influenced Koch's questions, research strategies, methods of investigation, and particularly the choice of his first subjects for investigation—anthrax and traumatic infections. This chapter is not meant to be a complete prehistory of medical bacteriology. Rather, the study uses a few examples from the time between the 1840s and the late 1870s to examine how an understanding of the relation between doctrines of disease and microbiology came into being in this period.[9]

In this chapter, then, I hope to open up the intellectual space in which Koch's early work developed. Its purpose is not to refute Koch's statement that there was no bacteriology at the time. Delivered as part of his inaugural lecture at the Prussian Academy of Science, this thesis was couched in a rhetoric appropriate to the occasion and reflects an understandable positive self-image. One could confirm or qualify it with equal ease.[10] More pertinent, however, than a revision of Koch's judgment is placing the discoveries and inventions that marked the beginnings of medical bacteriology into a historical context.[11] Going beyond the cliché of the heroic creation of a new science, we should see this history as comprising a few acts accomplished in a very limited space of time.

## II.2. Botany, Pathology, and Infectious Diseases

### II.2.1. Miasmata and Contagia

When trying to delineate the historical space where the origins of medical bacteriology must be sought, one is tempted to set out large segments of time and, for instance, date the prehistory of the pathogenic bacteria concept far in the past. After all, since the seventeenth century, if not earlier, human beings have used optical aids to see structures that the modern observer recognizes as bacteria.[12] The idea of a physical transmission of disease by so-called contagia—literally being touched by a disease—also dates far back in history. As early as the beginning of the sixteenth century, the Italian physician Girolamo Fracastoro had postulated different contagious substances, called *seminaria contagionis*—that is, germs of contagious diseases—for each specific disease, be it syphilis, plague, or leprosy. He then distinguished these diseases, which are passed from person to person, from the miasmatic ones, which resulted from the exhalations of unhealthy places.[13]

Yet the seventeenth-century microscopists, Fracastoro's contagia, and other seeming analogies have little to do with nineteenth-century medical bacteriology. These early modern objects only look like those of later centuries, and specific historical continuities can rarely be demonstrated.[14] We who accept the general premise that living organisms can cause diseases find that this premise was unknown to the early modern microscopists and contagionists. Antoni van Leeuwenhoek described his microcosm without giving any thought to the question of the causation of disease. He examined, for instance, dental plaque and the microorganisms it contained, but this did not lead him to the subject of caries and tooth decay.[15] Attempts to differentiate among these organisms were rare, and the inhabitants of Leeuwenhoek's microcosm were usually referred to in their entirety simply as *animalcula*.[16] Fracastoro's contagia were not only pure hypothesis—he was working without optical aids—they were above all inorganic and in that sense not even comparable to the living pathogens of modern science. Fracastoro's assumption that the mechanism by which his seminaria contagionis spread disease involves sympathy and antipathy to certain areas of the body. He was closer to the atomists of antiquity and their idea that a stream of particles imparts perceptions and effects than to the microbiology of the nineteenth century.

Aside from mere similarities, then, the history of medical bacteriology begins in the late eighteenth century. An important prerequisite for understanding bacteria as living pathogens was the botany of the years around 1800, which increasingly studied microorganisms.[17] Johann Gottfried Ehrenberg, for example, in the book he published in 1838 under the title *Die Infusionsthierchen als vollständige Organismen* (The infusion animals as complete organisms) classified the multiplicity of shapes he had observed under the microscope in a taxonomic system borrowed from Linnaeus. Other scientists, among them Theodor Ambrose Schwann, also attempted at that time to determine the role of microorganisms in putrefaction and fermentation and to develop a biological theory of these processes, noting that their course had similarities with that of contagious diseases. Early in the century Antonio Bassi observed that mortality in silkworm nurseries was accompanied by a fungal infection. In the 1830s Johann Lucas Schönlein described a fungus, later named *Achorion schönleini*, as the cause of a skin disease.[18]

Yet we also have to remember the fundamental differences between the microbiology of that time and conceptions that gained acceptance by the end of the nineteenth century. The idea that microbial species are fixed, which is self-evident to the modern observer, did not have much of a following at the time: such phenomena as the transformation and spontaneous generation of microorganisms were frequently reported and then used to account for pathological changes. As long as it could not be explained how germs had come into infected tissues, the assumption that they had developed in situ through spontaneous generation was perfectly plausible.[19]

And in any case, medical thinking had little use for microbiological observations. Without attempting a complete survey of pathological thinking here, I should point out that two particular schools of thought were incompatible with the notion of a causation of disease through exogenous germs.[20] One of these was contemporary physicians' pronounced interest in pathological anatomy, which, following in the footsteps of the Parisian school of clinical medicine, had gained acceptance in Germany as well.[21] Scientists had learned, for instance, to distinguish among different types of tissue and the specific transformations they underwent. Accordingly, the study of disease emphasized the endogenous changes of the diseased organism and paid little attention to possible exogenous causes. The second conflicting school of thought, particularly in Germany in the first third of the century, was caught up in the concept of parasites without a microbiological dimension. The protagonists of the so-called school of natural history viewed diseases as immaterial entities that take over the body. Strongly influenced by vitalism, this school assigned to pathological processes an autonomous and immaterial vital force comparable to what proponents saw as the driving force in human physiology.[22] Parasitism solved the problem of the causation of infectious diseases in a manner that did not need microbiological insights. Yet eventually it was this problem of causation that gave bacteriology its importance in medicine.

The situation began to change in the 1840s, as a new generation of medical researchers took an increasingly dim view of the vitalism underlying the concept of parasitism. By midcentury these scholars set out to understand the objects of medicine as nothing more than the sum of physico-chemical processes, thereby redefining the discipline as an experimental science without any reference to metaphysics.[23] Physiologists like Emil du Bois-Raymond and pathologists like Rudolf Virchow were united in the endeavor to anchor the knowledge of organic processes in observation, measurement, and experiment. This reorientation gave rise to such innovations as experimental electro-physiology, and new clinical diagnostic methods, among them the use of the thermometer for measuring fever and the recourse to cellular pathology.[24]

This so-called pathological medicine, its methods, techniques, and theories, also gave rise to a specific concept of disease.[25] Diseases were seen no longer as entities qualitatively different from health, but as processes subject to the same natural laws that governed the life of the healthy organism.[26] Seen in this manner, disease is life under different conditions, a phenomenon that does not need to be explained by a special ontology, for it can be seen conventionally in relation to normal organic processes. If disease is thus understood as a deviation from normalcy and as a process engendered by the organism itself, then the causes of disease do not have much importance. Jacob Henle (1809–85), an anatomist of whom more will be said later, distinguished between predispositions (*causae praedisponentes*) and opportunistic causes (*causae occasionales*). Neither of these explains much about the pathological processes they have set in motion, even though the ideas of unspecific preconditions and triggering incidents do have a certain significance.

It was Henle who, in a short study published in 1840 under the title *Von den Miasmen und Contagien und von den miasmatisch-contagiösen Krankheiten* (About miasmata and contagia and about the miasmatic-contagious diseases),[27] pointed medical pathologists in the direction of infectious diseases and caused them to consider the possible involvement of living pathogens. In examining this subject, he developed ways of thinking that medical bacteriologists of the next generation considered important prerequisites for their discipline. Today this study is considered an early step in the study of bacterial infections in the nineteenth century.[28] The significance of Henle's study lies not so much in its empirical content—it was a strictly theoretical paper—as in the fact that the author, albeit in the abstract, called for demonstrating the presence of microbiological pathogens. Many scholars believe that this idea of Henle's anticipated the criteria for documenting pathogens in infectious diseases that have become known as Koch's postulates.[29]

Koch did not think so. He felt it necessary to make the point that his academic teachers had not conveyed to him any specific knowledge of bacteriology. His time at the university, he said, had indeed awakened his "interest in scientific research," but no one should look for the origins of his later bacteriological work there, as he put it in 1909. And while on that occasion he did mention "the anatomist Henle," this characterization interestingly enough sidestepped the author of *Miasmen und Contagien*.[30]

It is indeed unlikely that Henle's ideas of 1840 inspired Koch to begin his studies of pathogenic microbiology thirty years later. Henle did not teach microbiology in Göttingen; he taught anatomy. According to Koch, his teachers there were the pathological anatomist Wilhelm Krause, the physiologist Georg Meissner, and the clinician Ewald Hasse.[31] His scientific achievements during his student days were in the area of microscopic anatomy and physiology.[32] Moreover, when he graduated from Göttingen in 1866, he did not embark on an academic career, which his family could hardly have afforded, but rather looked for a position as a general medical practitioner. In 1865, when his uncle Eduard Biewend wanted to encourage him to do research on a trichinosis epidemic near Magdeburg, he declined for a revealing reason: "Such an extensive undertaking would take me too far away from the practical direction I am pursuing now, and so I shall have to renounce this activity."[33]

On the other hand, Henle's ideas were by no means forgotten in the 1870s, when medical bacteriology came into its own as an academic field. In 1875 Felix Victor Birch-Hirschfeld, a pathologist in Leipzig and author of a widely used pathological anatomy textbook that gave considerable space to the causation of disease through bacteria,[34] showed how far this research had advanced in relation to Henle's standard for proving the involvement of pathogens in infectious diseases. "Such considerations, which suggest an analogy between the conditions of life of bacteria and those of the hypothetical infectious matter, could be carried much further, as was already done thirty years ago by Henle. But at this point we are not attempting a deductive demonstration; we must take the more arduous path of induction to treat this important question in pathology."[35]

Indeed, Birch-Hirschfeld described the (bacteriological) research of that period—in which, according to Koch, bacteriology did not exist—as extremely vigorous.[36] At the 1877 Naturalists' Congress, where the pathologist Edwin Klebs wanted to ring in the age of medical bacteriology, he too claimed Henle as an ancestor, pointing out that the latter, "in his under-appreciated and underused studies of the 1840s" had shown the way to new research.[37] Throughout the 1870s *Miasmen und Contagien* was considered a classic of the nascent discipline, which is why it should be examined here in some detail. What was the state of knowledge concerning infectious disease that Henle described, and which of his ideas made him—even in his lifetime—a kind of ancestor of medical bacteriology?

The anatomist Henle, along with Carl August Wunderlich, Rudolf Virchow, Ludwig Traube, Theodor Ambrose Schwann, Robert Remak, and others, had been a student of the physiologist Johannes Müller.[38] Like these men he was a protagonist of a medicine that consciously distanced itself from the natural-history school of the early nineteenth century and conceived of itself as pure science. Hence it is not surprising that in his *Miasmen und Contagien*, which served as the introductory chapter to his lengthy *Pathologische Untersuchungen*

(Pathological studies), he was not necessarily interested in reviving Fracastoro's concept of seminaria contagionis or in transmitting the concept of parasitical disease entities that had been postulated by the school of natural history. Rather, his treatise was intended to subject certain concepts of the historico-geographical pathology of his day—namely, its understanding of endemic and epidemic, miasmatic and contagious diseases on the one hand, and of parasitical disease entities on the other—to a revision grounded in the principles of natural science. His way of proceeding was partly speculative, in that he tested the internal coherence of existing theories, and partly empirical, in that he compared the older concepts with what had become known about putrefaction and fermentation and the diseases of plants and animals in his own time.

Henle began by examining the traditional distinction between miasmatic and contagious diseases and then added a third group of miasmatic-contagious diseases, since transitional forms between the two categories were frequently observed. The main purpose of this operation was to eliminate the purely miasmatic diseases from consideration. Conceding that for contagious and miasmatic-contagious diseases it was possible to demonstrate, or at least assume, some kind of contagious agent, Henle considered the idea of an immaterial miasma "purely hypothetical"[39] and as such not amenable to scientific examination. There was no possibility of testing the miasma by means of observation or experiment unless here too one postulated a contagium and tried to pinpoint its peculiarities. The scientific investigation of contagious diseases made sense only if one assumed some kind of contagious matter that, as Henle cautiously put it, when "excreted from the sick body (I do not say, as is usually said, generated in the sick body) and transmitted to healthy persons will cause the same disease in these persons." But since contagia are distinguished by their ability to "assimilate at the expense of other substances"—in other words, to multiply—and to produce effects with very small doses, and since the course of epidemic diseases suggests the involvement of a living pathogen, it seemed "very likely that their substance is organic." Following this observation, Henle formulated the hypothesis that "the substance of the contagia is not only organic, but also active, endowed with an individual life, whose relation to the sick body is that of a parasitical organism." From there he immediately moved on to comment on the "seeming congruence" between this concept and the romantic idea of disease as a parasite. Whereas in the latter view "not the contagium but the disease is seen as a parasitical organism or, even more ambiguously, as a parasitical living process," Henle postulated the exact opposite: since to him the organic contagium only transmitted the disease, it was "not the disease but the cause of the disease ... that propagates itself." He thus separated disease from the cause of disease and replaced parasitism as the essence of disease with a material and organic parasite. That parasite is not the disease, but it is causally related to it: "In our sense, then, the contagium is not the germ or seed of the disease, but that of the cause of the disease ... What is inoculated is not the disease, but the cause of the disease; that cause multiplies in the sick body and is excreted when the disease has run its course."[40]

On the basis of this hypothesis, Henle found contemporary studies on putrefaction and fermentation—topics that had been worked on, for instance, by Theodor Schwann, whom Henle had known when both were assistants to Müller in Berlin—to be the area most likely to furnish information about possible contagia.[41] He named a few candidates for identification as pathogenic microscopic fungi, among them the above-mentioned fungus of silkworm disease described by Bassi[42]—Schönlein's *Achorion schoenleini*, a fungus discovered in pustules that caused skin diseases.[43]

The achievement of replacing parasitism with the parasite makes it clear why Henle's book was considered the standard and the beginning of modern bacteriological research in the 1870s, but Henle actually added two other and equally important considerations: since, as he put it ironically, he did not "consider the assumption of a spontaneous generation of infectious matter any less justified than the belief in a *generatio aequivoca* in the realm of animals and plants,"[44] it followed that he rejected not only spontaneous generation—an idea that was still widespread in science at midcentury[45]—but even suspected that the formal constancy of certain infectious disease symptoms indicated the specificity of the pathogens involved. "Where the disease appears with such specific characteristics, we have the right to consider its cause as something constant and unchangeable, as a distinct species.[46]

Not convinced, however, that he had unequivocally demonstrated the presence of a *Contagium animatum* in even a single case, Henle almost casually added some remarks about the criteria for a possible experimental proof of such organisms. These criteria are often referred to in the historical literature as the source of what became known as Koch's postulates for the demonstration of pathogens.[47] While Henle's treatise does not call for the classic threefold process of isolating the germ, growing it as a pure culture, and testing it in animals, it does explicitly discuss the problem of causality. The mere presence of "living mobile animalcula" or "plants" in infected tissue could not be considered proof. These animalcula might well have appeared by chance, or they might be "parasitical though constant elements of the contagia." In the latter case they could be useful for the diagnosis, even if nothing was known about the effects they produced. Proof of an effect demanded more than observing these elements in infected tissues: "That they really are the effective cause would be empirically proved only if one could isolate both the germs and fluids of animalcula and the organisms and fluids of contagia, and then observe the effect of each of these separately, an experiment we will probably have to forego."[48]

Thus Henle emphasized the need to establish a material and causal connection between the pathogen and the pathological process. He also recognized that observation in isolation of the suspected pathogen was a practicable way to do this. However, it is important to see the limits of his reflections: he said nothing about cultivating such a pathogen, nor did he mention animal testing for control purposes. On the contrary, he expressly stated that such proof, which a generation later became the very center of the medical-bacteriological experiment, was impossible to obtain. Moreover, in interpreting his thought, one must consider Henle's views of pathology in general. As I pointed out, he attributed specificity only to the pathogen, not to the disease. To postulate causes for infectious diseases is not tantamount to attributing to them major importance in the disease process. Given his conventional and anti-ontological understanding of disease, whether Henle conceived of microorganisms as necessary or as sufficient causes of disease is hard to determine and moreover misses the salient point of this concept. If diseases are not understood as natural objects, the question of whether their causes are considered necessary or sufficient makes little sense. Henle's concept, in which only the causes of disease but not the diseases themselves are specific, is difficult to reconcile with a bacteriological understanding of infectious diseases that identifies pathogens as necessary causes of disease.[49]

Henle thus appealed to the early bacteriologists because of reflections that were useful in areas beyond his own pathological thought. Among these were his rejection of the idea of purely miasmatic diseases, his explicit separation of disease from its cause, his establishment of a connection between the cause and an organic parasite, and finally his conviction that the isolation and examination of such a pathogen might well be the prerequisite for the experimental study of the etiology of infectious diseases.

The further development of the concept of disease in German medicine between 1840 and 1870 clearly shows that for a time no one went much beyond

Henle.[50] Under the domination of the rising field of pathological anatomy and its prestigious representative Rudolf Virchow, German research continued to focus on internal pathological processes, showed no interest in external causes of disease, and rejected all ontological concepts of disease.[51] The classic example for this is the understanding of tuberculosis: early in the nineteenth century, Théophile Laënnec had proposed a unitary concept of this disease on the basis of a pathological-anatomical structure, namely, the tubercle, but in Germany this concept was abandoned in favor of a highly artificial doctrinal edifice in which the so-called caseous pneumonia of phthisis (pulmonary tuberculosis) was separated from other forms of the disease, and various tubercular processes were seen as transitional phases of other diseases.[52]

## II.2.2. Cholera Fungi

Between 1840 and 1860, then, German academic medicine showed little interest in the microbial causes of infectious diseases. More intense research activity did not begin until the 1860s. In 1867 Hermann Eberhard Richter, one of the editors of the *Schmidtsche Jahrbücher der gesammten in und ausländischen Medicin*, inaugurated a series of reports on the latest literature on this topic with a survey of the recent, rather active research.[53] As far as the microbiology of infectious diseases was concerned, the survey pointed out that while Bassi's and Schönlein's studies had begun to "transfer the hitherto very indefinite concept of parasitism in disease from an ideal to a real foundation,"[54] that theme had subsequently found little resonance. First applied to botany, a revival of microbiology had led to "the most extraordinary and surprising discoveries" by the 1860s.[55] Richter described these in contemporary terminology as a rising tide of research on "illness-producing parasitical fungi" and " preliminary phytophysiological concepts," with which he sought to familiarize his readers. This development had virtually assumed the character of a revolution: "Sometimes it seemed as if the newly discovered facts went far beyond anything an exuberant imagination might have invented, and that a new assertion was the more likely to be correct the more heretically it overturned the concepts grounded in the dogma of the traditional botanists (especially in the systematic-descriptive area), and made them completely untenable."[56]

Unmistakably, this upswing in research took the form of a major botanical controversy about the order of the microcosm, which in turn had its own preconditions. Although around midcentury Justus Liebig and Friedrich Wöhler had heaped ridicule on Theodor Schwann's fermentation experiments and postulated their own purely chemical theory of fermentation, the biological nature of these processes had been brought to public attention, mainly by Louis Pasteur.[57] Moreover, the proponents of spontaneous generation theories, which were still widely held at that time, had been so greatly outstripped by Pasteur's work on this subject that from the mid-1860s on, thanks in part to enormously improved testing equipment, the classification of microorganisms became the subject of the hour.[58] An additional factor that lent urgency to the subject of infectious diseases was the renewed and very violent outbreak of cholera that swept Europe in 1866–67.

Pauline Mazumdar has shown convincingly that German microbiology of the time no longer considered the spontaneous generation of schizophyta the decisive question, but rather concentrated on the constancy or variability of the organisms involved.[59] Such studies, in which the search for natural laws of microbiological development and the description of variability were paramount, were more in tune with the contemporary understanding of science than the classification of species, which was still redolent of a system and hence of deductive reasoning. Hermann Eberhard Richter summarized the latest development for his readers in this sense. If biologists had until now tried to classify the forms of life they had observed under the microscope into species according to their morphology, it was now "becoming increasingly clear that for the most part these forms are mere *temporary phases in a development*." The amount of assured knowledge about species of microscopic creatures had shrunk, and now, according to Richter, the most important thing was "for us physicians to keep our minds free from the older systematic view with its orthodox adherence to the fixity of species and genera."[60] Around the middle of the century Matthias Schleiden had attempted to redefine botany as an inductive science in this sense, and the leading botanist of subsequent years, Carl von Naegeli, was also convinced that species of microscopic creatures did not represent natural objects but could change and assume transitional forms.[61] This scientific botany, which inductively sought to discover the natural laws governing the development of the microcosm, was not easy to refute for systematicians in the Linnaean tradition like Ferdinand Julius Cohn or Anton de Bary. Someone who, like Cohn, "had come to the conclusion that bacteria fall into equally useful and distinct categories as other lower plants and animals"[62] came under the unflattering suspicion of valuing a system more than scientific observation. Such a person, moreover, held a view that for the time being was hard to prove. By contrast to the many instances available around 1860 that showed the morphological transformation of microorganisms, not a single microbial species could be demonstrated to be constant.⁶³ Although Louis Pasteur also held the view that each different fermentation process needed its own, constant species of microorganism, the culture broths he used to cultivate them were poorly suited to yield proof, and in any case Pasteur, who had little experience with microscopic work, did not make any particular effort in this direction.⁶⁴ There was no conclusive way to prove that an experimentally demonstrated change in the forms and characteristics of microorganisms was attributable to contaminated apparatuses. The answer largely depended on the trust one placed in the effect of certain measures of disinfection. The history of Pasteur's experiments to disprove the theory of spontaneous generation has shown that it was well-nigh impossible to decide whether the appearance of microorganisms in germ-free liquids was an effect of faulty technique or the result of spontaneous generation. Depending on the objective of an experiment, one and the same outcome could be seen as a case of spontaneous generation or as the result of contamination.⁶⁵ Unable to fully document the life cycle of even one microbe and reduced for the time being to using the rather unstable distinguishing marks furnished by morphology, the "specificists" remained in the minority.

This was the situation in the 1860s, a time when science was much more interested in the life cycles of yeast, fungi, and similar creatures than in categorizing the different species. Yet it was also the time when a first comprehensive effort to define the relation between microorganisms and the etiology of infectious diseases was offered. The subject of this endeavor, the living pathogen, now began to assume greater importance in the medical history of its time. This research was conducted by the botanist and microbiologist Ernst Hallier (1831–1904), a nephew of Schleiden, who for most of his career worked as a botanist at Jena University.⁶⁶ His research on the etiology of infectious diseases was founded on the phenomenon of pleomorphism, which commanded a great deal of attention at the time. Pleo-, poly-, and heteromorphism initially referred to the fact that certain microscopic fungi followed very different patterns of development under different growing conditions, whereas they had hitherto been considered separate species.⁶⁷ In observing this phenomenon, scientists did not stop with the few fungi in which they had actually seen this change in forms, but increasingly came to see it as a characteristic of life itself. This could be exemplified by the humble potato: "Who, unless he had known it all along, would recognize the potato tuber, the long shoots it grows in the storeroom, and the potato plant with its leaves and fruit as one and the same creation?"⁶⁸

Prominent scientists, such as the evolutionary biologist Thomas Henry Huxley and the surgeon Joseph Lister, were convinced of the significance of pleomorphism,[69] and Hermann Eberhard Richter recognized that seeing these phenomena meant an upheaval in microbiology. Hallier's success was due to his development of a dynamic system of morphological change in microbiology. This system not only brought together a multiplicity of observations and was experimentally verifiable, it also proposed an elegant and far-reaching thesis concerning the etiology of infectious diseases.[70]

In Hallier's view polymorphism was the natural law of microbiological life.[71] Fungi, yeasts, bacteria, and other forms of life were not genera that could be divided into species. Rather, they were developmental stages, so-called morphs, into which the original forms of fungal entities could develop, depending on the environmental or nutritional conditions they encountered. Hallier combined a relatively small number of fungal entities and a similarly limited number of morphs to produce a system of so-called morphic cycles that ordered the observable multiplicity of microbiological life and related it to existing knowledge about its requirements for life.[72]

Hallier's theory could be experimentally demonstrated and showed a thorough knowledge of the living conditions of microbes, things like their reaction to acidic or basic culture media or their needs for carbohydrates, nitrogen, oxygen, and so forth. The initial form of fungal entities—the so-called granular protoplasm—could turn into mold if deprived of air and into yeasts if aerated; but these yeasts in turn became differentiated depending on their food supply. There were transitional forms; different morphs of one and the same fungal entity could appear simultaneously, and it was understood that every development was reversible.[73] Appropriate experiments, which for the most part involved variations in environmental and nutritional conditions, allowed Hallier to identify a large number of microorganisms as developmental stages of a specific fungal entity. According to Hallier, about twenty of these were known. Hallier's microbiology can be characterized by a comparison with Pasteur's: whereas Pasteur defined a fermentation by the microorganisms involved, Hallier determined the character of a fungal morph by the milieu from which it emerged.

To demonstrate and observe the development of his cultures, Hallier used two especially developed apparatuses, one of them called isolation apparatus, the other culturing apparatus. The former, an obvious variant of the usual contemporary apparatuses, served as a control: it was not manipulated and it remained closed. According to Hallier it was not suited for experimentation.[74] What he used for the work was the culturing apparatus, for it allowed him to manipulate its contents during the experiment and to remove samples for examination. Both apparatuses made it possible to keep the air under the cultivation apparatus germ free and to exchange it by means of the methods used at the time, such as cotton wool, strong acids, and so on. This meant that he could observe experiments for extensive periods, sometimes several weeks.[75] Periodically examining the samples he took from the culturing apparatus under the microscope, Hallier was able to describe the morphic cycles and to document them with drawings. A series of drawings of some of the morphs of the common *Penicillium* (*Penicillium crustaceum*) illustrates his technique. Hallier documented some thirty of these morphs. The culture media he used, all of them sterilized by boiling, ranged from blackberry juice to boiled dog feces and black Köstritzer beer.

On the basis of his model, and starting with diseases of the skin, Hallier began to study infectious diseases and was able to identify a large number of them in short order.[76] Concretely this meant that, following the appropriate experiment, he assigned the structures found in the pathological material—that is, usually in infected tissue—to a known fungoid. In all the diseases that Hallier investigated, it turned out that however varied the fungal entities, a specific form of yeast, which Hallier called "micro-coccus," was etiologically significant.[77] This also provided an empirically plausible explanation for the spread of diseases and their pathogens: outside the body, mold and spores were most prevalent, and these were responsible for the spread of disease. Under the growing conditions within the body, they then changed into pathogenic cocci, which thanks to their small size could easily disperse throughout the body.

Hallier's popularity was greatest in the aftermath of the cholera epidemic of 1866.[78] In May 1867 he succeeded in cultivating a coccus out of the feces and fragments of intestines from the cholera victims of the previous year. He identified it as the morph of a fungus named *Urocystis cholerae* growing on Indian rice stalks. He had found the pathogen of cholera.[79] In a few short years Hallier published a considerable number of studies, founded his own journal, the *Archiv für Parasitenkunde*, and even his own institute at Jena University.

The public, and the medical community as well, initially reacted favorably to Hallier's system, but the botanical community soon voiced harsh criticism.[80] The morphic cycles of fungal entities could not, as especially Anton de Bary emphasized, be considered documentation for the developmental history of a fungus but had to be seen as the effect of inadequate technique. They could be explained as the results of impure preparations containing a wide variety of

*Lower Fungi and Diseases* 35

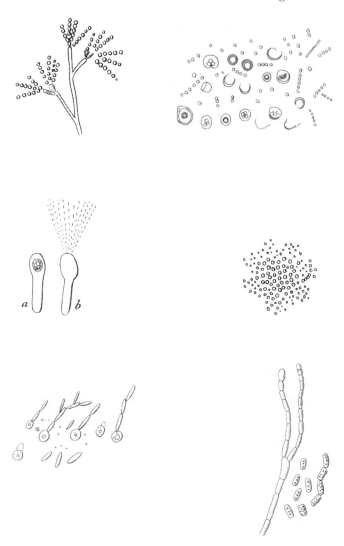

Some morphs of *Penicillium crustaceum*. From Hallier, *Gärungserscheinungen*.

microorganisms, along with the use of unsterile culture media. De Bary refuted the experiments in every detail and stressed that Hallier had failed to demonstrate the developmental history of his fungal entities. Rather than a temporal sequence of different forms, he had demonstrated nothing more than the presence side by side of the different organisms contained in his inoculate. What was lacking was the "direct observation of organic continuity, which at some point must exist between the two developmental stages or generations of a species." Hallier, without producing further proof, had understood sequentially or simultaneously appearing forms as belonging together. "A proper demonstration of a genetic connection of the organisms . . . is not even attempted in Hallier's paper."[81]

De Bary supported his critique with the example of the cholera fungus. He contrasted Hallier's drawing of the so-called cysts of the cholera fungus containing cocci about to swarm with drawings showing the subsequent development of these structures as germinating sprouts of a fungus not related to coccus forms at all. What Hallier had identified as a developmental stage of his cholera fungus *Urozystis cholerae* was shown by De Bary to be the sprout of a long-known fungus called *Urozystis occulta*.[82] He described the supposed micrococci as fat granules. In order to judge Hallier's findings, he suggested, one only had to realize that Hallier had worked with preparations that had been stored in containers for months, and that he had failed to compare his findings with others based on materials taken from the intestines of living persons.[83] Citing Hallier's disregard for the impurity of his inoculates, further compounded by extended storage, De Bary felt justified in his withering comment that Hallier's "tales of his cholera fungus and his fantasies about fungi and their developmental stages are altogether outside the realm of science."[84] And it was indeed remarkable that Hallier had always identified precisely the most common fungi species as pathogens for infectious diseases.

In the medical community, however, Hallier's experiments were less controversial, and not only because he approached the problem with the tools of botany. No one initially criticized the lack of pathological-anatomical investigations. The fact that Hallier had not tested his pathogens in animal trials was a valid objection only if the critic rejected the theory as a whole. Hallier was interested in showing the transformation of fungi into cocci, but there was a whole ocean of imponderables, from nutrition to disposition, between their presence in the body and the possible outbreak of a disease. That is why animal testing seemed long and involved at best, but all in all of rather limited usefulness.[85]

What had first appeared to be a breakthrough soon turned into a fiasco, which, according to Loeffler, "once again discredited . . . the view that the contagia of infectious diseases were of a vegetable nature."[86] But Hallier's work had other effects, which can definitely be counted among the preconditions for the development of medical bacteriology in subsequent years. First of all, its critical discussion generally strengthened the position of those who, like Cohn or De Bary, were staunch champions of the stability of microbial species. It also sensitized colleagues to the issues raised by these researchers concerning culturing techniques. The barb of a pioneer of the new culturing techniques, Oscar Brefeld, that in the absence of pure cultures "all we get is nonsense and *Penicillium glaucum*" seems to have been addressed directly to Hallier.[87] Brefeld not only caricatured Hallier's "favorite fungus" but also gave its name to the opposing program, the pure culture. Proposed by Brefeld and others, this procedure was designed to eliminate the problems arising from a descriptive-morphological ordering of microorganisms through the use of culturing experiments. The objective of these experiments with the smallest possible number of microorganisms, and ideally a single one, was to document their entire generational cycle, thereby demonstrating the stability of the species.[88] This procedure thus differed markedly from the pure culture developed by Koch in the early 1880s, which focused on bacterial colonies rather than on individual microbes. Beyond these questions, Hallier's work placed the theme of contagia on the scientific agenda and used it to illustrate the potential and the limits of a purely botanical approach. Its very weaknesses made it clear to such observers as the pathologist Birch-Hirschfeld that the problem of the contagia for infectious diseases could not be solved by microbiology and botany alone, but that pathological anatomy and animal experiments were indispensable for the purpose.[89]

This raises the question of whether the 1860s discussion about the doctrine of contagion was connected to the medical bacteriology of the following decade. Looking first at the work of Robert Koch after 1873, such a connection seems unlikely.[90] All that can be said is that in many different ways Koch's work was indebted to Hallier's critic, Ferdinand Julius Cohn. Nonetheless a closer look reveals some evidence for such a connection. Koch left Göttingen University without having studied the problem of living contagia in any depth. But it is likely that he made up for this omission soon thereafter, and that one of its points of entry was precisely Hallier's microbiology. Immediately after graduation he worked from June to September 1866 as a junior physician at the General Hospital in Hamburg.[91] There he witnessed the cholera epidemic of that year[92] and, encouraged by his uncle Eduard Biewend of Hamburg, began to work on a microscopic-pathological study of the disease.[93] The drawings of microscopic preparations found in the Koch papers led his biographer Heymann to suspect that Koch had become acquainted at this early date with the microscopic structure that would become world famous as the cholera bacillus.[94] Even though

## 38  Laboratory Disease

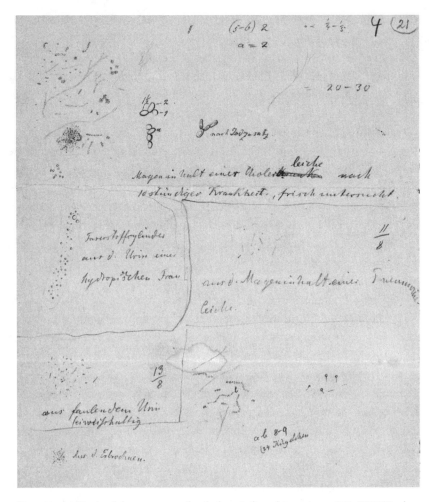

Drawing by Koch of the contents of a cholera-infected intestine, 1866. RKI/Koch papers, as/w1/o15.2. Note that the word "Cholerapatient" (cholera patient) has been changed to "Choleraleiche" (cholera cadaver)!

there is no evidence that Koch went beyond pathological anatomy and attempted to experiment with the microorganisms he had found, they did arouse his considerable interest, as one can surmise from the precise measurements he noted. On the other hand, the labels of the preparations, bearing simple descriptions of their origin, such as "from vomit," do not suggest that he had any thoughts about an etiological significance of these structures, which he called "globules." Koch drew these microorganisms as Hallier had drawn his, that is to say, as round or elongated forms, whose precise shape obviously was of less interest to the observer than their configuration.

Against the background of this episode, it seems plausible that when the young physician studied cholera, he came upon the book by Hallier, the most prominent cholera expert of his day and a proponent of the contagion theory. And it is unlikely that he missed the controversy about Hallier's work; there are some allusions to his reading of Hallier at later times: Koch cited the researcher, who by then had become hopelessly obsolete, in the protocols of his anthrax experiments and repeated his opinion in barely veiled form in the published version.[95] The three oldest bacteriological works in Koch's scientific library were all by Hallier.[96] Given the extremely critical discussions of these texts, it does not seem likely that Koch acquired them very long after their publication. He still cites the fungus researcher in the draft of a paper he presented in 1878.[97]

*II.2.3. Micrococci*

For all the havoc Hallier's theories had wrought, and all the negative effects of the collapse of his system, particularly on this scientist's career, his observations remained pertinent.[98] He and his supporters had been able to identify in diseased tissues large quantities of extraneous organic structures that were not usually found in healthy tissues. In so doing they had directed their contemporaries' attention to this subject. The ubiquitous presence of micrococci in infected tissues—and their count subsequently grew even higher, thanks to improvements in microscopic techniques—became one of the prime research topics of medical bacteriology in the 1870s. Hallier's theories were short-lived, but they did leave a legacy in the form of a clear recognition of contaminants in diseased tissues. These had not gained significance just because of their large quantity. The insight that these structures were organic matter also marked a qualitative turning point. As a result the view of foreign bodies in inflammations shifted from the realm of mere triggers, such as a splinter, toward a causation by living organisms: foreign bodies gradually became seen as pathogens.

The problem of living pathogens that was first broached by Hallier thus remained on the agenda. His work marks the beginning of medical bacteriology in the sense that it initiated the kind of research that was pursued with increasing intensity in the 1870s. Even his critic Cohn had to acknowledge that the Jena botanist "undoubtedly had the very real merit of having been the first to bring direct and continued microscopic observations to the question of the relation of ferments and contagia to bacteria, a question that had hitherto been discussed mostly on a theoretical level."⁹⁹ Even if, as Cohn continued, "the materials collected by [Hallier] himself were unusable because of the well-known defects in his methods of investigation," many of Hallier's observations could be meaningfully employed in other research contexts.¹⁰⁰

Specifically this was the case for the thesis of the coccus stage as the pathogenic morph of his fungal entities. The significance of certain spherical structures in connection with symptoms of disease was confirmed by many other researchers, among them Chaveau, Burden-Sanderson, Weigert, and even one of Hallier's foremost critics, Cohn.¹⁰¹ Different forms were also observed, but the research on spherical cocci occupied a special position in quantity and quality over many years. The reason for this was the large number of these structures and the difficulty in distinguishing among them. In the medical field they were usually found in infected wounds. By the 1870s, however, cocci were no longer what they had been for Hallier: it became possible to differentiate them by their configuration, for instance, in rows or clusters—in our words as strepto- or staphylococci—to the point that in 1878 Robert Koch was able to identify no fewer than six kinds.¹⁰² And even though quite a few researchers, among them Theodore Billroth, continued to believe that such forms had the ability to transform themselves, the speciesist point of view began to make considerable strides after the 1870s.¹⁰³ Here special mention must be made of the system of bacteria developed by Julius Cohn after 1870.¹⁰⁴ This system prevailed not so much because it divided bacteria into distinct groups on the basis of their morphology (globules, rods, threads, spirals), but because Breslau University, where Cohn had directed the Institute for Plant Physiology since 1866, was the birthplace of a number of innovative laboratory techniques that supported the system. Julius Schroeter, for instance, succeeded in using chromatogeny—that is, the production of pigments by certain bacteria—to identify distinct species with the help of the boiled potato, a culture medium of long standing that could be tested against reagents.¹⁰⁵ Another technique developed in Breslau would be of decisive importance for bacteriology: the staining of bacteria with aniline dyes invented by Carl Weigert. The use of synthetic dyes was a reliable and easily reproducible means of identifying and describing bacteria.¹⁰⁶ These techniques would be crucial for the research Koch undertook after his studies on anthrax. A major strength of the Breslau group, finally, was the fact that in personnel as well as ideas, it maintained relations with physicians such as the pathologist Cohnheim and the teacher of anatomy Prosector Weigert. This made it possible to link biological and medical questions. When Cohn strengthened his originally morphological system by adding the biological characteristics of bacteria as distinguishing marks, he demonstrated that the ability to cause disease was more than a species-related effect; it was the defining characteristic of the species.[107]

Indeed, the situation around 1870 was characterized by the fact that important observations now adopted a more medical perspective. In the heyday of Hallier's culturing experiments, it had briefly seemed as if botany had solved the problem of the contagia, but now pathological anatomy assumed greater importance. This was also reflected in terminology changes. Whereas in 1867 Hermann Eberhard Richter, writing in *Schmidtsche Jahrbücher*, had still reported on "the knowledge about disease-causing parasitical fungi and the basic phytophysiological concepts,"[108] Felix Victor Birch-Hirschfeld, who in 1872 began to publish a series of parallel reports to counterbalance Richter's more botanically oriented approach, first spoke of pathologic-anatomical studies of disease-causing parasitical fungi.[109] In his second report of 1875, he referred to these as bacteria. In 1872 he had characterized the present state of knowledge in the field as follows: "While the majority of earlier publications on the presence of fungi in the organism left much to be desired, especially with respect to pathologic-anatomical insights, this shortcoming has now been remedied for a certain group of diseases. Anyone who sets aside prejudice when considering the results of observations made by reputable scientists will hardly come to the conclusion that in such affections the presence of fungi in the body was merely a fortuitous and unimportant element."[110]

It is as if by the middle of the 1870s the microscopic image of pathological preparations had rapidly filled with extraneous organic structures. The initial result, however, was not the birth of medical bacteriology but in fact considerable chaos. This showed itself most clearly in one area which—as mentioned above—occupied a somewhat special position: wound infections, referred to at the time as septicemia or pyemia, and obviously related manifestations such as puerperal fever.[111] To begin with, one is struck by the sheer number of contributions to this theme published by bacteriologically interested physicians such as Billroth, Hueter, Recklinghausen, Weigert, and others.[112] No doubt this was in part related to the rise of Lister's antisepsis, whose newly achieved control directed increased attention to microorganisms in inflamed tissues.[113] And since that attention was essentially focused on cocci, a part of Hallier's system was also salvaged and integrated into the newer research. Theodor Billroth for instance frankly admitted that the impetus for his bacteriological study of *Coccobacteria septica* had come from Hallier's work.[114] Karl Hueter's investigations of the so-called disease monads of inflammation also showed clear connections to Hallier's system.[115]

And finally, there was a political impetus for this boom of medical biology: the military hospitals of the Franco-Prussian War provided expedient investigative opportunities for pathologists. Perhaps the most important work on the subject was produced on the basis of 115 autopsies performed between August and October 1870 at a military hospital installed in the Karlsruhe railroad station. Edwin Klebs's *Beiträge zur pathologischen Anatomie der Schusswunden* formulated the problem of the microbiological causation of disease manifestations with unprecedented clarity.[116] The research agenda that Klebs established as a follow-up to his work on gunshot wounds not only led to a fundamental attack on cellular pathology,[117] it was also, as K. Codell Carter has shown, perhaps the most important source for what we now call Koch's postulates.[118]

What was the subject of this study? First of all, Klebs demonstrated the presence of microorganisms in secretions from wounds. It had turned out that "in almost every case these contained the lower organisms that are referred to as bacteria, monads, etc."[119] The forms of these organisms were extremely varied: "I found rod-shaped bodies, so-called bacteria, which did not move and were frequently attached to others in rows, so that they formed long articulated threads; also numerous microspores, shiny and extremely small bodies whose diameter might be half a micromillimeter at most, and these lay unattached by themselves and then made oscillatory movements or else were clustered together in groups (zoogleiaforms) or lined up in rosary-like threads."[120] Klebs understood all these microorganisms as various developmental stages of a fungoid (Pilzwesen). Referring to it as *Microsporon septicum* harked back to Hallier's system.[121] Klebs's thesis indicated that "these parasitical organisms are the cause of severe manifestations,"[122] that they produce pus and fever, not only as a mechanical but also as a "physiological" reaction. Klebs bolstered this assumption by calling on both pathological anatomy and experimentation. Concerning the former, he showed that his *Microsporon septicum* could be detected in virtually every open wound. He provided a detailed description of its entry into the organism and its spread by way of the bloodstream as well as a theory about its effect. He attributed the formation of pus to mechanical action and the fever to metabolic secretions of the pathogens in the blood.[123] Klebs's view about the physiological effect of the *Microsporon septicum* in particular were based on experimental work that had been carried out in his institute in Bern. The animal experiments of his associates had succeeded in showing that when the animals were injected with fungus-free—that is to say, filtered—secretions from wounds, they developed fever, but not pus, and that in this case the animals usually survived.[124] Since it could also be shown that the *Microsporon septicum* was never present in the blood of healthy animals, Klebs felt justified in identifying it as the cause of wound infections. This was further corroborated by the course of the infections, which—including the relapses—could be seen as analogous to the growth cycle of a fungoid.[125]

As for the experimentation with cultures, Klebs had developed his own method he called "Fraktionierte Kultur," which worked with very small amounts of culturable materials. The material to be seeded was extracted from the infected tissue in a complex procedure, then purified and added to the culture medium. Klebs worked with very small quantities of material, but his method suffered from the same problems as Hallier's: even if he had an effective way of protecting his inoculates against subsequent contamination, they were likely to be of a heterogeneous nature in themselves—with the result that he always cultivated those microorganisms that thrived best under the given circumstances.[126]

Klebs, who in his first publication had already referred to similar observations by other researchers, among them Recklinghausen and Waldeyer-Harz,[127] subsequently endeavored to position his findings against Virchow's view of inflammation.[128] He defended it in 1872 at the meeting of German scientists and physicians in Leipzig, where he also supported Cohnheim's concept of inflammation in a vivid debate about this subject.[129] The linkage of the problem of wound infections with the matter of inflammation in general made this research exciting, for the problem of inflammation touched on one of the central elements of Virchow's cellular pathology. Once Virchow had—to put it briefly—committed himself to the causation of inflammations by mechanical irritants and defined them as internal processes, any view that gave a biological explanation for the same phenomenon and at the same time assigned an important part to an exogenous pathogen was bound to represent a clear opposition to the dominant doctrine of the age.[130]

Perhaps Klebs's culturing experiments found little resonance because his techniques were too reminiscent of Hallier's and because he had observed some unusual phenomena, such as the lengthwise division of bacteria. Birch-Hirschfeld did consider his pathologic-anatomical observations pathbreaking, but he too felt that the "botanical" part of the study had jumped to conclusions and did not meet the standards of modern microbiological technique: "But if the expression '*Microsporon septicum*' means that Klebs believes that the organisms peculiar to septicemia he described are lower morphs of a specific fungal species, then the very existence of the latter would first have to be demonstrated."[131] To arrive at a final assessment of Klebs's work in the 1870s, it is useful to direct attention to the difference between this work and Cohn's bacteriology. For Klebs the question of the specificity and constancy of bacterial species, which was decisive for Cohn, was not only open but above all not of the same importance. Klebs endeavored, of course, to gather the purest possible culturing material, but the changing and diverse forms of his samples did not prevent him, in his spirited polemic against Cohn's views, from regarding them as the contagium of wound infections and from steadfastly postulating their ability to change.[132] And even though a few years later Klebs came to hold the view that, in principle, every disease was related to a specific species of fungus,[133] he still believed that normally the etiology of infectious diseases could be immediately inferred from the pathological anatomy of the infected tissues. His work on gunshot wounds essentially consisted of descriptions of such tissues and, as far as microbiology was concerned, reported on the work of his students. Even if Klebs as coeditor of the *Archiv für experimentelle Pathologie* published a number of microbiological studies, he primarily treated the question of etiology as one of pathological anatomy. Only if anatomy did not allow for clear-cut conclusions would proof need to be produced by animal experiments: "Generally speaking, there are two ways to demonstrate the involvement of the organisms essential to the causation of infectious diseases. (1) The anatomical findings can be decisive in themselves in cases where fully characterized forms exclusively found in a specific disease are present, (2) in cases where an unsual form does not provide conclusive evidence, a transmission through the isolated organisms—cultivated if possible outside the body—can be decisive in determining the genetic involvement of these organisms."[134]

K. Codell Carter has shown that the requirement of experimental demonstrations formulated by Klebs was an important source for what we call Koch's postulates.[135] In the passage cited above and in others, the three steps of isolating, cultivating, and inoculating are clearly enunciated. However—and this marks a considerable difference with Koch—Klebs did not consider these steps as postulates, since in his view they were needed only when pathological anatomy, which he considered sufficient to establish an etiology, did not provide satisfactory information. For Klebs the examination of the isolated pathogen called for by Henle was simply a procedure for clarifying a few ambiguous cases.

## II.3. Medical Bacteriology
### II.3.1. Anthrax

Around the year 1875, the discussion of microbiological pathogens of infectious diseases had once again reached a new level. Although descriptions of individual diseases based on pathological anatomy did exist, there was controversy about each one of the bacterial etiologies proposed for these diseases.[136] Nonetheless the microbiology of infectious diseases now appeared in a different light than before. Whereas the weaknesses of Hallier's work had lain in the area of pathology, it now appeared that microbiological findings were less conclusive than those of pathological anatomy. To be sure, microbes could regularly be identified in many infectious diseases, and especially in sepsis, but no one was able to correlate their multiple forms to specific disease entities in a generally satisfactory manner. As Birch-Hirschfeld remarked in reviewing the work on wound infections of the years 1872–75, "there still is hardly a single fact that one might describe as [a sign of] undisputed progress."[137]

But medical researchers now increasingly saw the mono- and polymorphism of bacteria as alternative interpretations. Rudolf Virchow, who in a ceremonial address of 1874 voiced his opinion about the now much-discussed question of the bacterial etiology of infectious diseases, made it clear to the medical public that this was a matter of irreconcilable and mutually exclusive views: "Either the microorganisms of all the infectious diseases I named are identical, and this . . . points to special toxic substances that must be present in addition to the fungi or algi. Or else the microorganisms are different, despite their apparent uniformity and thus, as the carriers and stimulators of the most dangerous events in the body, the actual causes of the disease. A third view does not seem possible to me."[138] This rejected the concept of polymorphous bacteria on the theoretical grounds that they did not lead to a medically operative object. For Virchow microorganisms were significant only in the context of a "speciesist" understanding of disease that connected specific diseases to clearly defined species of bacteria. If bacteria were variable, they could not have any etiological significance. The decisive factor, in addition to the disease and the bacteria, was a third, still unknown entity that caused the variability of the microorganisms. The pointed form in which Virchow presented the problem made it clear that he sided with speciesists like Cohn or Birch-Hirschfeld. Accordingly, he gave the medical public Cohn's thesis that beyond their morphology, biological properties of microorganisms also had to be understood as species-dependent. "If the same forms produce completely different effects, they must be internally different. [. . .] Thus, if following an inoculation or by pathological chance, bacteria that look exactly like ordinary putrid infusoria produce anthrax, even though the bacteria of infusoria ordinarily do not produce it, we will always have to conclude that the bacteria of anthrax are as different from those of the infusoria as hemlock is from parsley."[139]

This shows that Virchow was by no means, as is too often assumed, opposed in principle to the bacteriological explanation of infectious diseases. In 1874 he saw the problem quite concretely when looking at the demonstration of varied species, which he considered, so far, unsuccessful.[140] It was a weakness in Klebs's argument that he did not pay enough attention to this point, believing that it would be possible to formulate definitive etiologies even on the basis of provisional definitions of bacterial species. That is why in his passionate plea for medical bacteriology delivered at the 1878 meeting of the Naturalist Society in Kassel, Klebs postulated an astonishingly large number of bacterial etiologies, thereby signaling his dissent, not only from Virchow but also from some of his fellow bacteriologists.[141]

All in all, then, the question of the pathogenic significance of bacteria was as popular as it was controversial. Edwin Klebs, who by 1878 already believed that he was identifying many infectious diseases,[142] was fairly isolated in his views and could not even count on the support of a medical researcher like Birch-Hirschfeld, who took his bearings from Cohn's system. Klebs also had to contend with positions like that of the influential Viennese physician Theodor Billroth, who considered all the bacteria appearing in connection with infections as variants of ubiquitous fungal entities and therefore dismissed them as noncausative side effects in such disorders as wound infections.[143] Not swayed by this view and holding fast to his thesis of specificity, Cohn formulated a critique of Billroth, but he did not prevail against the Austrian.[144] The research community was divided, and in 1878 Friedrich Sander, reviewing the situation for the newly founded *Deutsche Medizinische Wochenschrift*, raised an objection to medical bacteriology of the Cohn school by pointing to an observation that could be confirmed with the help of any microscope. This was the presence of microorganisms in bodies that were obviously not sick, or at least not suffering from infectious diseases. One plausible interpretation was to understand these bacteria as opportunists rather than as responsible for the diseases: "The second fact that is difficult to fit into the bacterial theory is the apearance of massive amounts of vegetation in the healthy human body but not in infectious diseases . . . Under these circumstances one can only agree with Billroth's conclusion that the living cell is protected against the corrosive effect of putrefaction-causing organisms, and that the development and proliferation of the latter can only take place in dead matter. Some have tried to counter this second fact with the excuse that there are two kinds of bacteria, the harmless and the dangerous ones."[145]

Sander's recourse to Billroth illustrates the preeminent position the latter occupied in the controversies of the day. In 1874 the Viennese professor had published a book about his research on the *Coccobacteria septica*, in which he described the round and rod-shaped structures as forms of an algaelike entity. Concerning the question of a connection between bacteria and infectious diseases, he was Cohn's most adamant opponent.[146] Billroth saw his microorganisms as polymorphous entities and explained them as a result, not a cause, of infectious processes. The organisms always present in small quantities in the body, he maintained, developed under the special condition of an inflammation and then—depending on the growing conditions they encountered—assumed the most diverse forms and constellations. The designations coined in this connection, such as mono-, diplo-, strepto-, or staphylococci, did not refer to different species, as they would later, but to transitional forms, whose most striking distinguishing mark was the constellation of these entities. The technical difficulties Cohn, Brefeld, and others were encountering in their pure cultures were caused, according to Billroth, by these researchers' unrelenting attempts to cultivate artifacts. Pure cultures—if indeed they could be produced at all—were more the result of identical experimental conditions than proof for the existence of constant bacterial species. Billroth made some ironic comments about the medical relevance of Cohn's bacteriology: "Out of tender solicitude for the medical profession Cohn established the genus 'pathogenic globular bacteria' with four species, supplementing it with one more, a kind of 'bacillus anthracis,' but to me this is nothing more than a concession to the modern current of our time . . . I cannot find any essential morphological differences between the manifold forms of vegetation of the so-called cocci and the bacteria. And we know even less about what chemical differences might exist."[147] This indicates that even the multifarious new culturing methods that had been developed in Cohn's Institute for Plant Physiology in Breslau had not yielded any real progress. The differentiation of morphologically identical bacteria by means of chromatogeny, for instance, presupposed that the production of the dye under study was constant. But since a generational change in identical organisms—the famous pure culture—had never been demonstrated to everyone's satisfaction, the matter itself remained unclear, even in the eyes of its defenders.

Concerning infected wounds, the multiplicity of the bacteria involved gave rise to several competing theories, none of which prevailed. But research was also conducted about other infectious diseases. Two of these diseases stand out. In the opinion of scientists, they seemed particularly apt for establishing unequivocal etiological connections between clearly defined microbes and specific infectious diseases. Both in relapsing fever and in anthrax, morphologically well characterized microorganisms had been known for some time. Here it seemed possible to avoid the confusions of the microscopical picture that plagued the study of infected wounds.[148]

As far as relapsing fever was concerned, Otto Obermeier had published his studies on the presence and the consistent demonstrability of microorganisms in the blood of relapsing fever patients.[149] These bacteria were not only highly characteristic in their form and their motility, they were also absent from the healthy organism. Obermeier's often confirmed observation made it clear that these microbes had a connection to recurrent fever. But it was not clear what kind of connection it was. Obermeier's early death is not the only reason his spirochetes did not come to be seen as the first truly and incontrovertibly identified bacterial pathogens, for two problems had not been solved. For one thing, Obermeier had not succeeded in finding the pathogen between two or several flareups of the disease—a time span during which its presence was assumed. For another, he was unable to produce the disease in animal experiments.[150] All in all, Obermeier's findings presented strong circumstantial evidence for the correctness of Cohn's theory, for they could easily be integrated into it—but they did not constitute proof.[151]

The critique of Obermeier's research illustrates that since the 1860s, research based on animal experiments had become increasingly important in the field of pathology.[152] In 1840 Henle had still spoken of the animal experiment as "something we will probably have to forego."[153] The views on inflammation that Virchow formulated in the 1850s were based on series of pathological preparations rather than on animal experiments.[154] Hallier had limited himself to cultivation experiments only with suspected pathogens. The absence of animal experiments in his studies was not cited as a shortcoming until years later.

The 1860s brought an increase in animal experiments for studies on infectious diseases. In 1863 the French microbiologist Casimir Joseph Davaine conducted experiments for propagating anthrax. His innovation was that he transmitted infected tissues from a sheep to guinea pigs so that he could study the disease under laboratory conditions.[155] In 1865 Jean Antoine Villemin's manipulation of infections in guinea pigs and rabbits definitively established animal experimentation as a method in tuberculosis research. The Breslau pathologist Julius Cohnheim introduced animal experiments into his work; and even though he, like Villemin, worked with infectious materials and not with bacteria, his method had considerable influence on Koch's early bacteriology.[156] In Breslau, Cohnheim and Carl Salomonsen found a way to increase the usefulness of the rabbit as an experimental tool by injecting tubercular material into the eye cavity. This made it possible to observe the pathological processes directly in a transparent medium.[157]

Animal experiments were also very important for the study of a second disease in which a morphologically well-characterized bacterium had been described, namely anthrax. Since the middle of the century, rod-shaped structures had repeatedly been found in the blood of animals affected by this epizootic, which was only rarely transmitted to humans. Animal experiments by Davaine and Otto Bollinger had finally confirmed the presence of these structures and the infectiousness of materials that contained them.[158] This time, however, the mechanism of transmission raised questions: sometimes the bacteria were difficult to detect, and sometimes infections had been produced even with bacteria-free blood.[159] This circumstance had prompted Bollinger, the leading German expert in this field around 1875, to postulate what he called "germs" of bacteria, which under certain conditions would develop the characteristic forms. Moreover, anthrax had a rather mysterious kinship with malaria; these diseases shared affinities for places, such as animal pastures, and warm seasons, which obviously provided favorable conditions for contagion. By 1875 Birch-Hirschfeld could designate anthrax, even in the absence of conclusive evidence, as "the one disease for which the relation between lower organisms and infection is most clearly established."[160]

On this basis Ferdinand Julius Cohn was able, in the years 1875 to 1878, to formulate another model that greatly strengthened the theory of bacterial species and became extremely important for the study of anthrax. In several cases he succeeded in demonstrating a long suspected spore phase in the life cycle of bacteria of the genus *Bacillus*.[161] Under certain conditions, among them aeration, humidity, and favorable temperatures, these entities, such as the *Bacillus subtilis* (also known as the hay bacillus) were able to form so-called durable spores.[162] These were resistant to drying out and high temperatures. When studied in themselves, spores yielded explanations compatible with Cohn's system for several conspicuous phenomena that were usually considered proof of transformation or even spontaneous generation. A case in point was the origin of light-refracting granules that had repeatedly been observed in connection with bacilli. These were now no longer seen as proof for the transformation of bacteria forms but as phases in the life cycle of bacteria. The spores also provided an explanation for the temporary disappearance of bacteria. And finally, they made it possible to understand the sometimes dramatic variations in the hardiness of cultures and infectious material as the result of sporulation.

The discovery of the spore phase may well have been the most important prerequisite for Koch's work on anthrax. Before him, Cohn had already postulated a great similarity between the anthrax bacterium and what was known of the sporulating *Bacillus subtilis*.[163] Koch's work, which he began in March 1873, was indeed "a brilliant confirmation of Cohn's teaching."[164] Such a confirmation was particularly significant because Bollinger, the leading authority in the area of anthrax, had offered an alternative view of the anthrax bacteria, one that rejected Cohn's classification of the bacterium and also had no use for spores. According to Bollinger, the rod-shaped structures were chains of cocci that at times were also found singly, in which case they looked like globules.[165]

Koch's first publication, which involved culturing experiments and animal experiments, confirmed Cohn's view.[166] His first step was to document the constancy of the shape and the infectiousness of the anthrax bacillus by means of the serial inoculation of bacteria-containing substances in animal tests on up to twenty mice—or rather generations of mice—over lengthy periods.[167] Koch was surprised to find that outside the organism, the bacilli were far less resistant than Davaine had claimed and than the etiology of the disease would lead one to believe.[168] The bacteria were highly sensitive, for instance, to dessication. This paradox cleared up when in early 1876 Koch attempted to observe the growth of the bacteria directly under the microscope in a transparent culture medium (liquid from the eye chamber of a cow). This allowed him to identify the spores postulated by Cohn:

> Fresh anthrax blood from a guinea pig was diluted with fluid of the eye cavity from the eye of one calf and placed in several batches in the incubator, with or without the magnifying glass for 10–12 hours at a temperature of 35 degrees ... In one preparation consisting of the serum of guinea pig blood on a concave slide, masses

of bacteria in long threads had developed . . . The threads that were located at the edge of the cover slip were longer and better developed and showed at their extremities regularly arranged strongly light-refracting grains that remained behind in formation when the threads fell apart and disappeared. (Perennial spores).[169]

Koch succeeded in precisely describing the development of bacilli into spores and the transformation of the latter back into bacilli and also uncovered the conditions necessary for sporulation (temperature, aeration, humidity). This botanical account, which completely clarified the conditions and circumstances of the life of the anthrax bacillus, allowed for a far-reaching understanding of the etiology and the epidemiology of the disease, yet did not demand a thorough knowledge of the pathology of the disease and the role played by the bacteria.[170] The knowledge of the conditions necessary for sporulation thus led to the insight that burying cadavers in shallow pits close to the grass roots was a source of future infections. Even if "we are not close to the construction of a seamless etiology,"[171] the bacterium had in principle been identified as the pathogen: "Given the fact that anthrax substances, regardless of whether they are relatively fresh, completely putrified, dried out, or years old, can only generate anthrax if they contain bacilli or spores of the *Bacillus anthracis* capable of development— in view of this fact all doubt as to whether the *Bacillus anthracis* is really the cause and contagium of anthrax must fall silent."[172] Although Koch now proudly claimed that he had succeeded "for the first time to shed light on the etiology of one of these strange diseases," he still remained cautious when it came to extending his conclusions from the etiology of anthrax to other infectious diseases.[173] He considered the animal experiment to be of limited evidential value and doubted its applicability in infectious diseases such as typhus or cholera, which, as far as was known, could not be transmitted to animals. At the theoretical level Koch followed Cohn in 1876, providing decisive evidence for the Cohn system of bacteria. This evidence consisted of the complete cycle of generations of a bacterium whose etiological significance Koch had documented in animal experiments.

## II.3.2. Bacteriology in Breslau

Against the background I just described, the enthusiastic reception Koch's work received in Cohn's Institute for Plant Physiology in Breslau is easy to understand. In April 1876 Koch had contacted Cohn, requesting the opportunity to demonstrate his findings, and had traveled to Breslau upon receiving an invitation. Bruno Heymann has described the circumstances and the course of this event in some detail.[174] For the topic of this chapter, two results of this trip deserve special attention. The strong relationship with Breslau that ensued brought Koch into close contact with colleagues in the field for the first time since his university days. Only the year before, in 1875, when in Munich for the Naturalists' Congress, he had to be satisfied with a simple visit to Pettenkofer's laboratory and otherwise was just another attendee of the congress and a tourist.[175] In April 1876, Cohn brought the bacteriology-minded country doctor to his laboratory in Breslau and introduced him to a great many botanists and pathologists. Koch had the opportunity to demonstrate his work and learned about the methods of Cohn's and Cohnheim's working groups. These encounters must have been extremely important to Koch, for his diary entries for April 1876 consisted largely of lists of persons he had met.[176] The acquaintance with Cohn's two assistants, Eduard Eidam and Carl Weigert, and with the pathological anatomist Cohnheim was evidently particularly profitable. Eidam, who taught private courses in bacteriology for physicians in Cohn's Institute, actually visited Koch in Wollstein in the early summer of 1876.[177] He familiarized himself with Koch's laboratory techniques so that he could teach them to those who took his courses.[178] Weigert, who remained Koch's friend for life, had much to teach him about the latest methods of experimental pathology, a field whose leading practitioner in Germany was Cohnheim.[179] Cohnheim used animal experiments, histology, microphotography, and cutting-edge techniques such as the use of microscopes with immersion mechanisms and especially the staining of bacteria with aniline dyes in tissue samples, a technique invented by Weigert himself. All this made it possible to achieve a sharper contrast that, along with improved microscopic techniques such as the use of the condenser, was a decisive prerequisite for the microscopy of bacteria that Koch would later develop.[180]

Koch's anthrax studies thus afforded him invaluable contacts. But the enthusiastic reactions in Breslau did not reflect a comparable resonance among the rest of the specialists in the field. Aside from the attention the work aroused in France—where it reached its full scope only years later in the controversy with Louis Pasteur[181]—its impact in Germany was rather slight. In the *Schmidtsche Jahrbücher*, normally a good reflection of medical research, the study (which after all had been published in a botanical journal) was not mentioned for many years. On the other hand, Bollinger, whose position in the matter of anthrax Koch had carefully examined, called the study "by far the most important study on the etiology of anthrax to appear in the year under consideration"[182] in a survey for Virchow and Hirsch's *Jahresbericht*. The reaction of Theodor Billroth, to whom Cohn had sent the relevant issue of the *Beiträge zur Biologie der Pflanze*, was positive—but based on a misunderstanding on the part of the Viennese professor. Billroth counted Koch's study among the confirmations for the existence of polymorphous fungal entities, which his assistant Gustav Frisch had described in connection with anthrax, and he replied to Cohn in that sense: "You will soon receive a study by Professor Frisch, who concerning the anthrax bacteria has reported the same findings as Dr. Koch. Surely, nothing confirms the correctness of such observations better than two researchers who, starting from two totally different points of view and working with different methods, nonetheless come to the same result."[183]

At this point, Koch did not continue to work on anthrax; his interest was rekindled only after 1881 in the context of the controversy with Pasteur. He now focused on improvements in experimental techniques, making use of what he had learned in Breslau. Best known among these were the new methods of microphotography, which he presented in 1877.[184] His use of aniline dyes had achieved such vivid contrasts in cross sections that Koch was able to apply Gustav Frisch's invention of microphotography to medical bacteriology.[185] The microphotography of bacteria then afforded him the possibility of publishing his findings quickly and in a form his contemporaries accepted as objective.[186] The question as to which medical problems could be examined by these methods was not immediately broached. In November 1876 Koch still stated as his general goal that in the future he hoped to distinguish even much smaller structures clearly. He wrote to Cohn: "All in all, my interest in the question of anthrax has diminished considerably ever since I have been working on a problem I consider much more important. For it is my goal to find a proceeding that makes it possible to recognize the most varied kinds of schizophytes, even the smallest and most inconspicuous ones, and in the most varied liquids, by their characteristic forms and groups with complete certainty."[187]

Koch considered the microbiological techniques of the contemporary medical researchers inadequate: "Almost every study coming from medical people indicates how far they still are from achieving reasonably adequate results. They barely know about the larger species, but below a certain length they can only talk of rods, micrococci, bacteria, or fungi."[188] With the help of microscopes, aniline dyes, and microphotography, Koch was now in a position to classify even these very small and varied cocci, bacteria, and so forth, and to bring medical bacteriology up to the level of experimental pathology. He was now able, he wrote to Cohn, "to document the smallest schizophytes in putrefying blood and other liquids with extreme sharpness and preserving them for later examination in the same manner as ordinary microscopic preparations."[189]

Scarcely six months later, in March 1877, when the innovations in microscopic techniques had come to an evidently successful conclusion,[190] Koch told Cohn about his next step, which, once the varied kinds of bacteria had been identified and photographed, did involve a specific medical problem: "In this way it must surely be possible to solve the question of the form of the septicemia-related bacteria, once the different species present in the putrefying blood have been identified, and then only one of these species is found in large quantitites in the cadaver of the animal that has died of artificial septicemia."[191] In autumn 1877 Koch once again spent some time in Breslau, where he conducted experiments, above all with Eidam and Weigert. This time he obviously focused on problems connected with questions of pathological anatomy and with animal experiments, a forte of the Breslau school.[192] Again it can be assumed that it was Carl Weigert who helped Koch learn and improve on the necessary techniques. Recall that Weigert had mastered the techniques of staining and preparing tissue samples that were already routine in pathological anatomy, as well as in the animal experiments in pathology propagated by Cohnheim. Indeed, Weigert's own writings introduced some of these techniques to the nascent science of medical bacteriology.[193] Koch's diary entries from these days report animal experiments in the Institute of Pathology and mention that Koch and Weigert did histological work together. The entry of October 17 reads: "From three o'clock on in the Institute of Pathology to find out about using the microtome, hardening etc. from Dr. Weigert."[194]

Not until his return from Breslau in December 1877 did Koch explicitly state the new object of his research, the crucial problem of medical bacteriology of his time, namely, the infection of wounds. "You may be interested, dear Professor," he wrote to Cohn, "to hear something about the work that now occupies me. Among the diseases presumably caused by bacteria, the one that has been studied most frequently, yet is still quite obscure in many respects, is septicemia."[195]

## II.3.3. The Etiology of Traumatic Infections

Koch thus used the methods he had learned through the newfound contacts with his colleagues in Breslau to take up the theme of traumatic infections, which began to dominate the research in the field of infectious diseases. In the wake of Klebs's 1872 study of gunshot wounds, large numbers of observations were published and new theories were advanced—but there was not even a hint of an answer to the question of the causation of septicemia or pyemia by microorganisms.[196]

Three aspects characterize the complexity of the situation prevailing in the mid-1870s. First, in 1874 Peter Ludwig Panum had succeeded in isolating an operative toxin in infected wounds. On this basis he pleaded for a strict distinction between bacterial septicemia and putrid intoxication.[197] To be sure, as Panum himself conceded, the documented existence of an independent operative chemical toxin did not in itself constitute an argument against the causation of disease by microorganisms, but it did complicate matters even further. As Birch-Hirschfeld remarked in 1877, "studies about the action of putrid masses are now complicated by the fact that one has to distinguish between the effects of a transmission of the ferment on the one hand and that of the fully developed toxin on the other—and this does not even envisage the possibility that a multiplicity of substances, each with its own effect, may be involved here."[198] Second, Klebs's *Microsporon septicum*, which had initially appeared to be an etiologically significant bacterium, soon encountered open opposition. The reason was not even that critics raised doubts about the techniques of Klebs's culturing experiments. More important, the work of Theodor Billroth and his students proposed an alternative and nonetiological interpretation of the structures observed by Klebs. Those who followed in this direction firmly held the view that such microorganisms could also be found in the healthy organism but were unable to multiply there. Their massive presence in infected wounds, they maintained, was the side effect of an inflammation, whose causes must be sought elsewhere. After 1874, Billroth's views received massive support from the work of Hiller,[199] who pointed out that, given the observed effects, entirely too few bacteria could be shown in the affected sites, and that, indeed, the blood in infected tissues would have to consist of virtually nothing but bacteria. He also demonstrated that in animal experiments, bacterial masses isolated from infected wounds did not consistently produce wound infections. Hiller, like Billroth, considered bacteria the effect, not the cause of the infection.

Third, the French microbiologist Casimir Joseph Davaine had reported remarkable septicemia experiments on rabbits in 1872.[200] He not only documented the transmissibility of the disease by means of bacteria-filled material, he also described the virulence of the infectious substance as its ability to remain effective after passage through several generations of rabbits in smaller, even minimal doses. The virulence of the first generation varied from that of the twentieth as 1:100,000,000![201]

All in all, the situation in the 1870s was highly perplexing. Nobody knew how to establish a plausible and provable connection between the uncontested presence of microbial life in infected wounds and a bacterial etiology for it. Those who, like Birch-Hirschfeld and Klebs, were convinced of the etiological significance of such microorganisms nonetheless had to admit "that with respect to the significance of lower organisms in the appearance of septicemia and pyemia, the various researchers have become embroiled in such contradictions that a survey of the studies in this field does not yield a single uncontested advance." "There is something depressing," Birch-Hirschfeld continued, "in this admission for one who is searching for the truth."[202]

In this situation, Koch's study of traumatic infections marked a turning point, if only because it defined the objective of the investigation in a new way. Eschewing further study of the pathogens of pyemia, septicemia, and so forth in humans, Koch examined five purely experimental infectious diseases in mice and rabbits, diseases that were clearly distinguishable by their symptoms and the microorganisms involved. This approach was made possible by Koch's innovative methods of investigation, which enabled him to accurately identify unprecedented numbers of bacteria in infected tissues. The first experiments in the area of sepsis can be dated from October 1877, when Koch, returning from Breslau, began to conduct a long series of animal trials in Wollstein.[203] The "culture media" for these innovations were Koch's newfound contacts. The enterprise as a whole clearly shows that the scientific isolation in Pomerania was a thing of the past: Koch was able to present a brief report of his findings at the 1878 Naturalists' Congress in Kassel and to publish them shortly thereafter in the *Deutsche Medizinische Wochenschrift*.[204] An extended version also appeared in book form that same year.[205] A first glance at this text reveals the changed working situation and the determination of the country doctor to make a name for himself as a medical bacteriologist. When writing the anthrax study of 1876, Koch had had only limited access to the literature in the field and had learned about important authors through reviews or even, as in the case of Bollinger, after completing his own investigations.[206] In 1877, he used an old professional diary in which he had occasionally noted indications, medications, minor interventions, and—though this was rare—relevant therapeutical literature in the early 1870s to make extensive excerpts from the literature on the microbiology of traumatic infections. The thoroughness of his reading is impressive and documents his will to master this new field of research. In addition to a wide selection of specialized writings on the subject, the book contains large excerpts from Pasteur, whose work Koch was evidently assimilating at this time.[207] Thus Koch's work on traumatic infections began with a thorough and extensive discussion of the ongoing research.

In the context of this chapter's endeavor to situate Koch's bacteriology in its time, it is most interesting to investigate to what extent his solution to the sepsis problem was not only technically innovative but also conceptually independent. A first indication is the fact that Koch, who had considered his findings about anthrax as pertaining to that case only, opened his study of traumatic infections with a general discussion of bacterial infections and their demonstration. Under the premise that "bacteria are not present in the blood and the tissues of the normal . . . organism,"[208] he formulated two conditions for making valid statements about their etiological significance. First, he demanded the constant demonstration of bacteria in relation to the site and the intensity of the symptoms, so that "in every case bacteria are demonstrated in such quantities that the symptoms of the disease in question find their complete explanation."[209] Second, he made it clear that these observations had to refer to clearly distinguishable species of bacteria, pointing out that if one and the same bacterium appeared in several different diseases, it could not be the cause and was nothing more than a side effect: "The morphological characteristics of the bacteria found in pyemia, diptheria, smallpox, and cholera are so similar that it is indeed easy to mistake them for identical forms. But this would mean that one could not assign any specific importance to these organisms. In this case they would be parasites of the diseases, not their cause."[210]

Focused on his etiological quest, Koch felt that it was plausible to assume that even if the bacteria involved look similar, different symptoms are produced by different bacteria, and that "micrococci that because of their smallness appear to us very similar or even identical in outward form nonetheless are internally different and able to give rise to such a variety of diseases."[211] This central hypothesis, which he supported with references to Birch-Hirschfeld, Cohn, and even Virchow,[212] also defined the most important goal of the study, which was to learn to differentiate among the bacteria found in infected tissues. At first blush this sounded like a technical problem, but it actually stood in for two new approaches to the concept of infectious diseases and the appropriate strategies for studying them.

Hitherto practically all researchers had taken their cue from pathological anatomy, that is to say, from the symptomatological boundaries—however controversial—of the different varieties of traumatic infectious diseases, and had then looked for the pathogen connected to each of these. Koch, for his part, focused on the question of the definition itself and proposed a reverse relation, one that started with the bacterial species. Provided that the pathological symptoms and the distribution of bacteria could be connected to each other, it now became possible to chart the terrain according to the systematically identified microorganisms involved. The task, then, was "to find a morphologically well-defined microorganism as the parasite for every individual traumatic infectious disease."[213] Doubts about the division of infectious processes into pyemia and septicemia had been widely voiced for some time.[214] But Koch went a step further from the very beginning when he proposed a mutuality of definitions between disease and bacterium in which nosology, that is, the classification of diseases, correlated with the classification of bacteria.[215]

As for the second point, his method, Koch's approach was based exclusively on animal experimentation. He did not study sepsis in humans and had little to say about it. Instead he produced artificial cases of sepsis in laboratory mice and rabbits and considered them analogous to pathological phenomena in humans. He did not care whether the experimental animals could contract such infections outside the laboratory; the important thing was that the animal model that functioned in the laboratory could be used to draw conclusions about the situation in humans.[216] Transferring the investigative process from the outside world to the laboratory then led him to answer the question "whether infectious diseases are of parasitical origin or not."[217]

Starting from these premises, Koch further developed the infection experiments that Davaine and others had carried out.[218] The point of the animal experiments was to observe typical pathological changes in connection with morphologically identical microorganisms. This usually demanded two steps, as can be seen in the following experiment:[219] On November 13, 1877, Koch poured ordinary ditch water over a piece of pork meat. On the very next day the meat had a strong putrid smell, and on November 15, it was inoculated as a solution under the skin of a rabbit. To speed up the disease process, a second inoculation was performed. Three days later the animal died. Although characteristic swellings were found in the area around the puncture, the microscopic findings concerning bacteria were initially unsatisfactory. "Samples from the skin on the back: innumerable bacteria (small, larger micrococci and short thin rods) spread in clouds throughout the subcutaneous tissue." But if the blood of the first rabbit was used to infect other rabbits, uniform bacterial masses were seen. "The fluid taken from the abdominal skin contains only micrococci in large quantities." These experiments led Koch to the fifth of the diseases he described, rabbit septicemia: "[We found] exclusively more or less spread-out single layer deposits of oval micrococci lining the inner walls of the capillaries in consecutive or side-by-side rows . . . this caused the micrococci colonies to look like bowls or rills . . . These micrococci thus differ greatly from the cocci of pyemia by their size. But also in most other respects. They never encapsulate blood corpuscles, but push them aside, even when they accumulate inside the blood vessels. They do not cause blood clotting and hence no embolic processes either."[220]

Characteristic changes in the location and behavior of the bacteria now provided information about the disease process. The example of the third disease, the progressive abscess formation in rabbits, clearly shows that this process is very similar to a culturing procedure. Koch first noted that "while no bacteria are found inside the abscess, its wall is formed on all sides by a thin layer of micrococci agglutinated in dense clusters of zoogloea."[221] At the same time the bacteria located on the outside of the wall appeared to be more vigorous, and therefore easy to stain. Going in the other direction, they became more difficult to stain, and the center of the abscess consisted of dead bacteria. The distribution and behavior of the bacteria thus turned out to be analogous to the pathological process, which by the same token unfolded like a culturing experiment when it consumed the tissue of the experimental animal.[222]

Altogether, as Koch stated at the end of his investigation, there were three criteria for proving a bacterial etiology: (1) the effect produced by very small doses must preclude a confusion with poisoning; (2) bacteria must be demonstrated in every case, "and there has to be a different and well-distinguishable form for each different disease"; and (3) the quantity and constellation of these bacteria must have a plausible relation to the symptoms and the course of the disease. Five of the six diseases Koch described met these criteria. As a result these diseases were no longer isolated cases but showed "the greatest similarities with human traumatic infectious diseases" and were to be understood as their model.[223] Koch felt that it was immaterial whether the same bacterial species caused such diseases in both humans and mice. The decisive breakthrough was that the animal experiment had provided a model that could furnish fundamental insights about traumatic infections. As Cohn had found out, the characteristics of bacteria were unchangeable. Following Koch's study, one could add to these characteristics the ability to produce disease. He himself summarized the possibly most important finding of his work as follows:

> Even in the small series of experiments I was able to carry out, one fact was so evident that I must regard it as constant, and, as it helps to remove most of the obstacles to the admission of the existence of a *contagium vivum* for traumatic infectious diseases, I look on it as the most important result of my work. I refer to the differences among pathogenic bacteria and to their unchangeability. A distinct form of bacterium corresponds, as we have seen, to each disease, and this form always remains the same, however often the disease is transmitted from one animal to another.[224]

To be sure, the diseases he studied were themselves artificial, being confined to the laboratory environment. The course of such infections in free-roaming mice was unknown, and of no interest to Koch anyway. The decisive aspect was the analogy to be drawn between the laboratory model and infections in humans. He felt that since the experimentally generated diseases "showed very great similarities to human traumatic infectious diseases," the conclusion was plausible that "if the same improved methods of investigation were applied, it was very likely that all of the traumatic infectious diseases in humans would turn out to be parasitic diseases."[225]

## II.4. The Pathology of the Laboratory

Koch's study on traumatic infections developed a bacteriologically based concept of disease. In the study on anthrax, the disease and the pathogen had really been known from the outset, so the main task had been to clarify the relation between the two. The clinical and the bacteriological diagnoses led to very similar demarcations of the disease. By contrast, the investigation of traumatic infections produced a number of completely new and experimentally generated diseases. In this sense, Koch's experiments represented the inauguration of a new process that defined infectious diseases by means of medical bacteriology, a process that would come into its own over the next decades. It now became possible to define diseases whose identity had hitherto been tied to pathological findings and clinical symptomatology through the presence or absence of the pathogenic agent. Every time a pathogen was identified, the disease in question was newly defined, and the decisive argument no longer came from the clinic or the autopsy room, but from the bacteriological laboratory.[226] This could lead to completely novel definitions of a disease, as it did in the cases of tuberculosis and diphtheria. The best method of investigation was now no longer the examination of a sick or deceased patient but an experimentally generated animal model of the disease. Even when the definition of the disease was linked to the taxonomy of the pathogens, it was only the combination of experimental pathology and bacteriology in the animal experiment that allowed for a general description of the disease that went beyond the individual case. In view of the theme Koch had chosen and the original solution he proposed—at the Naturalists' Congress in Kassel, no less!—it is not surprising that his second etiological study caused a greater stir in the medical public than his anthrax study of 1876.[227]

In the form in which it was developed in 1877–78, the model relied on two presuppositions: the stability of the bacterial species and the validity of the animal model. It is remarkable how much energy Koch expended on establishing the stability of the bacterial species. In fact, he attempted to apply a concept borrowed from botany to the medical field, where it was not easily reconciled with the observations of other researchers. Thus he was deeply suspicious of the phenomenon of variable virulence of pathogenic bacteria as described by Davaine and others. Even in outlining his research project, he made it clear to Cohn that his study would also serve to refute this observation, which was incompatible with Cohn's bacteriology. "I have always been extremely puzzled by the intensification of the effect of the septic toxin as shown by Davaine. This characteristic, which probably misled Naegeli into attributing the rapid transformation in the character of his schizomycetes to assimilation, seems to me so extraordinary and so incompatible with what I have learned so far about the life and behavior of bacteria that I am setting out to solve this puzzle."[228] The sharp attack against the botanist Carl von Naegeli and Naegeli's student Buchner that Koch published in 1878 was thus motivated in part by his strict insistence on bacterial stability.[229] This attitude misled him into condemning transformism and virulence as kindred fantasies produced "by the magic wand of adaptation and heredity."[230] "Some otherwise perfectly precise researchers have allowed themselves to be dazzled by this beguiling theory," he remarked in the published text, referring still quite politely to Davaine's theory of virulence.[231] But in a letter to Cohn he openly expressed his hostility to Naegeli: "Apparently Naegeli imagines . . . micrococci that come together in order to form today a spirochete, tomorrow a bacillus, and the day after a spirillum, but will later walk away from each other and then vanish into thin air as harmless micrococci."[232]

Today we know that Naegeli's transformism did not have much in common with the virulence of bacteria that Davaine and later Pasteur described and used in the production of vaccines. Yet Koch saw similarities, since both called his

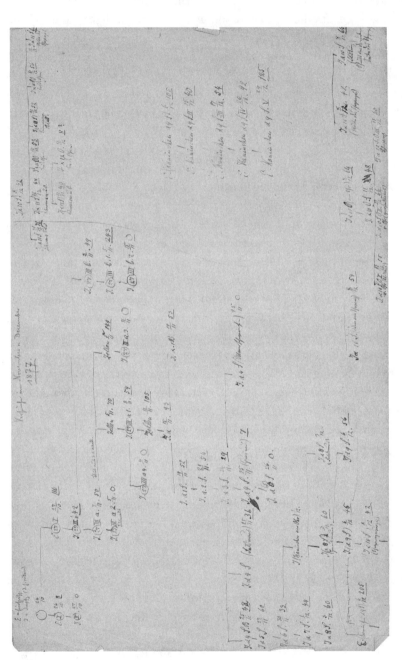

Listing of Koch's animal experiments on traumatic infections. RKI/Koch papers, as/w1/012.

fundamental assumption of the stability of species into question. It was years before he was able to understand virulence as anything more than the result of poor laboratory technique. As Andrew Mendelsohn has shown, it was no coincidence that Koch chose this very theme to set his microbiology apart from Pasteur's.[233]

The second presupposition of Koch's approach was the relevance of the animal model. The views he developed on this topic were as multilayered as they were original. They therefore deserve to be examined more closely. To begin with, animal experiments were helpful in solving one central technical problem, namely that of culturing. In the case of anthrax, Koch had been able to observe bacteria isolated outside the body in culturing experiments. Although he did not explicitly embrace the concept of the pure culture that was being propagated at the time, particularly by Oscar Brefeld, Koch had actually practiced it in that study.[234] In the case of traumatic infections, however, no one had yet succeeded in growing cultures of this kind, and the variability of Klebs's *Microsporon septicum* or Billroth's *Coccobacteria septica* seemed to indicate that this could not be done. In the discussion about the resulting problems, Koch voiced doubts about the practicability of Brefeld's approach for the study of medical questions. Brefeld's "demand is no doubt theoretically correct, yet it must not be presented as the sine qua non for every investigation of pathogenic bacteria."[235] It would not do to stop etiological investigations "until the botanists have succeeded in identifying the various bacterial species through pure cultures and cultivation."[236] Even if such information as nutritional requirements were used, Koch felt that a small number of bacteria—among them the anthrax pathogen—were morphologically so characteristic that a contamination of the culture would surely be recognized. "If pure cultures of very small bacteria—so small that they might not even be visible unless they were stained—were to be made, how could one detect a contamination of the culture?"[237]

Koch solved this problem by viewing the body of the experimental animal as the culturing apparatus where pure cultures were produced by the pathological process itself. By 1877 or 1878 this idea had assumed its full significance, and there is no indication that Koch's study on traumatic infections involved in vitro culturing experiments at all.[238] The animal model's effectiveness for this purpose was based on two assumptions: first, Koch was convinced that the healthy organism is free of pathogenic germs. It was therefore possible to see the organism as analogous to a culture medium. A second assumption was that bacterial infections always took the same course in members of the same species but varied considerably from species to species. That is why, for example, a change in experimental animals made it possible to filter an individual bacterial species out of a mass of mixed bacteria. "It is quite amazing how differently bacteria act in different animal species," Koch wrote to Carl Flügge in 1879.[239] The rabbit experiment described above shows that Koch—unlike Klebs and others—was not interested in starting with the most homogeneous bacterial material possible. Instead, he injected his experimental animals with heterogeneous substances, which in the course of several passages through a variety of animal species purified themselves. On November 18, 1877, for instance, he started an experiment by injecting some "skin liquid" of that rabbit, which had earlier been infected with putrid substances, into a domestic mouse.[240] On the next day blood was taken from that mouse and injected into a field mouse, and a series of similar experiments followed. The move from domestic mouse to field mouse to rabbit and so forth led to different pathological and bacteriological findings from case to case, and this could be understood as the result of a filtering action by the animal body. In addition to taking notes on individual experiments, Koch also summarized entire series in diagrams that resemble genealogical charts. Their most important graphic information was the documentation of the path taken by the infective substance. The illustration on the preceding page shows such a schema. Even though most of the laboratory notes to which it refers are lost, so that some parts lack a context, the figure is informative. In principle, Koch noted the kind of transmission (inoculation or injection), the survival time of the animal, and, whenever he found it interesting, the site of the injection (the root of the tail, behind the ear, and so forth), as well as the kind of animal used. The continuity marked by a line is that of the pathogen.

Even though Koch showed a certain predilection for guinea pigs and mice,[241] his animal experiments were marked above all by great attention to and skill in varying the experimental animals used. In the case of traumatic infections he had worked with rabbits, domestic and field mice, and later, when studying tuberculosis, he had recourse to two dozen different species, among them such unusual ones as goldfish, hedgehogs, and turtles.[242] To be sure, some of these animals were used for control purposes, but varying the animals starting with the same or a similar kind was a fundamental tenet of Koch's experimental strategy: "The experiment must not, however, stop with the transmission of the infective substance to individuals of the same or a similar kind: the next step is to examine the reaction of as many animal species as possible to the infective substance . . . Such experiments yield most informative observations about the extraordinary sensitivity of pathogenic bacteria to the nutrient medium on which they can thrive or which they reject."[243] Under the assumption that the healthy organism was free of pathogenic germs and that the unfolding of a disease was similar to a culturing process, this solution was logical. Koch was therefore delighted with the effectiveness of his procedure: "And yet we do have pure cultures of the smallest and most difficult to recognize bacteria. But these are not produced in the culturing apparatus but in an animal's body . . . There simply is no better culturing apparatus for pathogenic bacteria than the animal body."[244]

By combining different experimental animals, each susceptible to a different disease, it became possible to solve even difficult cases and to distinguish bacteria that were morphologically almost undistinguishable: "It is quite evident that my experiments have succeeded in producing true pure cultures if one examines the two diseases provoked . . . in the mice. The putrifying blood that gave rise to the two diseases contained a whole variety of bacterial species. Of all these only two found the conditions necessary to their existence in the body of the live mouse. All the others perished."[245] But since field mice were not susceptible to one of the diseases, it was possible to separate the two bacterial species by moving to a different experimental animal. The result was that "the inoculation of the two bacteria into field mice eliminated the bacilli, so that the micrococci could be further cultivated in pure form . . . In short, this makes it quite possible to cultivate several species of bacteria side by side, unmixed and in pure form, to separate them, and if need be recombine them."[246] The technique of achieving pure cultures on solid media that Koch developed a few years later by no means diminished the importance of the animal body as culturing apparatus, since the preparation of such cultures required the purest possible initial material.[247] This was far easier to obtain from an experimentally created infection than from material taken from human corpses, which was "always dirty on the surface and moreover often not quite fresh."[248] In this sense a preliminary passage through an animal by way of cleansing the infectious material was part of the technique of pure culture.[249]

By comparison to other laboratory sciences of the period, the creative variation in the use of experimental animals was a procedure specific to medical bacteriology. It is true that contemporary physiologists also worked with a variety of species, but it was usually done to enhance the general applicability of their findings.[250] By contrast, Koch's use of various animals was a heuristic process that, as described above, served such purposes as filtering out certain species of bacteria.

Experimental animals served not only as culturing apparatuses but also as models of the disease process. This raises the question of which animals were useful in such research, and why. Koch addressed this question, usually in a cursory fashion, in some writings he published in the years after the study on traumatic infectious disease.[251] In theory the choice of the experimental animal that would facilitate the observation of the pathological process was based on its phylogenetic proximity to the starting point of the investigation. In human pathology, for instance, primates were the most suitable experimental animals: "As for the choice of the experimental animal, it is expedient to start with animals of the same species as those from whom the infective material has been taken . . . In the case of human infectious diseases, one should have recourse to the animals closest to humans, the primates."[252] Differing from the modern view that explicitly conceives of animal pathology as a model—a view that became fundamental in the biology of the twentieth century—Koch and other bacteriologists of his time only demanded the closest possible proximity between the original pathogen and the experimental organism.[253] Hence one finds hardly any reflections on the differences between human and animal pathology. Indeed, the studies based on animal experiments starkly contrast in scope and even content with the scant systematic statements about this theme. In practice the requirement of experimenting with genetically closely related organisms was usually disregarded. The consistent reproduction of the disease process by culturing in a susceptible animal was more important (to Koch) than the creation of similar pathologies in closely related species, which is why primates were used only rarely and in small numbers. There are several reasons for this. First of all, it was a matter of cost and of space. Replacing the countless rabbits and guinea pigs that Koch used up in his tuberculosis studies of 1881–82 with larger animals would have gone beyond the spatial and financial possibilities of the Imperial Health Office.[254] However, there is no indication that the decision not to use certain experimental animals was related to or anticipated antivivisectionist criticism.[255]

It is also noteworthy that consistent susceptibility and genetic kinship between species did not necessarily have to correlate. In inconclusive cases Koch was indeed aware of the different reactions in differing species, but in his pathological experiments he generally chose to use the susceptible animal, as his writings about tuberculosis indicate: "The more susceptible an animal is to infection with the tubercle virus, the better suited it is to these infection experiments."[256] At the same time he increasingly emphasized the need to repeat experiments in long series. And indeed, this requirement is of major importance for the postulated species-specific constancy of the infections. While it is true that no preferred experimental animals appeared on paper, in practice rabbits, guinea pigs, rats, and mice were prevalent. These showed particularly good reactions, could be kept in the laboratory in requisite quantities, and were affordable. In the case of the mice it was initially sufficient to use the offspring of two specimens that a friend and colleague of Koch's had given to his daughter in 1876.[257] Moreover, Koch had long experience in working with these animals. Some of his experiments, to stay with the example of tuberculosis, also involved monkeys, but these were exceptions contrasting with the use of "many hundreds of rabbits and guinea pigs."[258]

To summarize, the animal model as developed in the research on traumatic infections had a twofold significance: Provided such a body was free of pathogenic germs, it served as a culturing apparatus and as a model of the pathological process. It secured the identity and stability of bacterial species and demonstrated their actions in the disease process. This combination made for an understanding of infectious diseases essentially as bacterial invasions that, once they had entered the body, spread there as they did in a culture medium.[259] Koch's description of the pathological process in the second disease he investigated—the progressive necrosis of tissues in mice—provides a clear example of this view: "And so it comes about that micrococci are always present in necrotic tissues and that in spreading they push a wall of cellular nuclei ahead of themselves. This wall constantly melts on the side turned toward them [the micrococci] and is replenished on the opposite side by lymph cells that constantly reattach themselves."[260]

While Koch was motivated by the contemporary debates to choose the subjects of his first studies, anthrax and traumatic infections, and while his methodology was greatly indebted to the experimental pathology of Cohnheim and Weigert, his solution to the problem of sepsis in particular was all his own. In terms of biology it was unusual, since he approached the question of pure culture differently by way of the medical phenomenon of infection. The medical approach itself was innovative in the sense that Koch did not define disease through pathological anatomy but rather through the pathogens, thereby clearly differentiating himself from such researchers as Klebs.[261] Koch's ability to use this foundation to construct a functional experimental system returned the specific ontology of infectious diseases—a concept that seemed to have died with romantic medicine—to the field of pathology.[262] The characteristic feature of the bacteriological concept of disease and its construction of what is pathological is not just its reduction to etiology. Above all it is the fact that the measure of the disease is no longer the patient's pathology, but is the successful reproduction of the pathogen in an animal experiment.[263] Such knowledge is laboratory science par excellence. It should be seen as the product of an artificial environment created for the purpose of observing and destroying pathogenic agents. Experimental animals as living apparatuses play a crucial role here.

Unlike in the assumptions of physiological medicine, the diseased body once again was seen as qualitatively different from the healthy one, the difference being the presence of an exogenous pathogen, without which the disease process could not be explained. Even though Koch's work at this early stage revolved around the theme of etiology, it is already clear why medical bacteriology must be seen as a break with the medical thinking of the time. The construction of bacterial etiologies led to a new understanding of infectious diseases that was based on a combination of nosology and taxonomy. Ideally, disease entities could now be tied to items in the taxonomy of bacterial species.[264] A related aim was to identify the manifestations of the disease as resulting from the life cycle of the pathogen. Although Koch's work on traumatic infections did not discuss matters of pathology at any great length, its effect on this field was nonetheless enormous: the process and the cause of the disease—which Virchow wanted to separate conceptually a few years later in a clear criticism of bacteriology[265]—coincided in Koch's view with the pathogenic agent, which confronted the body viewed as a passive culture medium. The pathogenic agent thus became what one might call a deterministic object.[266] The reductionist concept of infectious disease that in historical sight characterizes medical bacteriology is not often formulated as such, and methodical reflections are usually limited to the discussion of the best investigatory techniques at any given time. The bacteriological model of infectious diseases is hardly ever made explicit; it is implied in a concept of bacteria as pathogens that has led to far-reaching consequences for pathology.

CHAPTER III

# Tuberculosis and Tuberculin
History of a Research Program

## III.1. Tuberculosis as a Research Problem

Robert Koch's identification of the tuberculosis pathogen is considered his greatest scientific achievement. His friend and colleague Friedrich Loeffler said that the discovery of the tubercle bacillus, which Koch unveiled on March 24, 1882, in Berlin, "suddenly made him the greatest, most successful, and most meritorious scientist of all time."[1] In 1905 he received the Nobel Prize for his studies of the epidemiologically most important infectious disease of the nineteenth century. Later biographers also saw the identification of the pathogen as "the greatest and most significant success of his life."[2] Paul Ehrlich, who was present at the lecture to the Berlin Physiological Society, wrote in an obituary for Koch, "that evening has always remained in my memory as my greatest scientific experience."[3] The study was recognized as a breakthrough from the very time of its publication. As Albert Johne put it in the history of tuberculosis he published in 1883, "as a result of Koch's newest work, the pathogenic aspect of the tuberculosis question can now be considered essentially solved."[4] Even Koch's latest biographer so far, Thomas Brock, counted the identification of the tuberculosis pathogen as one of two steps on the way to fame: if the cholera expedition founded Koch's public reputation, the tubercle bacillus was emblematic of the scientific sensationalism that marked medical bacteriology in the early 1880s.[5]

Traditionalist arguments of this kind are insufficient for writing good history. Their understanding of the pathogen to be discovered is simplistic in its concreteness—as if it had hidden all along and only waited for its discoverer. Its discovery is depicted as an event whose significance is self-evident and whose position in the history of bacteriology and in Koch's scientific work does not need to be discussed. And yet there are good reasons, particularly in this case, to place this work within a larger context. The fact is that Koch's discovery marked not the end but the beginning of a long and intense preoccupation with tuberculosis. From 1882 until Koch's death in 1910, numerous articles, lectures, and expert reports testify to this preoccupation. The identification of the tuberculosis bacillus in 1882 was followed by a series of studies designed to develop further the newly gained knowledge and to apply it to other questions. The first item in this endeavor was the appearance in 1890–91 of tuberculin as a tuberculosis medication. The year 1897 brought an attempt to improve this therapeutic agent. After 1902 Koch published the studies that sought to prove the difference between human and bovine tuberculosis.

Unlike the identification of the pathogen in 1882, which was highly praised, Koch's later work gave rise to critical discussions among his contemporaries. Once the initial enthusiasm for tuberculin had passed, the number of its supporters remained small, a situation that not even improvements in the medication could change. Differentiating between the pathogens of human and bovine tuberculosis, as Koch did, was and would remain controversial.[6] This chapter investigates the nature of the connection between the writings on the etiology of tuberculosis and those on tuberculin within Koch's research program. Until now, the history of this medication has usually been depicted as an unmitigated disaster,[7] but tuberculin itself has sometimes been counted among the pioneering achievements of immunology.[8] Either way, little attention is paid to its connection with earlier studies, and hardly anyone has looked into the cognitive structures that, beyond success or failure, informed Koch's thinking and his work.

Precisely this connection between Koch's various studies within the framework of a research program is the subject of this chapter. It will focus on the road from tuberculosis to tuberculin, disregarding matters of success and failure. The point is to see the scientific conception of the therapeutic agent as part of a research program on tuberculosis that began with the etiological studies of the years 1881 to 1884 and was later expanded to include the area of therapeutics. I will start with two connected hypotheses concerning the development of a methodology and the choice of research topics.

Regarding the former, I am interested in the dynamic of a fully developed research program. The etiological studies of the years 1881 to 1884 had yielded a set of methods, means, and subjects for research that both created and limited the parameters of future work. Thus the development of the animal model in etiological research was the first step into the area of therapeutics. Although it was not clear whether this model could be made to work for therapeutic research, such a transformation of the research program would not have been neutral in relation to its origins, and it would have expanded on the original setup. In the new area of investigation, this origin may well have led to what the philosopher of science Ludwik Fleck has described as a stylistically consistent perception of observations and the staying power of fully established opinions. Once such a consistent system of perceptions is in place, it seems unthinkable to contest the established view, with the result that unsuitable observations are suppressed or forcefully fitted into the existing body of knowledge. It can even lead researchers to produce images of objects whose existence within the system is plausible or necessary, even in the absence of empirical observation. This chapter will examine to what extent this phenomenon—which Fleck dubbed "the harmony of illusions"—contributes to an understanding of Koch's procedure.[9]

Regarding the choice of research topics, this chapter advances the thesis that tuberculin as a medication and diagnostic tool can best be understood as a sequence from Koch's earlier studies on the etiology of tuberculosis. Developing tuberculin was not, in itself, a mistake. Conceiving of it as a medication remains incomprehensible if one is too quick to point out mistakes that were not recognized as such until much later. Historically speaking, tuberculin was embedded in a research program on tuberculosis that had started in the early 1880s in the area of etiology and was subsequently enlarged to include matters of therapeutics. Seen in this perspective, the conception of tuberculin is closely related to ideas about the pathogenesis of the disease that had been developed years earlier—the problems with tuberculin are in fact problems that characterized Koch's understanding of tuberculosis itself. Although the tuberculin reaction has later been studied as an immunological problem,[10] it would be anachronistic to interpret Koch's thinking as immunological. His ideas about tuberculin and its effect had nothing to do with the discovery of the "bacterial allergy";[11] indeed, they were not even located within the immunological context that we now consider a prerequisite for understanding the subject. This chapter also aims to shed light on Koch's understanding of tuberculosis by contrasting his views on the etiology and pathogenesis of the disease with what he regarded as a therapy.

Methodologically, this approach is biographical in the largest sense. In keeping with the reflections of Ludwik Fleck, it surmises that the notion of (self) deception may hold the key to reconstructing the path that led Koch from tuberculosis to tuberculin. Here, particular attention will have to be paid to his private, professional, and intellectual situation in the late 1880s—the period between the discovery of the pathogen and the introduction of the medication—for this may help us understand his proceeding in developing and introducing the medication. In this context one may also ask whether Koch's attitude toward his own work changed as a result of his earlier successes.[12] It will be interesting to see how Koch's image of tuberculosis was brought about and shaped by innovative techniques of investigation—from the Petri dish and the methylene blue dye to the guinea pig. It can be assumed that these techniques were among the more important aspects of continuity in the research program. In this sense, the application of certain dyeing techniques and the use of certain animal species in pathological experiments were vehicles for assumptions about the characteristics of bacteria and the pathogenesis of the disease that had originally been gained with these techniques.

And finally, this chapter will endeavor to place Koch's research program into a larger historical context. This will entail such matters as gauging the impact of the competition from Paris and defining the originality of Koch's concept of tuberculosis in relation to the speculative pathology of infectious diseases current in his time. Although Koch's work on tuberculosis ended only with his death,[13] the present discussion intends to trace the path from the tuberculosis studies of 1882–84 to the introduction of an antituberculosis medication, tuberculin, in 1890.

## III.2. The Etiology of Tuberculosis

Koch presented the identification of the tuberculosis bacterium in 1882–84 in four stages. The first was a lecture to the Physiological Society of Berlin, delivered on March 24, 1882. It has become a fixture in the literature of bacteriological memorabilia, and the site of its presentation in the former Physiological Institute at Berlin University is marked by a commemorative plaque.[14] A second, shorter lecture on the same topic followed that same year at the Internal Medicine Congress in Wiesbaden. In early 1883 Koch published a brief review of some critiques of his work, and finally, in 1884, he published his lengthy article "Die Ätiologie der Tuberkulose," in which he described and reflected on his proceeding in great detail.[15]

In trying to account for the resonance of these studies, one must first refer to the subject of the investigation. Early in his career Robert Koch had worked on anthrax, an epizootic that only rarely affected humans.[16] Later, when he had become interested in traumatic infections, he looked into disease processes whose infectious nature was obvious.[17] But then, in turning to tuberculosis, he encountered a subject of investigation that was as portentous as it was controversial. The disease, which most often took the form of pulmonary tuberculosis (phthisis), was considered the most important infectious disease of the nineteenth century. The characteristic features of the "white death" were its constant incidence within the population and a usually lengthy chronic disease process in individual victims.[18] No wonder, then, that tubercular affections were among the great themes of contemporary medical research.[19] And this was before its bacterial etiology had established that all of these affections were but different forms of tuberculosis. For the time being they were seen as a group of tubercular diseases that followed different clinical courses and showed diverse patterns of pathological anatomy. Back in 1819 the French clinician Théophile Laënnec had postulated the connection among certain diseases exhibiting very diverse clinical manifestations, such as lupus, phthisis, scrofula, and so forth. The evidence he cited was the presence in all of them of characteristic nodules, the tubercles.[20] But this view was not generally accepted. After midcentury, German medical scientists were more likely to look for tubercular forms of diverse disease processes than to search for a single cause of these processes, thereby integrating them into the concept of tuberculosis. In the wake of Rudolf Virchow's studies of the pathological anatomy of the tubercle, tubercular processes began to be seen as metamorphoses of other diseases, and phthisis was clearly separated from other forms of tuberculosis.[21]

With respect to causes, the situation was similarly complex. Aside from contagion by—as yet hypothetical—germs, such factors as disposition, age, environmental conditions, heredity, and an unexplained connection with cancer were thought to play a major role. But none of these factors could claim the status of a necessary cause of the disease, such as we would assign to a bacterial etiology.[22] This is not surprising, considering that researchers were much more interested in transformation than in causation. In 1863 the pathologist Felix

Niemeyer was critical: "It is a confusing use of language that the term *tuberculosis* applies both to a peculiar form of something emerging and to a peculiar form of transformation."[23] Niemeyer, of course, was interested in the latter case, for instance, when diseases like cancer became tubercular.

Nonetheless, anyone who embarked on etiological research could refer to some existing studies that had sought to prove the infectious nature of the disease. In 1843 and 1863, respectively, Friedrich Klenke and Jean Antoine Villemin had shown that tuberculosis could be produced in animal experiments with the help of diseased tissue and hence had to be classified as contagious.[24] In 1877 Edwin Klebs had proposed considering the contagium of tuberculosis as a bacterium yet to be identified.[25] At Breslau University, a place of major importance in Koch's career, a whole team of researchers was working on this topic. In 1879, Carl Weigert had launched the idea that the unity of various forms of tuberculosis should not be attributed to the pathological anatomy of the diseased tissue but to their etiology.[26] Julius Cohnheim and Carl Salomonsen had confirmed Villemin's experiments and proposed that the etiological demonstration be conducted on the basis of animal experiments.[27] Thus someone who, like Koch, attempted to demonstrate a bacterial etiology for tuberculosis could draw on a rough outline of such a demonstration. Nonetheless the demonstration of a bacterium as a necessary cause of the disease would have dramatic consequences for the complicated edifice of the various tubercular processes—even if in itself the idea of tuberculosis as an infectious disease was not new. This demonstration would be the first instance of a proven bacterial etiology of a human infectious disease. It would therefore be the model of a new concept of infectious disease in which the disease was identified by a bacterial etiology reconstructed in the laboratory rather than by clinical observation.[28] And finally, one could expect—given the major impact of tuberculosis itself—that such a demonstration would command widespread attention.

Koch did not feel that the significance of the discovery lay in the fact that he had used entirely new methods to hunt down the tubercle bacillus. To be sure, he emphasized new problems that had cropped up, such as the fact that the "tried and tested staining methods" had failed in this case.[29] All in all, however, "it was best to use the same path of investigation that had proved to be the most expedient on earlier occasions."[30] The identification of the tubercle bacillus was a variation and expansion of "tried and tested methods" applied to a new problem. In his publications, Koch constantly referred to tried and tested ideas, as when he explained his method of identifying pathogenic germs by citing the case of anthrax, on which he had worked.[31] The speed with which this work was carried out is indeed impressive. Barely eight months passed between the beginning of the investigations in August 1881 and the lecture in March of the following year.[32] It should be noted, however, that Koch was no longer working by himself, as he had done in his earlier work, for in 1880 he had become head of a rapidly growing working group at the Imperial Health Office (IHO).[33] In the two years since then, the group had developed such fundamental tools as pure cultures and solid culture media, which were used in the investigation.

In his writings Koch had the rhetorical skill to match the awe-inspiring dimension and complexity of his subject with the extraordinary difficulties of his investigation. "My investigations," he wrote, "initially also used the traditional methods, but these did not give me any information about the essence of the disease."[34] He then went on to describe new difficulties that arose, particularly in the identification, but also in the culturing of the pathogenic agent and in animal experiments, difficulties for which new solutions had to be found in every case.

To begin with, the tubercle bacilli proved to be considerably smaller than other known pathogenic agents—even under the microscope they were not necessarily visible without appropriate staining. Originally, in his anthrax studies, Koch had worked without staining. While working on traumatic infections starting in 1877, he had used the staining techniques he had learned from Carl Weigert to differentiate bacteria from surrounding tissues and to prepare specimens for microphotography.[35] The microscopic examination of tubercular tissues, however, was a different matter: at first glance one did not see anything that one could recognize as a bacterium and then stain.[36] Only the use of methylene blue brought out something. "After the cover slip preparation had been treated with this dye solution for 24 hours, one could begin to see in the tubercle mass very delicate rod-shaped structures."[37] The next step was to make these structures stand out from the surrounding tissue, which required a further development of the staining technique. A second, brown coloring agent called vesuvin affected only the tissue. The "rods" now appeared blue, the rest of the tissue brown. Because of their blue color, these structures could be differentiated from most other bacteria. "Under the microscope all the components of animal tissues, particularly the cell nuclei and the decomposition products showed in brown, whereas the tubercle bacilli were clearly blue. All the other bacteria that I [Koch] have tested in this connection, except the leprosy bacilli, also take on a brown coloring when treated with this staining technique."[38] This procedure, which Paul Ehrlich was soon to replace with a much more effective one,[39] made it possible to find all the rods in tubercular tissue and to describe their typical disposition. The bacteria were arranged in a characteristic fashion, usually forming "small, tightly packed groups, often occurring in bunches." Their appearance mirrored the disease process: "Where the tubercular process has just started and is progressing rapidly, bacilli occur in large quantities." Once "the climax of the tubercle eruption has passed, the bacilli become more rare . . . In very slowly progressing tubercular processes, the interior of the giant cells is usually the only place where the bacilli are found."[40]

The fact that Koch could make the tubercle bacilli visible only by staining them not only validates, as his biographer Thomas Brock pointed out, his "firm belief"[41] in the parasitical nature of tuberculosis, it also obviated the need to compare his bacteria to those of other researchers. Since so far no one had used a similar staining technique, and since, as we just saw, the bacteria remained invisible unless they were stained, these researchers must have seen something else.[42]

Somewhat in contradiction with the description of the methods, which emphasized their continuity, double staining was thus more than a simple technical innovation. Even more than other microorganisms Koch had studied, the tubercle bacilli were artifacts of the investigatory process. A faulty application of the staining procedure could, as Koch stressed, cause other components of the preparation to appear stained in blue.[43]

Given the thoroughness with which Koch discussed the staining technique, it is surprising that he passed over two related problems without comment. One was that the Königsberg medical researcher Paul Baumgarten had almost simultaneously observed the same forms under the microscope and described their connection to pathological-anatomical changes—without using a staining procedure.[44] Koch's scant notes about his own work indicate that his investigations were still in full swing in March 1882. A whole series of experiments that Koch described in his detailed account of 1884 were only carried out after March 1882.[45] Inoculations involving pure cultures had been successfully tried only a few weeks before March 1882.[46] In May 1882 a confrontation between the experiments of Koch and Baumgarten seems to have taken place in Berlin; Baumgarten acknowledged that although the structures identified by both of them were identical, Koch had demonstrated their etiology.[47]

Koch also passed over the problem of being unable to photograph his preparations. After all, he saw himself as a pioneer of microphotography and had only

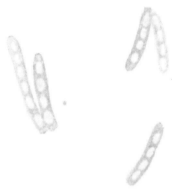

Koch's tuberculosis spores. From
Koch, *Die Ätiologie der Tuberkulose*
(1912 [1884]), plate 29, figure 47.

recently contrasted this technique—which he called the only "purely objective vision, uncontaminated by any preconceptions"[48]—with the subjective views that were expressed in drawings. The fact that his preparations could only be published as drawings was a problem with which he dealt only indirectly.[49] Not that he had not made every effort in the area of photography. Loeffler recalled that double staining was developed in connection with attempts to photograph microorganisms stained with methylene blue.[50]

All this assumes particular significance in light of Koch's 1884 article, in which he postulated spores—resistant permanent forms of the tubercle bacillus. According to today's state of knowledge, these spores do not exist. But in tuberculosis research they brought back an object that had had major importance in the early history of medical bacteriology.[51] The demonstration of this stage in the life cycle of certain bacteria had been a decisive step toward proving the stability of bacterial species. Ferdinand Julius Cohn had introduced this concept into bacteriology in the 1870s. In Koch's anthrax study, for instance, which closely adhered to Cohn's microbiology, the demonstration of enduring germs had served to explain the (apparent) temporary disappearance of these bacteria.[52] The spores of the tubercle bacillus met the same need: Koch described them as "the lasting form needed for the conservation of the species"[53] and used them to explain the long-lasting infectious quality of dried sputum.[54] The spores also provided an explanation for the infectiousness of the caseous mass inside the tubercles, where bacilli frequently could no longer be found.[55] But then, according to Koch, spores could not be stained and were hence practically invisible: "Since so far there are no means of staining the spores of the tubercle bacilli in any way, their presence after the disappearance of the bacilli only manifests itself in the infectious quality of the caseous substance in which they are embedded."[56] Koch provided a rather unenlightening illustration of these spores, describing them as "of oval shape" and distributed in clusters of two to four over the length of a bacterium.[57] In the face of these contradictory statements, one must focus on the considerable importance of the spores for the argumentation itself. Spores made it possible to postulate the presence of bacteria precisely in places where bacteria could not be demonstrated. Moreover, the development of the bacteria into spores mirrored the disease process. The transition to a spore stage, expressed in the relative quiescence of caseation within the tubercles, stood in for the completion of the disease process.[58] Koch's pointed references to the tried-and-true methods at his disposal thus also glossed over problems. The spores might have gotten him into considerable trouble. But the fact was that the complete life cycle of the bacterium could only be documented under the assumption that spores existed.

Most of the problems, however, were related to the identification of the pathogen. Difficulties related to cultivation and inoculation, by contrast, could actually be dealt with by modifying existing methods. Koch's first success with animal experiments was the demonstration that only the transmission of bacteria-containing tissues could cause tuberculosis in experimental animals. These animal experiments also demonstrated the identical nature of the different forms of tuberculosis and of the experimental tuberculosis obtained by inoculation, as well as the sameness of tuberculosis in humans and in susceptible animals.[59] To make these points, Koch reproduced the experiments of Villemin and of his Breslau colleagues Cohnheim and Weigert. Here guinea pigs proved to be the ideal experimental animals: outside the laboratory they did not come down with spontaneous tuberculosis, but unlike mice, they turned out to be highly susceptible to inoculated tuberculosis.[60] Moreover, the course of the disease in guinea pigs was rapid and typical. The Koch papers include a list of relevant experiments and documents the transition from the use of diseased tissue to cultures in the spring of 1882.

| Guinea pigs inoculated with: | | | Number of days until death: |
|---|---|---|---|
| 1/11/81 | Tub. of | meninges (& diaphragm) | 51, 52, 52, 62, 65, <u>76</u> |
| 2/11/81 | " " | lungs, liver meninges | 40, 44, 47, <u>75</u>, <u>75</u>, <u>75</u> |

| | | | | |
|---|---|---|---|---|
| 14/11/81 | " | " | Caseous pneumonia, meningitis | 53, 93, 94 |
| 19/11/81 | " | " | Lungs, liver, spleen, kidneys | 45, 52, 58, 60, 70, 70 |
| 2/12/81 | " | " | Miliary tub. of the lungs | 43, 43, 54 |
| 18/10/81 | " | " | Caseous infiltr. of the lungs | 39, 52, 61, 78 |
| 26/11/81 | | | II gen. | 39, 65, 83, 83, 83 |
| 7/11/81 | | | Child, caseous pneum. intestinal tub. | 89, 89, 89, 92, 92, 92 |
| 14/11/81 | | | Phthisis, intestinal abscesses, peritonitis | 65, 75, 88 |
| 2/12/81 | | | Caseous pneumonia | 61, 61, 61 |
| 10/1/82 | | | Caseous bronchitis, intestinal abscesses | 34, 56, 62, 66, 67 |
| 24/1/82 | | | Caseous pneumonia, caverns | 49, 57, 60, 102 |
| 6/1/82 | | | Sputum | 57, 74, 78 |
| 12/1/82 | | | Sputum | 33, 37, 63, 71 |
| 24/1/82 | | | Sputum | 60, 104, 146 |
| 8/3/82 | | | Dried sputum from 6/1 | 33, 33, 33 |
| 18/11/81 | | | Tubercul. of the uterus | 58, 82, 83, 83, 83, 83 |
| 30/1/82 | | | Elbow joint | 106, 156 |
| 27/1/82 | | | Scrofulous gland (not casefied) | 119, 128, 179, 184 |
| 27/2/82 | | | Scrofulous gland (caseous pus) | 60, 70, 87 89 |
| 7/4/82 | | | Lupus (hypertrophicus) | 44, 48, 58, 60 |
| 20/2/82 | | | Culture from lupus XIII gen. | 33, 36, 41, 46, 46 |
| 9/6/82 | | | Culture M 1/11 (intraperitoneal) | 10, 10, 11, 13, 14, 15 |
| 2/8/82 | | | Culture M 6/2 (intraperitoneal) | 13[61] |

To prove that the tuberculosis bacteria and not other possible components of the tubercular material caused the disease, it was necessary to grow pure cultures and to use them to trigger the disease in animal experiments. The production of such pure cultures faced two difficulties. One involved the special growing conditions of the bacteria, for their cultivation required temperatures of over 30 Centigrade. This caused Koch's solid, gelatin-based culture media to liquefy.

The other difficulty was that the extremely slow growth of the bacteria entailed the danger of contamination and uncontrolled growth in the cultures. For the first problem, a culture medium made from coagulated blood serum eventually provided the answer. For the second, the solution was stringent hygiene in conjunction with the technique of the animal passage with which Koch had become familiar in his work on traumatic infectious diseases: given that preparations from human cadavers were usually contaminated, the team inserted a passage through guinea pigs between that preparation and the culturing. Thanks to the rapid course of the disease in these animals, this step made for much better, that is, more homogeneous, seed material for the culturing experiments.[62]

Once the tubercular material was placed on the culturing nutrient, it had to stay on the surface of that transparent substance. Unless it showed signs of growth earlier than ten to fifteen days after its preparation, it could then be further cultivated as a pure culture. Earlier growth or sinking into the nutrient were signs of contamination.[63] There were several other ways of checking the identity of the cultures: a macroscopic examination showed the bacteria on the culturing substance as "very small dots and dry-looking little scales";[64] the microscopic one revealed "the peculiar, very delicate" form of the cultures, whose "manifold, snake-like convolutions . . . are reminiscent of intricate lettering."[65] These characteristics referred both to the nutrient prepared by Koch and to the bacteria themselves.[66] But the staining method he had developed for the identification of bacteria provided an independent means of verification.

In the final stage of the investigation, Koch conducted infection experiments with pure cultures. These involved inoculating large numbers of animal species in a wide variety of ways, as well as experiments with inhalation and feeding.[67] In animals susceptible to tuberculosis, the pure cultures brought about tuberculosis.[68] Koch described the pure cultures he used for this purpose with obvious pride: "It is therefore no exaggeration to claim that in most experiments cultures of absolutely pure bacilli were used."[69] The important aspect of these experiments was not so much that symptoms of tuberculosis appeared at all, that bacteria were found in autopsies. The decisive point was that the tuberculosis produced with pure cultures manifested itself in the same way "as if fresh tubercular substances had been inoculated."[70] Koch had reproduced the inoculated tuberculosis of Villemin, Cohnheim, and other researchers with his pure cultures. The pathogen, which had been the missing link in these investigators' experiments, had thereby been identified.[71] In 1882 Koch summarized his study in a sentence that has become famous: "All these facts justify the statement that the bacilli occurring in tubercular substances are not just side effects of the tubercular process but their cause, and that we must recognize the bacilli as the actual tubercle virus."[72]

In the presentation of his work in 1882, Koch placed particular emphasis on the bacterial etiology of tuberculosis, a thesis that aroused few objections. His extremely thorough procedure seemed to exclude any doubt. Koch did not encounter sharp criticism because he was careful not to contradict other, nonbacterial factors in the etiology of tuberculosis too clearly. Thus he expressly acknowledged the importance of disposition, heredity, and social conditions.[73] Even Rudolf Virchow, who took a dim view of the validity of medical bacteriology, could not deny that the tubercle bacterium was etiologically significant.[74] Koch's reply to his critics in 1883 therefore seemed to be more a matter of setting them straight than of discussion. He dismissed their objections in an ironic tone and reproached them for faulty technique and even ignorance: "Sternberg could not find the bacteria and therefore felt duty-bound to deny their existence. I hope that by now he realizes his error."[75] For Koch the bacteria, once he had found them, were self-evident. In 1882 he therefore still recommended bacteriological testing as a simple diagnostic tool, since "anyone who has once seen the manipulation can easily perform the staining of bacilli."[76]

The relative lack of objections also had to do with the fact that Koch had, after all, redefined only the pathogen of tuberculosis, not the disease itself. His statement that whereas "physicians [see] phthisis as a noninfectious disease arising from constitutional anomalies,"[77] now "it has become possible to draw the boundaries of the diseases that can be designated as tuberculosis, which could not be done reliably in the past,"[78] is somewhat exaggerated. What he had actually done was add a bacterial pathogen to a concept of tuberculosis as an infectious disease that had been developed by thinkers ranging from Bayle by way of Laënnec and Villemin to Cohnheim. The boundaries of the disease that could be drawn with the help of a bacteriological diagnosis were not very different from those based on the biological and anatomical marker of the tubercles that had been in use ever since Laënnec. And the identical nature of human and animal tuberculosis had already been the subject of Villemin's experiments.[79] The consensus even extended to unsolved questions: concerning the disease of the glands (scrofula) that Villemin hesitated to characterize as tubercular, for instance, Koch too was initially unable to make a firm determination.[80] In opposition to the views of the pathological-anatomical school—exemplified by Virchow, who continued to consider phthisis an independent entity—Koch helped promote an already formed concept of tuberculosis as an infectious disease.[81] At the core of this concept was the definite rejection of clinical appearances and a new reliance on bacteriological findings. These, rather than pathological anatomy, now furnished the crucial evidence: "As for ... the confusion between nontubercular nodules and true tubercles, nothing is easier than to avoid it: the true tubercles are infectious and contain tubercle bacilli, the false ones do not."[82]

Koch presented this first demonstration of the bacterial etiology of a human disease as the result of the new methods developed by him and other bacteriologists. Given this scenario, his reference to the use of tried-and-true methods and the extended discussion of these methods in the 1884 article, which is considered the most important source for Koch's postulates, are part of an enhanced rhetoric.[83] Koch did not present his discovery as the result of an eight-month investigation of the subject—he gave no details about this work[84]—but referred back to his anthrax studies when he summarized his findings: "As for recognizing its etiology, tuberculosis is very similar to anthrax. The tubercle bacilli have the same connection to tuberculosis as the anthrax bacilli have to anthrax."[85] Although Koch is said to have told Loeffler that he did not count on a rapid acceptance of his findings, the rhetoric of his texts betrays a far more canonical view of his own work. It is obvious that he intended to bring together his findings about individual infectious diseases in a general bacterial theory of infectious disease. If, as modern authors still do, one sees medical bacteriology as marking the transition between traditional and modern medicine,[86] it becomes clear that Koch already thought of his work on tuberculosis as a turning point. He fully expected that the newly gained knowledge about the etiology of tuberculosis would also yield new ways of looking at all infectious diseases, and that the methods that had proved their worth in the investigation of tuberculosis would also be useful in the work on other infectious diseases.[87]

## III.3. Bacteria and Disease

Beyond the straightforward praise bestowed by some of Koch's biographers, the more recent history of medicine has particularly appreciated two aspects of his work on the etiology of tuberculosis. One is that it marks the beginning of the triumphant progress of bacteriological hygiene in public health and the medicalization of entire societies in the last third of the century.[88] Once the bacteriologists had defined the "unpolitical reason"[89] of the bacteria, it became possible to take politics out of the great epidemics and to expand medically legitimized social controls with the help of physicians, health authorities, insurance companies, and so forth. In this respect, the convergence in the relation between French society and Pasteur's microbiology as analyzed by Bruno Latour is comparable to the dominant influence that bacteriological hygiene began to exert in Germany by the mid-1880s.[90] Placing Koch into this context can surely be justified by his own assessment of his tuberculosis studies, for the author insistently advocated their use by public health services.[91] From this perspective, the tuberculosis studies of 1882–84 appeared to be more closely related to the cholera studies begun in 1884 than to his later work on tuberculosis. Together, tuberculosis and cholera studies marked the beginning of the era of bacteriological hygiene.

The second aspect addresses the theory of medicine and is related to the concept of Koch's postulates. Historians have described how Koch's bacteriology, which was committed to the concept of necessary causes for infectious diseases, could win out over an older concept committed to the findings of pathological anatomy.[92] This victory also entailed a shift of emphasis in the understanding of disease, since the place of internal organic processes was now taken by their external causes. In its bacterial variant, etiology, a concept that had hitherto referred to any number of disease-causing factors, ranging from climate and heredity to pathogenic germs, became a central concept of medicine in the late nineteenth century. Here Koch's tuberculosis studies marked the breakthrough to an understanding of human infectious diseases oriented toward bacterial etiologies. K. Codell Carter, the foremost authority in this area, considers the 1884 article to be the most detailed and conclusive formulation of Koch's postulates.[93] In this interpretation the tuberculosis studies that Koch published in the years 1882–84 had little connection with later work on the same subject. Instead, they appear to be closely related to Koch's other etiological investigations, such as those on anthrax and traumatic infections. These studies conducted between 1876 and 1884 thus add up to a first complete model of infectious disease.[94]

However helpful this approach may be for understanding medical bacteriology, it does raise a question. Does it not reduce the dimension of Koch's work on tuberculosis to focus on the etiological level and dismiss Koch's discussion of pathogenesis as the description of barely sufficient causes of disease?[95] Koch's demonstration that tubercular processes were identical in different animal species was more than just a description of the necessary causes of the disease. Rather, Koch explicitly linked it to the thesis that there is an analogy between the behavior of the bacteria and the course of the disease. He did not consider bacteria only as necessary causes of disease, believing that their number, movements, and constellations also accounted for the course of the disease. Koch had stated this as early as 1878 and repeated it in 1884: "Moreover, it was necessary to find out about their [the bacteria] environment, their appearance in the different stages of the disease, and similar circumstances, all of which already pointed, more or less convincingly, to a causative connection between these structures and the disease."[96] The premise was that bacteria, regardless of the infected species, always produce the same pathological and microscopically verifiable structures. In the case of tuberculosis, these were mainly the tubercles. This concept is related to the importance of bacterial specificity in Koch's thinking. The ability to reproduce a specific disease by means of experiments with specific bacteria founded a relation of mutual definition, in which the stability of bacterial species was demonstrated by their constant pathogenic effects, just as the diagnosis of a malady could be arrived at by identifying the bacteria in the diseased tissue.[97]

However, Koch treated questions of pathogenesis differently than problems of etiology. Especially in the 1884 study, he described invariable connections between the behavior of the bacteria and the course of the disease in addition to the causal chains that produced the necessary causes of the disease. These descriptions develop, albeit in an unsystematic and slightly metaphorical form, Koch's views on the pathogenesis of tuberculosis. Here the quantity, distribution, and constellation of the bacteria function as the analog of the disease process.

A fundamental feature of Koch's view is the assumption that the healthy body is completely free of bacteria, so that "the appearance of the tubercle bacillus marks the beginning of the tubercular process."[98] And although the number of bacteria present is not without importance, in principle a single one is sufficient. Thus Koch explains a certain form of miliary tuberculosis by claiming that "a single germ of infection, i.e., a single bacillus, had been deposited by the bloodstream in that particular site."[99] In this view of the matter, invasion, infection, and outbreak of the disease practically occur simultaneously. Koch accounts for differences in the course of the disease by the character of the infected tissue, which might, for instance, lead to the formation of caverns in the lungs. He describes the outbreak of phthisis as follows: "Originally only individual or small clusters of bacteria reach the lungs, and because of their slow growth, these will soon be enclosed in a cell infiltration [. . . however, they] do not perish in the cell infiltration but rather cause necrosis and caseation in the center of the cell mass, just as they do in the miliar-tubercle."[100] What Koch formulates here is clearly a simple invasion model of infectious disease, as can also be found in Klebs. This model became dominant among bacteriologists and in the contemporary popular imagination.[101] Within the bacteriological style of thinking of the 1880s, it seemed inconceivable that pathogenic germs were present in the healthy organism.[102] Accordingly, the defense against disease was a matter of fighting the pathogen at the margins of the body. Koch himself devoted a great deal of space to the intrusion of bacteria into the body through the respiratory organs, the digestive tract, or wounds, as well as to the question of their dissemination through dust, sputum, or food.[103] Comparable questions were soon among the most intensely studied, with the result that, beginning with Koch himself, the sick were often seen as transmitters of infection and, increasingly, as a danger to their fellow citizens.[104]

The body, conceived as the opponent of such an invasion, is essentially passive and actually serves as a kind of culture medium to the bacteria, which play the active part in the disease. In a striking analogy to the cultures in his laboratory, Koch analyzed the caseation inside the tubercles as resulting from the exhaustion of a nutrient substance. Initially the bacteria find ample nourishment: "The smaller and younger the nodules, the more numerous the bacteria that could be found; they were particularly dense around the center."[105] When the cells inside the tubercles dissolve, the bacteria, now deprived of nourishment, perish as well. "What is left is a uniform mass, which no longer responds to nucleus staining and in which all the originally present cells have died off. This mass forms what used to be considered the essential part of the tubercle, the carrier of the infective substance, namely, its caseous center. However, this caseous substance usually contains very few tubercle bacilli . . . very soon the bacteria too undergo further changes; either dying off or entering the sporulation phase."[106] Fresh, bacteria-laden tubercles are much more likely to disseminate the disease than caseated and almost bacteria-free ones. Indeed, since bacteria have no motility of their own, their dissemination throughout the body must occur passively through the growth of the colonies or by being carried along in the bloodstream. Yet the set of metaphors Koch uses to describe their dissemination throughout the body tends to characterizes them as the active part. "It appears very likely that the increase in the number of bacteria also causes their behavior vis-à-vis the nuclei to become more active." The bacteria "crowd in" at the edge of the cell, "push themselves between the nuclei," and in facing the cell nuclei, assume a positively military formation, so that cell nuclei and bacteria "keep each other in check."[107] Depending on the intensity of the tubercular process, giant cells are finally "exploded" by the bacteria or left in place as ruins of their former selves, resembling the "extinct craters of volcanoes."[108]

Once the infection has taken place, the cells have no chance to escape necrosis, and the body has no prospect of healing. An organism infected with tuberculosis will perish from it. This assumption of Koch's is clearly visible in his interpretation of an animal experiment: 0.5 cc. of pure culture tubercles were injected into a dog's abdominal cavity. Contrary to expectations, the animal recovered after showing initial symptoms. "This is the only case of tuberculosis in animals where I have seen a transition to healing."[109] But in fact Koch had killed a large number of his experimental animals and thus could not really make such a statement on the basis of his material. As a rule, animals that had survived the infection for some time were killed and autopsied, whereupon information about the presence of the disease was furnished by pathological anatomy.[110] That Koch thought of the case of this dog as the only one in which healing had ever been observed seems to indicate that for him, the lethal outcome of the disease was a foregone conclusion. His text never discussed such phenomena as spontaneous healing or delayed disease processes—and for guinea pigs, which usually succumbed to the disease in short order, such questions were indeed irrelevant. The body appeared to have defenses only against the entry of the pathogen.

Koch considered his understanding of the disease as bacterial activity, which I have outlined here, to be perfectly adequate. He did not explicitly state this, but his discussion of predisposition and heredity in the 1884 article points in that direction. In 1882 he had definitely acknowledged the significance of these factors. Now he treated them as residual categories for phenomena that had not (yet) found a place in a bacterial etiology and pathogenesis of the disease—they were available to deal with "facts which, difficult or impossible to interpret, compel us to stay with the assumption of a predisposition for the time being."[111]

Koch's views on the etiology of tuberculosis, then, were related to an understanding of the pathogenesis of the disease, and were indeed embedded in far-reaching assumptions concerning the nature of infectious disease itself. As a medical bacteriologist he saw infectious diseases as the expression of the life cycle of the pathogens rather than as a form of life in the human organism under changed conditions, as the leading pathologist of the time, Rudolf Virchow, believed.[112] To this extent, medical bacteriology was also speculative pathology,

*Tuberculosis and Tuberculin*    87

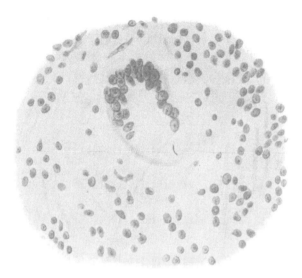

Giant cell with cell nuclei and tubercle bacterium. From Koch, *Die Ätiologie der Tuberkulose* (1912 [1884]), plate 25, figure 29.

for it combined experimental information about the behavior of bacteria with the ontological concept of an autonomous disease entity—which was embodied in the pathogen.[113] However, this thesis remained implicit in Koch's publications of the years 1882–84, for at that point he related the pathogenesis of the disease exclusively to the characteristics of the pathogen. The affected organism added nothing essential, except for the notoriously unreliable clinical picture.

It now became possible to look at the disease in isolation from the patient, and its reproduction by animal experiment took the place of the clinical picture. Patients appear in Koch's work only as corpses that are autopsied and furnish the initial material for culturing experiments.[114] This is clearly visible in the summary of the experiments Koch wrote down as he finished the etiology study. The top of the list shows the experiments that had reproduced a wide variety of human forms of tuberculosis in guinea pigs. One of these experiments, in which a patient who had died of miliary tuberculosis had furnished the initial material, is recorded as follows: "19/11/81 (tub. of lungs, liver, spleen, kidneys, A. Homann, inoculated by Dr. L.). 6 guinea pigs (cage LII) died 3/1, 10/1, 18/1, 28/1."[115]

The material harvested from the infected animals could then be used in further experiments. In the summary of the experiments reproduced below, the

## 88 Laboratory Disease

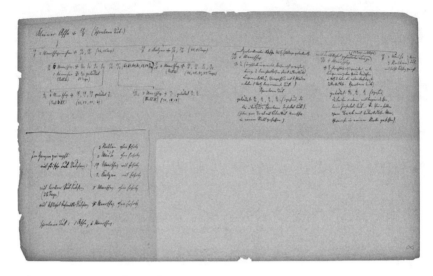

Schematic summary of animal experiments with tuberculosis, 1881–82 (transcription on opposite page). The groups of figures noted in decimal style in the original refer to the date of death. The lines indicate from which cadaver the material for the following infection was taken. The experiments designed to examine the effectiveness of different kinds of tubercular material and the susceptibility of different species are summarized under the dividing line. It is interesting to see the mention of additional laboratory infections. RKI/Koch papers, "versch. Tuberkulose— S. 1881/82" as/w2/001).

transmission of the pathogen from culture medium to culture medium is marked in a fashion reminiscent of a genealogical chart, regardless of whether it referred to experimental animals or to pure cultures. Clinical symptoms, indeed the patients, have almost completely disappeared from this depiction of the disease, replaced by a series of culturing processes of the always identical pathogen. This process shows the surrender of explicit and systematic discussion, compounded by a rationalistic simplification of the problem of disease as a whole.[116]

In drawing conclusions from the cause of the disease to the disease itself—as Koch did in his studies—he drew not only from his experiments but also from his analogies of the behavior of the bacteria that had microscopic symptoms of the disease. Koch was quite aware of the pathogenetic implications of his etiological thinking and emphasized that it "cuts deeply into the existing systems and causes us to break with old and cherished traditions."[117]

Edwin Klebs, who stated such claims explicitly and was given to uncompromising judgments, is today regarded as a kind of extremist in bacteriological

Kleiner Affe † 29/8 (spontane Tub.)

29/8 2 Meerschweinchen † 14/10, 18/10 (46,48 Tage)

14/10 6 Meerschw. † 3/12 3/12 6/12 6/12 13/13 15/12 (50,53,60,62,54 Tage)

6/12 4 Meerschweinchen † 28/1, 28/1, 30/1 getötet 4/2<br>(Stall XX) (43, 53, 55, 60)

15/12 3 Meerschw. †27/1 getötet 4/2 4/2<br>(Stall IV) (38, 50, 50)

Mit getrockneter Milz (56 Tage getrocknet)<br>
26/10 5 Meerschw<br>
† 2/2 (Impfstelle unverändert, Drüsen nicht vergrößert,<br>
Lunge [unlerserlich] mit der Pleura<br>
in Leber und Milz beginnende Tub)<br>
Sputum Tub<br>
getödtet: 4/2, 4/2, 4/2, 4/2 (1 gesund, die<br>
übrigen beginnende Inhal.tub).<br>
(Haben vom 2/12 ab mit tuberkul Meerschw.<br>
In einem Stall gestanden),

Mit in Alkohol getränkter Lunge (57 Tage in Alkohol)

27/10 4 Meerschw.<br>
† 8/1 (Lymphdrüsen nicht verändert, in<br>
den Lungen eingelegen kaum Knötchen<br>
in Milz u. Leber ausserd. Anfänge d.<br>
Tuberkulose. Sputum Tub.)<br>
getödtet: 6/2, 6/2 6/2 (1 gesund, die<br>
beiden anderen mit beginnender [unlerserlich] Inhal.tub.<br>
Die Tiere hatten vom 2/12 ab mit tuberkulösen Meer-<br>
schweinchen zusammen in einem Stall geschlafen).

29/8 3 Mäuse } keine<br>
3 Ratten } Tub<br>
Mit frischem<br>
Sputum<br>
geimpft

---

| | | | | |
|---|---|---|---|---|
| Im Ganzen geimpft: | 3 Ratten | ohne Erfolg | mit trockener tub. Substanz | 5 Meerschw. ohne Erfolg (56 Tage) |
| | 3 Mäuse | ohne Erfolg | mit Alkohol behandelter Substanz | 4 Meerschw. ohne Erfolg |
| Mit frischer InhSubstanzen | 19 Meerschw. | mit Erfolg | spontane Tub | 1 Affe, 6 Meerschw. |
| | 2 Katzen | mit Erfolg | | |

thinking.[118] And while Robert Koch did not engage in verbal attacks as Klebs did, it should be clear by now that his implicit assumptions betrayed an equally radical bacterial reductionism.

## III.4. The Road to Tuberculin

Historians have thoroughly investigated the consequences of Koch's work for medicine and public health. By contrast, the consequences of this work for its author are still largely unexplored. This is particularly the case for the connection between the tuberculosis studies of the years 1882 to 1884 and the research on tuberculin of later years. Koch had not published anything on tuberculosis in six years when he announced his antituberculosis medication at the Tenth International Congress of Medicine in Berlin.[119] The period in between was not necessarily one of productive research activity. On the contrary, it shows signs of a personal and professional crisis. Following the pursuit of the cholera pathogen, on which he embarked immediately after the tuberculosis study, there was a lull in Koch's publishing activity. Indeed, there are indications that for a time, he stopped doing research.[120]

To some extent this was due to external circumstances. In 1885 Koch, with the support of the Prussian Ministry of Culture, had hoped to leave the IHO and become director of an imperial bacteriological institute.[121] This plan did not materialize, however, and the scientist found himself occupying a professorship for hygiene and bacteriology at Berlin University, a position that had been created over the objections of the medical faculty.[122] The large-scale training and postdoctoral program pursued there, along with the *Zeitschrift für Hygiene*, which Koch founded with Carl Flügge, did much to promote bacteriology as a discipline,[123] but for Koch it was an unhappy situation. His position within the faculty was and remained precarious, and the daily obligations of teaching, giving examinations, and so forth were onerous. Added to this were personal problems, which in 1890 finally led to Koch's separation from his first wife.[124]

In a letter to Flügge of December 26, 1888, he lamented his fate: "In the last month I was occupied every day from morning to evening with training courses for government and school administrators, and more of these courses are scheduled to begin on January 3. Our institute has now become so overburdened with these kinds of courses that there is practically no time for other pursuits."[125] At the same time Koch was also wrestling with a problem of research strategy. Until then his successes had been related to spectacular demonstrations of pathogen involvement, which—through the control of the identified pathogens—seemed to open up a therapeutic perspective. For the time being, however, the practical applications of his work were limited to nonspecific measures of preventive hygiene, such as disinfection. Specific antibacterial therapies simply were not in the picture. Koch had produced nothing that could compare to Pasteur's preventive inoculations, which could indeed be considered specific applications of microbiological knowledge.[126] Koch's work in the 1890s did not continue to pursue the topic of the 1880s; the period 1885–90 does not appear as a simple interruption but as a decided break with previous work. Some historians have interpreted this as a downright tragic element in the career of a scientist who could never find his way back to the spectacular successes of the years 1875 to 1885.[127] Others have stressed the originality of the later studies with their more epidemiological than etiological orientation.[128]

Beyond the obvious breaks in Koch's biography, there were continuities across this period that still remain to be explained. In view of the striking fact that Koch worked on tuberculosis both before and after the crisis described above, one has to ask whether and how tuberculin was connected to his earlier work. Was it an attempt to put the knowledge of the disease that he had gained in his etiological research to therapeutic use? Koch himself was convinced that this could be done, and he believed that the discovery of the pathogen would almost immediately lead to the prospect of a therapy: "But in the future the fight against this dreadful scourge of human kind will no longer be against an indefinable something but against a tangible parasite," was how he put it in 1882, when he also saw "particularly favorable prospects for success."[129] He was certain that "by the time everyone is convinced that tuberculosis is a highly sensitive infectious disease, the question of how it can best be combated is bound to be discussed."[130]

The detailed study of 1884 already contains several reflections about antituberculosis specifics. The first idea that suggested itself was the attenuation of tubercle bacilli, even though Koch was skeptical about the phenomenon of virulence and was also involved in a controversy with Pasteur concerning the Frenchman's anthrax vaccine.[131] To be sure, a vaccine produced according to the Pasteur method would not have been a therapeutic; however, it would have been a specific medicine that demonstrated the importance of identifying the pathogen. But there were several arguments against this proceeding. The outcome of the experiment on the dog described above seemed to demonstrate that surviving an infection did not confer immunity. A second inoculation of the recovered animal with a fourfold dose yielded the expected result and killed the dog.[132] Moreover, the pure cultures Koch had successively grown in his laboratory over two years "did not show the slightest variation in their properties, and especially their virulence."[133] And finally, Koch could also cite failed attenuation experiments conducted at the IHO.[134]

A more promising path seemed to lead to the work on the life cycle of bacteria conducted in Koch's laboratory and the elaboration of his own studies of disinfection.[135] Immediately following the completion of Koch's etiology study, two collaborators at the IHO had studied the disinfection of discharged phthisic phlegm.[136] Koch conducted similar studies with Georg Gaffky. Substances that impeded the development of bacteria in vitro might offer the possibility of striking the disease itself in the body.[137] To be sure, the substances used in this endeavor, such as arsenic, had been part of the innumerable medicinal tuberculosis therapies that had been in use for some time. As expected, they proved ineffective in animal experiments.[138] Additional experimental therapies that Koch and Gaffky had tried out in patients in collaboration with the clinician Oscar Fräntzel of the (Berlin) Charité showed no results and were discontinued. Twenty-seven patients had been treated with such substances as camphor, creosote, or carbolic acid, all of which had been effective in vitro. Known and used in medicinal therapies for some time, these substances now proved completely ineffective.[139] Contrary to his earlier announcement, Koch did not publish the results of these experiments. In 1886 the IHO did report on the end of Koch and Gaffky's unsuccessful experiments.[140]

In light of the subsequent events, it is reasonable to assume that Koch continued his search for a therapeutic or supervised such a search.[141] In 1887 Georg Cornet, one of his collaborators at the Institute for Hygiene at Berlin University, conducted animal experiments to review several commonly used antituberculosis therapies.[142] In keeping with Koch's research strategy, he defined the desired effect of modern medications as a kind of internal disinfection designed to "impregnate the tissue itself with the antiseptic substances, changing it into an unsuitable culture medium—sterilizing it, as it were."[143] In the animal model of the guinea pig, however, commonly used substances, such as different forms of arsenic and the tar derivate creosote, proved ineffective. Aside from these chemicals, which were often used for therapeutic or disinfective purposes and had in part already been examined in Koch's experiments of 1881–82, Cornet also tested the effectiveness of garlic and actually sent three infected guinea pigs to Davos in an animal experiment that documented the ineffectiveness of climate therapy.[144]

Cornet's series of experiments had proved the ineffectiveness of traditional and popular antituberculosis medications, even in animal experiments. In keeping with the model of internal disinfection, however, the search for effective substances could now concentrate on other chemicals and on synthetic dyes and similar substances that were becoming increasingly important in the pharmaceutical chemistry of the period. A letter of May 1888 shows Koch embarking on a broad-based investigation of the bactericidal effect of aniline dyes. "My investigations propose to examine a number of substances in the aromatic series with respect to their influence on certain pathogens, and I began with dyes [because] these are most easily obtained. But soon I shall have to work with other substances as well, and then I will take the liberty of accepting your kind offer and ask you for certain preparations that are not commercially available."[145] Not many notes about these experiments have survived; they obviously had an exploratory character.[146] Koch tried out various approaches and tested them in a variety of diseases. Thus already infected animals were given chemicals, but at the same time attempts were made to infect animals that had been pretreated with such substances. The following can be considered the prototype of a large number of these experiments over the next one and a half years: Koch mixed different concentrations of a yellow dye named auramine with nutrient gelatin, which he inoculated with cultures of, respectively, anthrax and typhus.[147] Against the background of the pathogenetic thinking of medical bacteriology—which conceived of infectious diseases in analogy with culturing processes—this experiment provided an in vitro model of internal disinfection. The inoculated culturing medium would show the result of the internal disinfection, thereby making it possible to examine its effectiveness against bacterial invasions.

But in fact, this experiment did not yield any result, since the "antiseptic" culture medium had no effect on the growth of the cultures. Nonetheless, in September 1889, Koch used this very model as the point of departure for a whole series of experiments on the subject of tuberculosis. The surviving documents indicate that he tested the effects of chemically altered nutrient media on tuberculosis cultures. It turned out that nutrient media mixed with chemicals had about the same effect on the cultures as these substances had alone. Sulphuric acid, for instance, could barely slow the growth of the notoriously acid-proof tuberculosis bacteria, whereas alcohol in higher concentrations was effec-

Notes on a therapeutic experiment with auramine, 1888. RKI/Koch papers, as/w2/003.

tive. In effective substances, concentration usually played a crucial role. However, not many of these substances could be used in animal experiments, since most of them—including sulphuric acid—were hardly less harmful to the host organism than to the bacteria. But in cases where it could be assumed that the substance was more toxic to the bacteria than to the host organism, Koch tried to test the pathogenic effect of the substances in animal experiments. After it had turned out, for instance, that the addition of more than 12 cm$^3$ glycerin to 100 cm$^3$ agar had slowed down or stopped the growth of the cultures, cultures growing on glycerin-agar were inoculated into guinea pigs.[148] By February 1890, no success whatsoever had been documented, but Koch nonetheless decided to embark on a massive expansion of these experiments. Having tested no more than nine different substances since the autumn of 1889, over the next four weeks he used more than one hundred chemicals and synthetic dyes to prepare nutrient media on which he placed bacteria cultures. The figure opposite shows that at this time Koch's notes assumed the character of lists. Nothing is known about in vivo experiments at this time, but it can be assumed that Koch occasionally tested effective chemicals or dyes. However, it is not known that any of the tested substances stood out as promising in animal experiments. Instead, cultures that had proved viable in a chemically altered nutrient medium seem to have been inoculated onto other such culturing media. Thus, cultures grown on glycerin-agar would be transferred to another antiseptic nutrient.

There is no reason to believe that until the spring of 1889 the disinfection experiments had yielded any therapeutically useful results. Nonetheless, Koch was able, barely six months later, to introduce a medication he had developed in animal experiments and tested clinically. Announced in August in a dramatic public statement at the Tenth International Congress of Medicine in Berlin, the medication could be purchased from Koch's associate Arnold Libbertz starting on November 13.[149]

But then Koch presented the medication—tuberculin—as a secret remedy and held back information about the research that had led to it and its composition, testing, and expected effects to an extent that borders on disinformation. It became clear later that some of his statements were misleading. In the first announcement in August, he related the medication to the disinfection experiments described above. He said that he had "begun, very soon after the discovery of the tubercle bacilli, to search for substances that could be used therapeutically against tuberculosis [and had] over time examined a very large number of substances as to their effect on tubercle bacilli grown as pure cultures."[150] The

Notes on tests with chemicals, February 1890. RKI/Koch papers, as/w2/008.

objective of his experiments had been to find in the test tube substances that would prevent the growth of bacteria in the animal organism as well. In August 1890 Koch claimed to have found such a substance: "I can now announce that guinea pigs—which, as we know, are extremely susceptible to tuberculosis—no longer react to inoculation with tubercular virus when exposed to the effects of such a substance, and that in guinea pigs that have . . . fallen ill, the disease process can be brought to a complete standstill."[151]

However, it became known only later that tuberculin was something completely different. In terms of its origin and its intended effect, it was not comparable to any of the acids, alkalines, toxins, or dyes with which Koch had experimented. Rhetorically placing it into the series of his own work on disinfection was misleading, but it did mobilize that work as a legitimizing tradition of the new medication. Koch covered up the fact that his investigations, with their orientation toward the model of internal disinfection, had failed to yield results.[152] Against this background, it is not surprising that he subsequently released further information only hesitatingly and in part under pressure. Thus, when Koch released the medication to treating physicians in November 1890, he reported in detail about its effects on test animals and humans but did not say a word about its composition.[153] It was only around the end of the year, when euphoria had given way to a critical assessment, that pressure from the public and from the Prussian Ministry of Culture caused him to divulge the origin of the medication.[154] Until that point, "the world's physicians experimented with a completely unknown substance, a 'secret remedy' in which they believed solely because of the scientific reputation of a Robert Koch."[155] Tuberculin, it now turned out, was an extract of cultured tubercle bacilli suspended in glycerin. Suddenly the self-assurance the researcher had exhibited in the autumn came into stark contrast with the dubiousness of the object on which it was based: Koch clearly did not know exactly which elements composed his medication, nor which of the substances should be credited with its claimed effect.[156]

Here one must ask what—aside from his professional and financial ambition, of which more will be said later—had prompted Koch to proceed in this manner? Which reflections justified assigning a curative effect to tuberculin? The answer to this question cannot be found where one would expect to find it, namely in the tuberculin itself. Koch knew only where it came from and how it was produced; the composition of the final product was unknown to him. This did not put him far ahead of his critics and competitors: by the turn of the year 1890–91 Ferdinand Hueppe had already guessed, on the basis of a few of Koch's scant indications, that tuberculin came from tuberculosis pure cultures.[157] And as early as autumn 1891, Edwin Klebs was able to produce what he considered an improved tuberculin.[158] Tuberculin was a heterogeneous substance.[159] The conviction that Koch had a therapeutic agent was thus not based on the knowl-

Notes on therapeutic animal experiments with tuberculin, April–June 1890.
RKI/Koch papers, as/w2/003.

edge of its elements but on the observation of its effect in the organism, which he obviously interpreted as curative. The basis on which Koch made such claims is not clearly visible in his publications. The scientist himself kept largely silent about his experiments, and in the course of the year 1891 it became clear that few colleagues were able to replicate Koch's successful cures in animal experiments. Whereas Edwin Klebs's experiments on guinea pigs and rabbits showed at least a slowing of the disease process following the administration of tuberculin,[160] Paul Baumgarten and Ernst Grawitz obtained negative results with rabbits and guinea pigs, and even found cases of accelerated disease processes.[161] Only Koch's son-in-law Eduard Pfuhl published, in 1892, a series of experiments in which he was able to reproduce the curative effects observed by Koch.[162] Koch's partially preserved laboratory notes—supplemented by Pfuhl's reproduction of his experiments—make it possible to follow the path that led Koch from the experiments on internal disinfection to an entirely different conception and to see which observations he used for this purpose.[163] The first point to make is that the in vitro experiments on internal disinfection undertaken in March 1890 apparently were not continued. After a last series, begun on March 12 and 13, there is no sign that they were continued.

The further course of events was decisively shaped by a series of animal experiments that Koch probably started in early April. On April 11 he infected several guinea pigs with tuberculosis and began by observing the usual course of the disease, as described by his associate Pfuhl: poorly healing inoculation wounds and swelling of the glands, followed about four weeks later by weight loss and low-grade fever, with death ensuing after a total of six to eight weeks. The most important pathological finding was that the animal always showed—aside from tubercles in the tissues—a greatly enlarged spleen and a slightly enlarged liver shot through with necrotic tissue.[164]

But then a second inoculation of the diseased animals brought an unusual reaction: instead of speeding up the disease process, it unexpectedly prolonged their lives. The tissue around the inoculation wounds necrotized and was sloughed off. And finally, when the guinea pigs were killed at the end of July, the usual pathological findings were absent. The tell-tale signs of tuberculosis, enlargement of the spleen, necrosis of the liver, and so forth were not present, and the infected tissue had become caseated.[165]

Instead of the expected inoculated tuberculosis, then, Koch found a necrosis of the tubercular tissue. This striking effect of a renewed infection of diseased animals became the basis of Koch's understanding of the tuberculin reaction. As he emphasized in November 1890, tuberculin did not affect the bacteria, but did affect the diseased tissue, which it caused to necrotize.[166] This interpretation of his animal experiments mobilized the understanding of the disease pathogenesis that Koch had developed in the early 1880s: at that time he had considered the transition from an early, intensive, and bacteria-rich stage of the disease to the relatively quiescent state of necrosis and caseation as one of the central identifying marks of that process. This necrosis had apparently occurred in the tubercular tissue of the infected and reinoculated animals in 1890. Now deprived of nutrients, the pathological process had come to a standstill. In other words, the presumed curative effect of tuberculin was a bacteriological variant of the scorched-earth tactic, preventing bacteria from spreading throughout the body. "In the necrotized tissue the bacillus will find such poor nutritional conditions that it will be unable to grow further and most probably die."[167]

But what had actually produced this effect? Koch interpreted it as the result of the bacterial metabolism or a necrosis-inducing substance. Accordingly, he attempted to isolate that substance from the bacteria. Interestingly, he wrote, this effect "is not related exclusively to living tubercle bacilli, but is also found in the destroyed ones."[168] Thus it was not the bacteria themselves but some of their components that were responsible. A list drawn up in June and July 1890 shows Koch engaged in testing a wide variety of sterilizing procedures by means of heat, chemicals, and so forth.[169] Apparently all these procedures failed to cancel the observed effect. At the same time, a reaction occurred only in previously infected animals. This was not only proof of a specific reaction, it also provided the rationale for the diagnostic and therapeutic effect of the substance. Whereas tuberculin in higher concentration killed the infected animals, proper dosing led to a cure:

> A dose that is not quite sufficient to kill the animal can cause a large-scale necrosis of the skin in the area of the injection wound. If the solution is diluted further . . . the animals stay alive and, with injections continued in one- to two-day intervals, a pronounced improvement in their condition occurs; the ulcerating injection wound shrinks and finally closes, which is never the case without this treatment, the swollen lymph glands become smaller; feeding improves, and the disease process—unless it is too far advanced and the animal perishes from general exhaustion—comes to a halt.[170]

Since the solid elements of the cultured preparations were not assimilated by the body of the experimental animals, it was unlikely that the effect originated with them. Koch therefore felt justified in assuming that the effective substance was soluble and attempted to separate it from the solid elements. And indeed there were several ways of obtaining an effective extract from the material by means of alcohol or glycerin. After filtration, this extract was ready to be used as tuberculin.[171]

In March 1890, then, a profound shift in his experimental strategy made it possible for Koch to conceive tuberculin, but it was a shift he concealed in his sensational announcement in August of that year. Rather than disinfecting the nutrient medium with an effective medication, the new aim was to deplete it, and starving the bacteria took the place of destroying them. Moreover, as the animal experiments document, Koch was pursuing not a general disinfection but a specific effect produced only in infected animals, and only in the affected areas of the body. "All we know for sure," Koch stressed in November 1890, "is that this does not kill the bacteria present in the tissues, but that the medication only works on the tissue containing the bacteria."[172]

It was interesting to see that humans reacted to the medication in an "extraordinarily more sensitive manner"[173] than guinea pigs. Whereas among the experimental animals only the tubercular ones developed any reaction at all, humans almost always showed a general reaction with fever, painful limbs, nausea, and so forth, such as Koch had first observed in himself. In addition, infected humans showed a local reaction in the tubercular tissue. If the dose was reduced to 0.01 $cm^3$, only tubercular subjects showed a reaction, whereas healthy ones reacted weakly or not at all. The local reaction could be seen particularly clearly in lupus, that is, tuberculosis of the skin. After the injection "the lupous sites begin to turn red, usually even before the onset of violent shivering." Subsequently the tissue turns "brownish-red and necrotic," the tubercular foci "turn into crusts that fall off after 2–3 weeks and sometimes leave a smooth red scar after only one injection."[174] Koch also saw the pronounced reaction of tubercular humans to tuberculin as a diagnostic tool and strongly advocated its adoption. His contemporaries judged this use much more favorably than its purely curative effect, which was destined for a brief carreer.[175]

## III.5. Intoxication and Hangover

Against this background, Koch's announcement of August 1890 looks like a release from a professional and private dead end. After a quest of many years, his medication finally came on the market by mid-November. It carried both laypeople and specialists away in a euphoria reminiscent of the spectacular successes of bacteriology in earlier years.[176] Koch seemed to be in a position to outshine Pasteur's vaccines with his therapy for an infectious disease.[177] Under these circumstances it was reasonable to think about resigning his unsatisfactory position at the Institute for Hygiene and to revive the failed plans of 1885 for an independent institute that would be of comparable standing to the recently founded Institut Pasteur.[178]

Koch lost no time in trying to seize these opportunities. Even while the medication was still being tested, he took a leave of absence from his professorship and on October 31, 1890, asked the Prussian government for an institute that would be charged with studying and producing the tuberculin.[179] The medication seemed to open enormous economic perspectives to its inventor. The plan was to have its sales initially benefit him; only after six years would the rights be transferred to the government—in exchange for the financing of the institute.[180] The inventor laid out the potential gain in a letter to Friedrich Althoff, the relevant official in the Ministry of Culture. On the basis of a daily production of 500 doses of tuberculin, he estimated the profit for an institute at 4.5 million marks per year. Concerning the reliability of this prognosis, he dryly remarked, "As for the prospects for sufficient sales of the quantities produced, I take the liberty of pointing out the following fact: Among one million persons, one can on average count 6–8,000 who suffer from pulmonary tuberculosis. A country with 30 million inhabitants will thus have at least 180,000 phthisics."[181] Although the public was not aware of it—just then the Prussian parliament, for instance, celebrated Koch as a benefactor of humanity[182]—some officials in the Prussian Ministry of Culture were concerned that the scientist would take advantage of the subsidy he had been offered early on for the production of the medication. The Prussian authorities therefore tried to persuade Koch to renounce the direct exploitation of the medication in exchange for a grant. The scientist resisted and delayed any action when he bargained—citing lucrative offers from the United States, for instance—for a larger grant.[183]

But then the officials in the Ministry of Culture also played for time. In early December they worked out an agreement with Koch,[184] making sure at the same time that it would be nullified by the veto of Chancellor Caprivi.[185] And indeed, by the end of December Caprivi prevented the implementation of the contract signed by Koch and the ministerial officials. He justified this step by pointing to the pernicious consequences of allowing Koch to enrich himself—in any way—with the medication. "We would do ourselves great harm if, in an area where hitherto Germany enjoyed the uncontested reputation of pursuing ideal goals, we allowed the German thinker to appear greedy for profit. If Herr Dr. Koch is unable to adopt this point of view, it is, in my humble opinion, up to the government not to let this become known publicly and, as it were, be presented as justified."[186]

Beyond Koch's financial and professional plans, tuberculin seemed to fulfill the therapeutic hopes he had kindled in himself and in others. In 1882 he had solved the enigma of tuberculosis; now he was going after the disease itself. It was, in his own words, the decisive day "in the fight against the smallest but most dangerous enemies of the human race."[187]

But then the euphoria about the medication died down by the end of the year. Physicians gradually became aware of the dangers of using it, and therapeutic successes became increasingly questionable. The turning point of the scientific discussion may well have been Rudolf Virchow's presentation of pathological findings in early January 1891. When he documented the presence of new tubercles at the margins of the necrosed tissue, Koch's conception of the healing process, in which necrosis had played a central role, appeared increasingly questionable. Virchow did not make it clear whether the necrosis did not prevent or actually promoted the spread of the disease, but either way Koch's conception was damaged. The secrecy that had initially shrouded the medication as part of the media sensation now struck back at its author.[188]

In this situation Koch could no longer ignore the widespread demands for information about the composition of his medication. On January 15, he published a generally worded description of tuberculin.[189] And then he fled the scene. On the very day of the article's publication, he applied to the Ministry of Culture for a leave of absence from January 25 to April 30. In addition to a private crisis related to the separation from his first wife, the looming failure of the tuberculin was surely involved. Just when his presence in Germany would undoubtedly have been helpful, he ran off all the way to Egypt.[190] In his absence it became increasingly clear that tuberculin did not have the effect he had claimed for it,[191] and that he was obviously unable to produce the guinea pigs he had cured with tuberculin.[192] At the same time, more and more cases of worsening conditions, and even fatalities, after the use of tuberculin were becoming known.[193]

Koch had developed tuberculin together with two fairly undistinguished associates, Eduard Pfuhl and Arnold Libbertz, respectively his son-in-law and childhood friend, with whom he was personally comfortable. None of the other associates knew the composition of tuberculin. The first testing of humans seems to have been limited to himself and his seventeen-year-old mistress![194] It remained unclear whether tuberculin had no effect on the spread of the disease or whether it actually promoted it.[195] In late 1891 the tuberculosis researcher Paul Baumgarten, codiscoverer of the bacterium, rather maliciously summarized the latest findings based on animal experiments by saying that "large doses are harmful in fully developed inoculated tuberculosis, while small and medium doses don't help." All in all, then, tuberculin now came to look like the result of fishing in murky waters. Not much was left of Koch's discovery, once he had to disclose his secret. The publication also made it clear that his discovery was by no means as exceptional as had been assumed at first: the search for antibacterial substances in the blood had been an established area of research for a decade.[196] Two Frenchmen, Héricourt and Richet, had worked on a tuberculosis serum, although they had not gone beyond animal experiments.[197] The eventual discoverer of the plague bacillus, Alexandre Yersin, had experimented with a tuberculosis vaccine as early as 1888.[198]

The impending failure of tuberculin as a medication concluded the matter of a grant. In 1890 officials in the Prussian Ministry of Culture had wisely decided to "put off the negotiations [regarding a grant] until public opinion has a clearer idea as to the value of the medication."[199] By spring Koch was in a predicament: fully expecting to become the head of a tuberculin institute, he had asked for a leave of absence from his professorial duties in October, but to date he had not come to an agreement with the ministry about the structure of such an institute or about the matter of a grant. Now he faced the embarrassment of having to return to his unloved university position and accordingly had to give in on both points. On April, 20, 1891, he officially renounced the grant and abandoned the idea of a center for tuberculin research, accepting instead the directorship of the Institute for Infectious Diseases that was to open in the summer of 1891 and bears his name to this day.[200] At the founding of this institute, the Prussian officials took precautions against future financial adventures on the part of Koch, whose property rights to tuberculin were legally unassailable.[201] They demanded a written guarantee that "inventions and discoveries that would be made by him [Koch] at the Institute for Infectious Diseases or with the means of the Institute would be placed unconditionally and without any compensation whatsoever at the government's disposal."[202] This effectively meant that he had to sign away his right to apply for patents in exchange for the directorship. In addition, Koch's request to hold outpatient clinics in the new institute was denied.[203]

Indeed, he had to declare his complete renunciation of a private practice—a point that puts his exorbitant salary of 20,000 marks in perspective[204]—and promise not to accept another position without the approval of the Ministry of Culture for five years![205]

And finally, officials of the Prussian Ministry of Culture also decided in May to ask Koch for a detailed description of the production method and to have the medication chemically analyzed. They actually made disclosing the composition of tuberculin the condition for Koch's appointment to the directorship of the new institute. The minister of culture proposed to "place Koch at the head of that institute, but not before he has completely uncovered the secret and provided the government with the opportunity of examining the medication."[206] In April Friedrich Althoff had already contacted his friend Eduard Külz, a professor of physiology at Marburg University.[207] Külz went to work on this analysis, which presented major technical difficulties. After all, this was not, as Külz pointed out in an April letter to Althoff, a matter of simply identifying the chemical components in a small amount of tuberculin, but rather one of determining the chemical structure of a presumptive effective substance: "A so-called analysis, such as it can be carried out with one or two vials, would be worthless and downright unscientific, because we don't even know the effective components . . . What I had in mind was not an analysis, but the the isolation of the effective substance or rather effective substances."[208] But then, attempting to isolate the effective substance went beyond the limits of Koch's knowledge. For in the course of the year 1891, it became clear that tuberculin was an enigma, not only for the medical world and the general public but also for its inventor: Koch's own attempts to isolate an effective substance had obviously failed. The substance Koch labeled "pure tuberculin" was simply the one that was effective in the smallest dose in animal experiments.[209] He felt that he only had to know about its production and its effect, whereas Külz was focused on the isolation and chemical analysis of an effective substance.

Koch did not think that tuberculin could or should be improved, and he considered unnecessary the attempts to improve the medication by purifying it. He claimed that his publication of October 1891 had furnished information about the production of the medication that "could actually be used as a recipe."[210] Yet the Marburg team considered Koch's information imprecise and the tuberculin impure: "Surely, there is no need to elaborate on the statement that Koch's latest publication has failed to lay this question to rest."[211] The medication, they wrote, more than clearly showed its impurity by its ash content, which could be as high as 20 percent. In fact, they warned that there was no way of knowing whether what one produced according to Koch's instructions was really tuberculin.[212] Given the failure of the tests in Marburg and those of Koch's team in Berlin, it is worth noting that the two groups saw the importance of a pure form of tuberculin in very different ways. Külz expected that the identification of effective substances would lead to improved dosing: "If that could be achieved, a precise dosing of the medication, which is not possible now, would become a possibility."[213] Koch felt that the tuberculin reaction was sufficiently controllable and that the attempt to produce a pure form of tuberculin was really unnecessary. A year later he described the state of this question without much regret: "On the basis of my experiences so far, I do not think it likely that it will be possible to use chemical differentiation of the so-called raw tuberculin to isolate substances that will prove more advantageous in treating disease than the raw substance."[214] Instead he pleaded for improving the medication through changes in production techniques and methods of application, projects he had placed in the hands of the chemist Bernhard Proskauer at the Institute for Infectious Diseases.

### III.6. Deceptions

A historical analysis of Koch's concept cannot evade the question of whether he really believed that he had a medication. Allegations of some kind of "tuberculin swindle" were already raised during the events of 1890–91.[215] Some aspects of Koch's behavior are indeed hard to explain if one completely excludes the assumption of deliberate deception. As we saw above, he furnished only insufficient and in part misleading information about tuberculin and was—for whatever reason—unable to produce the guinea pigs he had cured with the medication.[216] Koch's attempts to earn a fortune with tuberculin are also suspect. This did not work out only because the officials in the Ministry of Culture put off the negotiations for the sale of the medication to the Prussian State until the euphoria about it began to turn into dismay.

Nonetheless there are some indications that Koch was firmly convinced of the curative effect of the medication. How else could one explain that he believed, practically to the end of his life, that he could improve it? In 1897 he introduced a new tuberculin, still giving advice on its proper use in 1901. Elias Metchnikov related that he had found Koch working on the optimization of the medication in 1909, a year before Koch's death.[217]

It is therefore more appropriate to interpret Koch's thinking and behavior as the result of a self-delusion brought about by the success and suggestive power of the ideas about tuberculosis he had developed in the early 1880s. This would mean that it was precisely his conviction that he had solved the enigma of tuberculosis in 1882 that prevented him from asking questions that might have allowed for a different interpretation of the tuberculin reaction. One of these questions was whether the guinea pig, which had proven an appropriate experimental animal in the reconstruction of the etiology, was also appropriate for the development of therapeutic agents.

It would of course be easy to point out a rather considerable number of errors in Koch's understanding of tuberculosis and in the resulting tuberculin. Doing so, however, is historically and scientifically problematic, since it would pass judgment without considering the contemporary state of knowledge. Thus Koch's supposition that the bacteria produced necrotizing substances is erroneous according to today's state of knowledge: tuberculosis bacteria do not produce any toxic substances and therefore cannot secrete any.[218] By the same token it seems plausible nowadays to understand the prolongation of the experimental animals' life as a strong immune reaction—in the sense of an auto-vaccination—triggered by the tuberculin injection.[219]

Koch, however, assumed that the products of the bacterial metabolism were responsible for the effect. His reflections lacked an immunological context. The thesis of self-delusion is supported by his interpretations of some available information in a most idiosyncratic way and in his expression of views that he could hardly back up with his own experiments and observations. Koch's bacterial spores are a fine example of this: visible or not, stainable or not, they occupied a central position in his argument. The partly invisible spores allowed him to assert the ubiquity of bacteria in tubercular tissue. At the same time the fact that these structures could not be stained furnished a plausible explanation for their partial invisibility.[220]

The spores offer a good example of what Ludwik Fleck calls "the harmony of illusions," that is, the persistence of once-formed views in the face of contradicting or uncertain evidence. Koch's understanding of disease as the activity of a pathogen falls into that category. For Fleck the entire bacteriological theory of infectious diseases was an outstanding illustration of his view: "The classical theory of the infectious diseases: it ascribed to each infectious disease a cause in the shape of tiny living 'germs' and did not see, could not see, that this 'germ' was also present in healthy subjects."[221] Koch did not look for tuberculosis bacteria in healthy subjects; he was satisfied with finding them in diseased tissue and in pathological preparations—and even that required the help of the spores.

"Whatever does not fit into the system will remain unseen."[222] Nowhere is this aspect of the "harmony of illusions" more palpable than in Koch's interpretation of the tuberculin reaction. As we saw, Koch considered the local tuberculin reaction as a sign of tuberculosis. He explained the striking fact that his guinea pigs showed practically no reaction while healthy humans definitely had a general reaction to tuberculin with the humans' greater susceptibility to tuberculin, which he estimated to be 500 times higher than that of guinea pigs.[223] This was not just speculation but also a bold statement, considering his own finding that the experimental animals were highly susceptible to living bacteria. The observation that the guinea pigs did not contract spontaneous tuberculosis, but that humans clearly did, placed another and more accurate interpretation right at his fingertips. But then Clemens von Pirquet's understanding of the effect of tuberculin as a delayed hypersensitivity reaction triggered by a possibly long-passed primary infection of a healthy person was hardly within Koch's grasp.[224] This understanding postulated a clear distinction between infection and illness, and Koch did not make this distinction; for him it was sufficient to know that the substance had a specific effect on tubercular processes, "of whatever kind they might be."[225] Moreover, Pirquet and others saw the reaction as an activity of the body's immune system, not as one of the bacteria, as Koch did. The violent reaction he observed when he experimented on himself could have been quite upsetting to him, but in fact he never asked himself whether or not he was tubercular because he did not use the reaction to diagnose an infection that might have occurred long ago. He only wished to gain a closer understanding of an acute illness. Accordingly he lowered the diagnostic dosing to the point where—in his opinion—the reaction occurred only in acute cases.

Koch's contemporaries who experimented with tuberculin and came up with results that, from a modern perspective, contradicted Koch's theory of its effectiveness failed to draw these conclusions on their own. Julius Schreiber, for instance, experimented with healthy infants and toddlers in the winter of 1890–91. The fact that none of these showed any reaction to tuberculin did not prompt him, who considered the medication a "reliable reagent to tubercular illnesses,"[226] to carry on with further observations. Erich Peiper as well, who conducted a whole series of experiments with healthy—or at least nontubercular—

adults, all of whom exhibited vigorous reactions, only expressed his doubts about the medication's diagnostic qualities.[227]

The consequences of this focus on the pathogen showed themselves in another area as well. Koch essentially understood the disease as an activity of the bacteria and paid little attention to the difference in pathologies between humans and guinea pigs.[228] In this sense it was consistent to interpret the absence of a tuberculin reaction in his guinea pigs as a lower grade of susceptibility. But in fact his experimental animals never lived long enough to develop a general tuberculin reaction. Indeed it was precisely because of the rapid course of the disease in guinea pigs that Koch had chosen them! In the pathology of tuberculosis in these animals, there was no room for a primary infection as a possible explanation.

## III.7. Tuberculosis and Tuberculin

Contemplating the road on which Koch traveled from tuberculosis to tuberculin yields a number of interesting insights into his thinking on infectious diseases in general and on tuberculosis in particular. Koch's initial etiological research in this area in the early 1880s was conducted in keeping with a model of infectious disease as an invasion of bacteria. This understanding characterizes Koch's early work in general.[229] But whereas Koch systematically and explicitly developed the etiological arguments on the subject of tuberculosis, the earlier studies lacked a comparable discussion of the pathogenesis of the disease. Here Koch simply described the characteristics of the bacteria and their behavior in the infected tissue. He thus did not formulate his reflections on pathogenesis directly, but they did inform his descriptions of the characteristics of disease agents. In this way he implicitly developed, but did not explicitly state, the reductionist model of infectious diseases as manifestations of the life of invading bacteria. In comparing this understanding of disease with the ideas of the so-called romantic medicine of the early nineteenth century, one realizes that in both instances the disease autonomously confronts the body; in other words, that it is a separate entity. The difference, and hence the modernity, of bacteriology was that it replaced a theory without practice, such as the romantic parasitism, with a practice without theory, a technology-based model of infectious disease. The pathological investigation of infectious diseases had been almost completely absorbed into the technology of the medical-bacteriological laboratory.[230]

Koch's research on the bacterial causes of disease was linked from the very beginning with the goal of finding a therapeutic agent. However, the set of methods and means he developed for dealing with etiological questions proved to be of little use for therapeutical purposes. The appearance of tuberculin in the winter of 1890–91 was predicated on several factors: his hopes of finding ways to combat tuberculosis, now that its etiology was known; professional competition, especially from the French microbiologists; seductive commercial prospects; and his refusal to reopen questions that seemed to have been solved. In explaining the reaction as a cure, Koch went back to the conception of pathogenesis he had developed in 1882. Its central tenet was his understanding of the caseous necrosis as a state of quiescence to be produced by tuberculin. His views were based on an unreflected transfer of the results of animal experiments and on an understanding of disease as an invasion of bacteria that—by conceiving of the body as a passive culture medium for the disease—made no distinction between invasion, infection, and illness. Koch's self-delusion thus resulted first and foremost from his unshakable adherence to the explanatory model he had developed. This attitude was reinforced by his competition with the French microbiologists, an intense desire to improve his professional situation, and also the prospect of financial gain.

It is also worthwhile to look at the consequences of new developments in Koch's research program. His contemporaries' critical assessment of tuberculin had consequences not only for his life but also for the perception of bacteriology as a discipline. Concerning these general consequences, it would appear that the discussion about the medication also raised questions about the prevailing bacteriological reductionism, that is, the bacteria-centered view of infectious disease. This change of attitude first became evident in small ways. Hans Buchner, a student of Pettenkofer, succeeded in producing Koch's allegedly specific tuberculin reaction with extracts of entirely different bacteria.[231] But most important, a number of fundamental critiques of medical bacteriology were published in the 1890s. The first, published in 1890, was by Heinrich Lahmann, a naturopathic practitioner who felt that the science of bacteriology was leading medicine in the wrong direction and advocated a "natural" cure for tuberculosis.[232] On the other hand, Koch was also attacked for being insufficiently scientific. Ferdinand Hueppe and Friedrich Martius criticized Koch's bacteriological reductionism, citing once again the supposedly obsolete categories of heredity, disposition, and constitution—in short, the reaction of the human organism.[233] And Ottomar Rosenbach, in his 1891 book *Grundlagen, Aufgaben und Grenzen der Therapie* (The principles, tasks, and limits of therapy), traced the emerging lines of conflict between medical bacteriology and clinical medicine.[234]

The tuberculin episode marked the beginning of the end of bacterial reductionism in medicine, and it is noteworthy that in the 1890s holistic concepts, such as the patient's constitution, fostered lively debates in connection with infectious diseases.[235] This was the period when the bacteriologists themselves finally said goodbye to the models dating from the heroic days of their discipline. They began to replace simple invasion models with complex epidemiological theories and studied such phenomena as healthy carriers, subclinical infections, and immunological questions in general.[236]

As for Koch personally, it is striking that he partook of these developments only to the extent that they did not touch the subject of tuberculosis. Even though in other areas, such as cholera and typhoid, he presented decidedly innovative research,[237] he continued to cling stubbornly to tuberculin as an antituberculosis medication. As mentioned, he brought out an improved tuberculin in 1897 and seems to have been faithful to its concept to the end of his life. The controversy about the nonidentity of human and bovine tuberculosis in which he allowed himself to become involved after 1902 still showed traces of the original research program. Based on his experimental research, Koch contended that human and bovine tuberculosis were two different diseases. But this idea proved to be untenable in the face of clinical criticism.[238] It is true that Koch's clinging to tuberculin was facilitated by considerable numbers of his contemporaries who shared his attitude. After the turn of the century there seems to have been a modest renaissance of the medication, which was still recommended in a widely used handbook of internal medicine after World War I.[239]

The tuberculin issue had a negative influence on Koch's professional situation. After the failure of this undertaking, which was to free him from the control of the Prussian Ministry of Culture, his position actually became even less favorable than before. The harsh conditions he had to accept when he took over the directorship of the institute put his salary of 20,000 marks, which was high for that period, in perspective. Koch's subservience to the Prussian science policy under the Althoff System had thus increased rather than diminished. Moreover, as far as the further development of tuberculin was concerned, his position was weak: while there was no question that the original tuberculin had been his property, there was also no question that this would no longer be the case in the future. By agreeing in June 1890 to make his inventions available to the government "unconditionally and without compensation,"[240] as was mentioned above,

Koch lost the right to apply for patents. At a time when patent protection for pharmaceuticals and their manufacturers came into being, this was a decided disadvantage.[241] It meant that Koch would fall behind in the competition with others who were also looking to improve tuberculin or generally working on sera and similar therapeutic agents. The original tuberculin had been protected by secrecy, not by patents. After Koch published the details of its manufacture in 1891, it could be copied. By 1898 the Berlin Institute for Serum Research and Serum Testing headed by Paul Ehrlich reported that it had evaluated no fewer than fifteen tuberculins from all kinds of sources.[242] In 1897, when Koch presented "his" improved therapeutic agent called tuberculin T.R.,[243] he found to his dismay that his former student Emil von Behring was applying for a patent to a process designed to extract a tuberculosis toxin that seemed to be similar to Koch's own and no more effective. Koch contested Behring's patent application in order to, as he put it, "protect the plaintiff's scientific research against exploitation by unauthorized parties."[244] Expanding his factual objections, he added a request for better patent protection: "I would humbly add the request that the results of both my research and that of my associates at the Institute under my direction be allowed to receive patent protection to the extent they appear to have practical applications."[245] Aside from the merits of the case, this kind of argument made it possible for Behring to fight Koch's opposition as personally motivated. He pointed out that the contestation was based not so much on factual objections as on the special legal situation of the plaintiff.

> The letter of Geheimrath Koch seems to me to indicate that
> 1. the latter is prohibited from claiming the benefits of the current patent legislation for himself and his associates,
> 2. that he seems to consider this prohibition unjust, and
> 3. that in principle he would be inclined to use the current German patent legislation for his own ends after this prohibition has been lifted.[246]

In the end, Koch's objection was rejected.[247] At the conceptual level, Behring had brought Koch's ominous necrosis-causing substance into his antitoxin research and thereby into a promising immunological context. In developing this train of thought, he had expressly referred to Koch in 1895 and also vehemently defended tuberculin against its critics.[248] But when it came to tuberculosis, Koch was evidently incapable of reacting positively to any widening, much less correction, of his position. After Behring had sided with Koch's critics in the controversy about human and bovine tuberculosis, Koch wrote a letter to Carl

Weigert, his colleague in the heroic Breslau days. His words illustrate the full measure of humiliation he felt as he watched developments in the field of tuberculosis research:

> Your letter most vividly takes me back to the old days, when we could freely and without interference pursue our studies. Over time, this has greatly changed, alas. Whatever I tackle and undertake nowadays, a crowd of jealous and resentful individuals always turns up and jumps onto the same subject in order to start a controversy or, if that can't be done, to try to discourage me ... Today every little yapping dog will inoculate a single calf, write half a dozen articles about it afterward and carry on as if this case of his had solved the whole question, and of course in his favor. And now I have to see what my former students are doing, as when v. Behring denies the existence of inhalation tuberculosis and tells us that bacilli contracted in the intestine just simply lie around, remain latent for decades and then almost always appear in the apex of the lung. Yes indeed, research would be a fine thing if everyone could think logically, but unfortunately this is the case for a small fraction only. Besides I believe that in my work I have been particularly unlucky, and that I have encountered more opposition, and unjustified opposition at that, than anyone else.[249]

An analysis of Koch's research program on tuberculosis, then, allows us to reflect on his intellectual goals and on the resulting limits of his version of medical bacteriology.[250] For all the wide-ranging knowledge of bacteria and disease that Koch's research program organized and produced, it was fundamentally anchored in a genuinely medical interest in understanding and combating diseases. In this sense it was limited. Additional factors of stability were Koch's special investigative techniques and the existence of a team of researchers around him, especially at the Imperial Health Office, the Institute for Hygiene at Berlin University, and later the Institute for Infectious Diseases. The frequently heard categorization of medical bacteriology as a medical science refers both to its subject matter and to a problem that became increasingly important as bacteriological hygiene came into its own. Connecting biological and medical concepts was a fundamental proceeding of medical bacteriology, which called bacteria pathogens. This created a body of knowledge that was heterogeneous in a very special way and practically invited criticism. The following chapters will examine, among other things, how and why after its initial successes in the microbe hunts of the 1880s, the discipline increasingly came to compete with fields such as internal medicine and parasitology.

A good example of a limited vision is Koch's lack of interest in the physiology of bacteria. For all his encompassing knowledge about bacteria before 1890, he later focused essentially on matters of identifying, staining, culturing, and destroying them, paying scant attention to other questions. Another example of this attitude (see chapter II) is Koch's refusal until the late 1880s to see the phenomenon of virulence as anything but the result of poor laboratory technique. It is obvious, however, that this limited focus was initially very helpful, since it justified Koch's and Cohn's insistence on bacterial specificity and was useful in the debates about Naegeli's transformist bacteriology and Pasteur's anthrax research. In the context of the tuberculin affair, this attitude accounts for Koch's disinterest in chemical analysis of the medication. Indeed, it would seem that he never reacted to any of the alternative theories about the tuberculin reaction that were formulated after 1891.[251]

Even though Koch did not in principle trust the clinical experience and as a rule developed his insights in the laboratory rather than at the patient's bedside, his work was undoubtedly motivated by what one might call a therapeutic impetus. In the early years of bacteriology, this hardly restricted the scope of his occupations, for this was the time when applied and basic research converged. The problems with tuberculin, which arose from Koch's attempt to force new problems into preexisting explanatory models, show that his research program had cast too wide a net. This not only made it difficult to reach the research goals he had set for himself, it would also, as will be shown in the next chapters, be the source of conflicts between medical bacteriology and other disciplines.

CHAPTER IV

# Of Men and Mice
Medical Bacteriology and Experimental Therapy, 1890–1908

"The least reliable results were obtained from experiences gathered at the sickbed."[1] Skepticism about the value of clinical knowledge was one of the mainstays of Koch's medical bacteriology. Rather than developing concepts of disease through the observation of patients, he focused on the experimental investigation of disease in the laboratory. His decision to experiment "not with humans but with the parasite alone in pure culture,"[2] was more than a physical move from the hospital ward to the laboratory, for it changed the very object of the investigation. New knowledge about diseases was now based not so much on the observation of sick humans as on the reproduction of pathological processes in experimental animals. To be sure, the patient—if the investigation concerned human pathology—was indispensable as the starting and end points of the research, but he disappeared from the experimental process itself. In the laboratory the researcher dealt only with the pathogens that, within the functional context of a biotechnical setup of microscopes, Petri dishes, culture media, experimental animals, and so forth, embodied the disease.

Even though the field of bacteriological hygiene considered itself an applied science, it decidedly kept its distance from clinical medicine when—to put it briefly—it disregarded clusters of symptoms and defined disease in terms of a concrete organic pathogen. The proverbial success of this model goes back to the "microbe hunting" in which Koch and other bacteriologists had engaged in the 1880s. The public saw the pathogen as the embodiment of the disease. With its connotations of invisibility, ubiquity, and extreme dangerousness, this representation of pathogenic germs became one of the essential cultural certainties of modern society.[3] If people were captivated by the sensational character of this knowledge, it was not only because of the reality of the research but also largely because such research justified hopes for future consequences. In scientific terms, defining a disease through a necessary pathogen provided a logical point of intervention that, once the pathogen was identified, made it possible to control the disease.[4] Reifying the disease in its pathogen seemed to hold out the possibility of annihilating any disease defined in this manner. In 1882 Koch had formulated such hopes when he wrote that "in the fight against this scourge of humankind [i.e., tuberculosis] we are no longer facing a vague something but, rather, a tangible parasite,"[5] adding that the means to combat it were bound to be developed before long. This belief clearly was widely shared. As Virchow once rather petulantly put it, in the mid-1880s bacteria "dominated not only the thinking but also the dreams of many older and almost all young physicians."[6] At a medical convention in 1886, a song called "War to Bacteria" proclaimed: "Once we know the enemy face to face/surely we'll find means to win the race."[7]

To this extent the prestige of the bacteriologists was strangely intertwined with that of the subjects of their research, which were yet to be fully understood. It was not the existence of bacteria that brought fame to the discipline, it was the dangerous character ascribed to them and above all the credible assertion that bacteriologists would be successful in combating these dangers. As we know today, decades would pass before therapies of this kind became possible, but scientists of the era felt that the identification of the pathogens had brought these therapies within reach.[8] In 1890 *Ulk*, the humor supplement of the *Vossische Zeitung*, published an ironically overdrawn report from the year 2000, a world without doctors or diseases, where "on the site of the former Institute for Hygiene loomed an over-life-size statue of Robert Koch." Future medical students would learn that "after Koch's discovery infectious diseases had become extinct."[9] A cartoon from that period presents a graphic illustration of this attitude. The bacilli as the sole representatives of Disease are facing victorious science in the person of a cook named Koch.

Cartoon from a supplement to the satiric magazine *Kladderadatsch*, November 23, 1890. The caption reads: Chorus of bacilli: "Hey kids, you look awful! What happened to you?" The tubercle bacilli: "Alas, our cook [Koch] fixed us some kind of a broth. It made us very sick."

The first attempts to develop and apply a bacteriological medicine brought these hopes face to face with reality, and this is the subject of this chapter. In the decades after the great microbe hunts, the relation between bacteriology and clinical medicine changed once again. Whereas hitherto the input of bacteriology had been limited to matters of prevention and diagnostics, the development and testing of therapeutic agents led to a direct confrontation between clinical and experimental medicine. This situation was mirrored, for example, in the promotion at that time of the self-image of clinical medicine as the physicians' art of healing in contrast to the bacteriologists' laboratory medicine.[10] As for the bacteriologists, it is hardly surprising that they encountered multiple and unexpected problems as they tried to make their way back from the laboratory to the clinic. To be sure, they had succeeded even before the mid-1880s in introducing effective means of disinfection and sterilization.[11] They had also been able to propagate the culturing methods they had developed as the basis for bacteriological diagnostics—although it is not clear when and to what extent these became part of day-to-day clinical practice.[12] Pasteur's rabies serum was enthusiastically greeted by the public and became the benchmark of what was possible, even if it involved, on the one hand, an achievement of the French competition and, on the other, a modernization of the traditional preventive vaccination.[13] In the end it proved vastly more difficult to make use of the newly acquired knowledge in the treatment of acutely ill patients: the introduction of Behring's diphtheria antitoxin after 1891 was a long drawn-out process;[14] the first truly promising therapeutic agents, such as salvarsan, became available only after the turn of the century;[15] and as everyone knows, the history of antibiotics began only after World War II. In retrospect, Paul Ehrlich's concept of the magic bullet refers both to the prospect of an antibacterial chemotherapy and to the long and initially fruitless search for such chemicals.[16]

As dreams of antibacterial therapies continued to elude bacteriologists, the established practice of research, development, and testing in this area began to be seen as a problem in medical ethics.[17] Transferring conclusions from mice to men had become an accepted practice in etiological research. If no suitable animal model was available, contemporary bacteriologists were perfectly willing—for instance, in the case of gonorrhea—to experiment on humans.[18] Although they were definitely aware of questions of possible harm and the matter of consent, they did not consider these as fundamental issues. They tried, for instance, to circumvent the question of doing harm by experimenting on terminally ill patients[19]—a proceeding that had already been advocated by Claude Bernard.[20] Although it is impossible to quantify this exactly, experimental bacteriology thus led to an expansion of the practice of human experimentation, which "unlike about thirty years earlier . . . had become perfectly acceptable in research on the etiology and pathogenesis of infectious diseases in the age of bacteriology."[21] The breakthrough of medical bacteriology in the 1880s was also favored by the temporary decline of the antivivisectionist movement.[22] Sustained by promises of therapy, the approval of bacteriology depended on the plausibility of such promises. Critical objections were therefore not likely to be raised, at least initially, should antibacterial therapies be introduced. The public reaction to Pasteur's decidedly problematic experiments with his rabies serum illustrates this attitude: in 1885 Pasteur had tested this serum in an adolescent patient who may or may not have needed this protection. Amidst the widespread enthusiasm about Pasteur's achievement, critical voices that expressed doubt about the ethical acceptability of such a proceeding could hardly be heard. To that extent the popular enthusiasm provided a foretaste of what could be expected when antibacterial therapies were introduced.[23]

Eventually, however, the therapeutic use of experimentally generated knowledge, however enthusiastically it may have been greeted initially, began to be seen in a dubious light. This theme was publicly vented in the German Empire during the Neisser scandal. Beginning in 1899, the experiments Albert Neisser had carried out years earlier—when he treated patients who neither had syphilis nor had given their consent with a cell-free syphilis serum—became a scandal. In its wake the clinical application of laboratory medicine was increasingly viewed as an unwarranted expansion of human experimentation.[24] Scientists had of course been able to distinguish between human experimentation and therapy all along but had usually discussed matters of information and consent only in the context of surgical therapy.[25] As a result of the therapeutic efforts of experimental physicians and the highly effective medications that were developed after 1890, the boundaries between therapy and experiment, science and art, laboratory and clinic seemed to be shifting. The Prussian state promulgated a decree regulating therapeutic trials, and the newly coined expression "experimental therapy" exemplifies the researchers' attempts to conceptualize the effect of laboratory sciences on clinical practice.

If, after 1890, Robert Koch tried twice to enter the field of therapeutics, his choice of this endeavor was hardly fortuitous. He felt obliged to give the world the miraculous remedies that seemed to be the logical consequence of demystifying infectious diseases. As already mentioned, the pressure he experienced in the late 1880s resulted in part from the fact that he had promised the world and himself these remedies, even though he was as yet unable to deliver them. This pressure was compounded when the rabies serum of his competitor Louis Pasteur gave the world, if not a therapeutic, then at least an undoubted application of microbiological science. In a lecture at the Tenth International Congress of Medicine in Berlin, where he was to announce his tuberculin, Koch formulated the problem with great clarity. Having spoken about the successes of prevention and hygiene, he acknowledged the difficulties involved in developing therapies: "But all these advantages can only be used indirectly in the fight against bacteria. To date we have no directly effective, that is to say, therapeutic means with which to complement these other, indirect ones. All that can be cited in this direction is the success achieved with . . . the preventive vaccinations developed by Pasteur and others."[26] The complex relationship between wishful thinking and reality in Koch's own research has been analyzed in chapter III in the context of his work on tuberculosis. The present chapter will examine the changes occurring in medical bacteriology and the way they were perceived by contemporaries at a time when the discipline began to use "bacteriological medicine" to redeem the promises made. Koch himself made two moves toward therapy, one being the development of tuberculin after 1890, and the other connected with sleeping sickness, namely the use of atoxyl beginning in 1905. These two examples will demonstrate the transfer of knowledge between the laboratory and the clinic.

As far as tuberculin was concerned, the clinical trials to which it was subjected beginning in the summer of 1890 were the first major test for the clinical promises of medical bacteriology. In analyzing the clinical trials of tuberculin in 1890 and 1891, one must start with the assumption that the faith in its miraculous qualities described above initially stood in the way of a critical evaluation of the medication, its effectiveness, and its possible dangers. Hence one must ask how, when, and to what extent contemporaries became aware of the problems with tuberculin treatment with respect to its effectiveness, its dangers, and indeed the experimental character of the enterprise as a whole. One must also ask about the image of antibacterial chemotherapy that subsequently came into being, and whether this development might have changed the attitude toward drug therapy in general.[27]

Thus the present study will also look into issues of medical ethics raised and possibly transformed by this process. This is not so much a matter of trite indignation at the unethical behavior of the era's physicians as an attempt to describe the ethical reasoning practiced in a time when there was practically no such thing as medical ethics. The period's more or less critical evaluations of novel experimental therapies give us access to the implicit ethical standards of the time.[28] This analysis is based on the finding that, as we saw, scientists were able to distinguish between human experimentation and therapy, even in the absence of a formal code of medical ethics. However, the concept of the therapeutic experiment was unknown to them. As a starting hypothesis, one can assume that faith in the miraculous qualities of medical bacteriology initially inhibited any attention to the experimental character of the tuberculin therapy and any ethical criticism of the use of this medication. One must therefore ask whether the increasingly critical evaluation of tuberculin brought a shift from a therapeutic to an experimental focus in the character of its use and how people judged this shift in terms of ethics.

The second illustrative case, from the years 1906 to 1908, evolved in a totally different context. Fifteen years after the tuberculin affair—when a highly developed chemical-pharmaceutical industry and its products were in place, and when the public had become aware of new issues in science—Koch's laboratory at the Institute for Infectious Diseases in Berlin experienced a crisis with far-reaching consequences: one of the research assistants had become infected with trypanosomes, the pathogens of sleeping sickness. This sudden confrontation between animal experimentation and clinical work resulted in a two-year therapy experiment on the person concerned. Facing a human infection, a team of researchers who specialized in laboratory research on sleeping sickness and syphilis had to find a new interpretation of the situation and modify their approach accordingly. The state of the discussion about therapeutic and nontherapeutic human experimentation around 1906 makes it interesting to analyze whether the situation was approached as a research or therapeutic problem, and to find out which ethical concerns were—or were not—raised. What were the objectives of the different protagonists? How did contemporary medical bacteriology deal with the patient, and was it able to cure him?

## IV.2. The Clinical Testing of Tuberculin, 1890–1891
### IV.2.1. Preparing the Miracle

"I have not shied away from searching for means of inhibiting the development [of bacteria] and eventually encountered substances capable of stopping the growth of tubercle bacilli, not only in the test tube but also in the animal body."[29] Until the early summer of 1890, Robert Koch's research on a tuberculosis medication was almost exclusively based on animal experiments. Even though there was no lack of drug-based therapies for the disease,[30] Koch's own research, begun in 1882, had not yet reached the stage of clinical testing.[31] This does not necessarily mean that such experimentation was without significance, but it is an indication of the pervasive trust in animal experimentation. By July 1890 Koch still had no more than one medication capable of curing guinea pigs.[32] It was not a serious problem to him that these animals' pathology differed from that of humans; he considered the difference a matter of degree. Since the experimental animals were "notoriously highly susceptible to tuberculosis," and since in those "that have already come down with severe general tuberculosis the disease process can be completely arrested," there was some reason to be optimistic.

The end of the experimental animal trials in the summer of 1890 brought a new dimension to Koch's work. He now attempted to transfer knowledge of animal pathology to human therapy, which meant that he had to engage in clinical studies to an extent probably unknown in his career until then. To be sure, this shift from the laboratory to the sickbed was already under way at the time of his Berlin lecture. Not counting an experiment Koch carried out on himself and his seventeen-year-old girlfriend—it is impossible to date, but it was to demonstrate the harmlessness of the reaction in healthy subjects in the requisite heroic manner of the time[33]—this shift began in June with experiments on healthy subjects: two of Koch's assistants, Shibasaburo Kitasato and August von Wassermann received 2 and 3 milligrams each on and June 24 and 25, and on July 13 two more injections of respectively 4 and 5 milligrams were given, one to a Dr. Maas and one to Paul Guttman.[34] Koch had worked with Guttman, head of Internal Medicine at the Municipal Hospital of Moabit, as early as 1882–83, and his hospital would later become the most important center of tuberculin treatment.[35] All four of these persons showed the symptoms that Koch had also observed in himself: "3 to 4 hours after the injection: aching limbs, fatigue, tendency to cough, rapidly increasing breathing difficulties; in the 5th hour unusually violent shivering set in, lasting for about an hour. At the same time there was nausea and vomiting as the body temperature rose to 39.6 Centigrade; after about 12 hours all these complaints receded."[36] Higher doses provoked more pronounced reactions in the subjects treated: if Kitasato reacted to 2 milligrams with a temperature of 38.2, Guttmann's fever rose to 39.2 after he received 5 milligrams. While the first two subjects did not experience shivering fits, this symptom was observed in Maas, and in severe form in Guttmann.[37] The rise in pulse rates showed a similar pattern. In all four subjects the reaction had completely disappeared within 24 hours.[38]

These rather scanty trials were obviously intended to prove the harmlessness of the medication, to ascertain the regular pattern of the reaction, and to establish a minimum dose (of 1 mg). By September, full-scale trials of the medication began in two departments of the Charité hospital, the III Medical Clinic directed by Professor Hermann Senator[39] and the auxiliary Internal Medicine Clinic for Men under the direction of Professor Oscar Fräntzel.[40] Expanded in subsequent weeks, this series of injections was supervised by Koch's son-in-law Eduard Pfuhl. It was designed to test the effect of the medication in various forms of tuberculosis and to verify its suspected diagnostic value.

The number of patients treated at this stage was small. By November 13, when he was finally able to buy tuberculin from Koch's associate Libbertz, Fräntzel had, aside from control injections, treated thirteen patients. He began his trials with four moribund patients, "far advanced cases in which the *exitus letalis* was only a matter of time."[41] He subsequently treated patients for whom a cure was possible and came to the conclusion that the course of the reaction was typical in every case.[42] In patients with slight cases of phthisis, improvement was measured in terms of weight gain, bacteria count in the expectorate, and general well-being. The twenty-six-year-old shoemaker August Weigt, for instance, had been complaining about symptoms like coughing and expectorating phlegm for fourteen weeks. He was described by Fräntzel as a man "in poor nutritional condition. Somewhat muffled on the right, to the second rib, intermittent rattling sounds, abundant expectorate, tubercle bacilli Nr. 7."[43] After fifty-six days of tuberculin treatment, the quantity of expectorate had diminished and been free of tubercle bacilli for several days, and the patient had also gained five pounds. In the case of the moribund patients on whom Fräntzel had conducted his first trials, "the injection of Koch's medication had not, to be sure, stopped the process," as Fräntzel put it in November, although the typical reaction had been observed in these patients as well.[44] Fräntzel subsequently provided the detailed description of a tuberculin treatment with its classic proceeding of gradually increased dosage and regular checks on fever and bacteriological status. He concluded "that this medication is effective in specific circumstances. The effects we produce with injections of small doses of the medication only appear in phthisics; furthermore, the medication influences the quantity and quality of the bacilli in formerly unknown ways."[45]

Some of the other physicians who had participated in the testing before tuberculin came on the market made even more favorable observations. Whereas Fräntzel had worked with cases of pulmonary tuberculosis, where prospects for improvement existed only if the case was slight, other forms of tuberculosis showed much better results. William Levy, in whose private Berlin clinic Koch worked as a consulting physician after September 22,[46] observed dramatic healing processes specifically in the skin tuberculosis known as lupus. Given the externally visible symptoms of this disease, "the medication's effect on the diseased tissue is revealed in a most convincing and concrete manner."[47] Moreover, lupus patients reacted promptly, and were therefore well suited to demonstrate the effectiveness of tuberculin. As described earlier, spectacular healing processes were observed after a small number of injections. After initially turning red, the diseased parts of the skin formed a scab, which eventually dropped off, leaving fresh, reddish scar tissue. Particularly impressive was the case of a "Fräulein K.," who had been suffering from lupus for eighteen years.[48] After the first injection, the patient ran a fever of 40.1, which receded the very next day. The diseased sites turned red; in addition, "her entire torso was covered by an exanthema resembling the eruption of scarlet fever," and "her gums disintegrated." After only the fourth injection, the improvement was unmistakable. "Reaction still visible, but lessened. General tendency to scarification." At the same time,

Levy noted, the local tuberculin reaction had "added tremendously to the reliability" of the diagnosis.[49] In doubtful cases, he felt, the reaction of the diseased tissue could now "provide a reliable indication." The use of Koch's medication appeared to be the basis of all future diagnoses and therapies, for the way it worked in lupus made visible a process that was assumed to take place in other forms of the disease as well. "We can now expect our treatment to achieve the outcome that satisfies us fully—the complete recovery of our patients."[50]

It is not easy to establish the exact number of tests to which the medication was subjected before it became commercially available in mid-November. In addition to continued trials in university clinics, whose results led to scientific publications, the medication seems to have been administered privately, for instance by Georg Cornet, a tuberculosis specialist in Brehmer's sanatorium and one of Koch's assistants.[51] In other clinical trials, such as those conducted in a department for diseases of the skin of the Charité Hospital, the reaction was found to be stable and the medication therapeutically and diagnostically effective. These findings were similar to those from Fräntzel's department.[52] The total number of injections for therapeutic purposes could not have been more than fifty. One of the recipients was probably Paul Ehrlich. Having diagnosed himself with tuberculosis in 1888, Ehrlich had set out for Egypt in 1889, intending to cure himself by means of the climate there. Following this clearly successful treatment, Ehrlich received nine injections of tuberculin between November 3 and 14, but his reaction was slight, causing fever of no more than 38.6.[53]

It should be said that until mid-November 1890, all the experiences gathered until then remained unpublished and were integrated in general terms into Koch's publication of November 13.[54] A week later they were published in the *Deutsche Medizinische Wochenschrift*—with the exception of one critical contribution that contested the use of the medication as a diagnostic tool.[55] In this journal's special issue on the topic, tuberculin and its use in humans were treated as largely accepted. Although Koch characterized his own contribution as a mere "orientation and survey,"[56] he did present his medication—which so far had been tested in fewer than fifty patients—as safe. He was firmly convinced that, thanks to the localized reactions of varying strength it provoked, it would become an "indispensable diagnostic tool." And he was equally convinced that not only lupus but also "incipient phthisis could unquestionably be cured with this medication."[57] Koch established the lowest (diagnostic) dose of the medication, to be injected subcutaneously, at 0.01 cc.[58] At this dosage, most people showed a general reaction, whereas tuberculosis patients experienced a strong and localized reaction that could be used for diagnostic purposes. In the context of this chapter, I am less interested in Koch's peculiar concept of the effect produced by the medication[59] than in such matters as the rules he established for its use, the importance he attached to clinical trials compared with that of his animal experiments, and the general significance he attributed to his therapeutic proceeding.

Except in the case of control injections, tuberculin was to be administered in increasing doses, starting with the minimum.[60] A weakening reaction to the identical doses indicated a healing process. The explanation Koch proposed for this was that due to the necrosis that the medication had caused in the affected tissue, not as much reaction-producing tissue was available: "Since each injection causes a certain amount of the reaction-producing tissue to disappear, proportionally ever larger doses are needed to bring about the same effect as before."[61] A tuberculin therapy thus consisted of injections of the medication in increasing doses, depending on the patient's reactions, over several days or weeks. Increasing the dose to 500 times the original (diagnostic) dose over a span of three weeks was a distinct possibility! As we saw, lupus was considered curable, as was incipient phthisis. In the advanced stages of pulmonary tuberculosis, the recommended course—probably inspired by the experiences in Dr. Levy's clinic—was a combination therapy in which a preparatory series of injections was followed by the surgical removal of the mortified but still bacteria-containing tissue.

As for Koch's animal experiments, their usefulness—aside from allowing him to study the mechanics of the tuberculin effect—was severely limited by the extreme difference in susceptibility between guinea pigs and humans. An important result of the trials with humans carried out between September and November was therefore the formulation of a human model of the tuberculin reaction, which set out the minimum dose, a typical reaction, its interpretation, and a schedule of applications to the point where further exploration could be entrusted to medical practitioners.

As for the importance of the procedure, in the eyes of its inventor it was more than a contribution of bacteriological hygiene to the therapy of tuberculosis. Koch felt that when it came to the diagnosis and therapy of the disease, the medication entitled him to set standards for clinical medicine based on bacteriological know-how.

> Until now, demonstrating the presence of tubercle bacilli in the sputum was considered a rather interesting but not essential activity, which, while it made for a sure diagnosis, had no other benefits for the patient and was therefore all too often omitted... This will have to change now. A physician who fails to use all available means, particularly the testing of the sputum for tubercle bacilli, to diagnose phthisis as early as possible will be guilty of severe neglect of his patient... In doubtful cases, the physician should use a test injection to obtain certain knowledge about the presence or absence of tuberculosis.[62]

Statements of this kind make one wonder whether it was really public pressure that caused Koch to begin selling the medication prematurely on November 13.[63] The course of events can be equally well interpreted as a skillfully orchestrated marketing campaign: Koch's publication of November 13 coincided with the extremely favorable trial reports of Fräntzel, Levy, and Köhler, which thus served to test and to advertise the medication. At the same time, as I pointed out above, skeptical observations coming from the Department of Internal Medicine, where the medication had been tested beginning on September 11, were not included in the special issue of the *Deutsche Medizinische Wochenschrift*.[64]

The climax of the introductory campaign was a public demonstration of tuberculin injections on November 16 at the clinic of the surgeon Ernst von Bergmann. The special importance of this event was signaled by the presence of the minister of culture Gossler, the medical officer of the general staff of the Prussian army, von Coler, Friedrich Althoff, and others.[65] Bergmann, the director of the surgical clinic of the Charité Hospital and one of the most prominent medical men of his time,[66] had treated a few patients with tuberculin since November 6. On the evening of November 16 he presented these patients, along with a group of others who had been injected that morning, and even carried out some injections during the demonstration. The public thus had the opportunity to observe the technique, the course, and the result of a tuberculin treatment firsthand. Koch had expressed the wish for such a demonstration to Bergmann.[67] And indeed, it was a good way to use a small quantity of tuberculin to acquaint a large number of physicians with the use and the effect of the medication. The patients reacted in a positively exemplary fashion: the five who had been injected that morning showed the usual symptoms, with fever between 40.1 and 41.2 Centigrade, violent shivering, and nausea.[68] Since all of them were suffering from lupus, they provided a vivid demonstration of the effect of tuberculin. It consisted, above all, of a high fever and the reddening of the affected parts of the skin and was strongest in the recently affected areas.[69] Five more patients were injected in the presence of the invited guests. The audience was also invited to observe all these patients' development over the next few days by visiting the clinic between 10 A.M. and noon. That evening, Bergmann intimated that, as far as he was concerned, ten days of testing would have been sufficient to convince anyone of the value of this medication: "We harbor the hope that the discoverer of the tubercle bacillus has also found the means of inhibiting and eventually banishing this scourge . . . Today we leave here with a salute to the greatness of this scientist and with gratitude to our famous and generous colleague."[70]

The "tuberculin high" to which the medical world succumbed for a time starting in mid-November 1890[71] was thus decidedly produced—in addition to general prerequisites, such as the prestige of its inventor and his methods—by several clinical tests that, especially in the case of Bergmann, were almost propagandistic in nature. Generating publicity for an important event by staging a semipublic demonstration before an audience of colleagues and prominent figures was frequently done in the nineteenth century—one only has to think of Thomas Morton's demonstrations of laughing gas in 1846 or Pasteur's demonstration of his anthrax vaccine at Pouilly-le Fort in 1881.[72] Here the result was the public acceptance of the tuberculin reaction as perfectly predictable, salutary, and—with the proper precautions—safe. Testing in humans, for which Koch had called in his publication of November 13, no longer seemed to require special justification. Experiments of the kind that Fräntzel had carried out in moribund patients, where therapeutic success was practically impossible, subsequently became more and more prevalent.

### IV.2.2. A New St. George

Considering that by November 13, 1890, tuberculin had probably been tested in hardly more than fifty patients and that the medical public was completely uninformed about the composition and production of the medication or about the animal experiments on which it was based, one realizes that the enthusiasm with which the medication was greeted had assumed the character of blind euphoria. While the scientific publications published reports of hitherto inconceivable cures,[73] the daily press overflowed with hymns to the scientist. The leading newspapers had daily columns under such titles as "Dr. Koch," or "Koch's therapeutic procedure."[74] Berlin became, in the words of the *Vossische Zeitung*, a "place of pilgrimage for physicians from every country."[75] To judge by the numbers of "unofficial" clinics and similar institutions that suddenly sprang up, the patients too went along: a coffee house in Berlin, for instance, could turn into a tuberculin sanatorium overnight.[76] The *Vossische Zeitung* reported on rumors circulating a week before the meeting and published the complete text of Koch's article in the *Deutsche Medizinische Wochenschrift* as early as November 14. It was accompanied by an allegory of the doctor as a "new knight St. George," who, seated on a horse named Research, carries a microscope as the sword with which to slay "the hydra Tubercul.Bacil."[77] By November 16, a definitive judgment was in place: "He [Robert Koch] has given the world a safe means with which to combat the angel of death named Consumption, for if that dreadful illness is recognized in time and treated properly, it is curable."[78]

And indeed, tuberculin—at that point still called by such names as "Koch's medication," or "Koch's lymph"—was very widely used immediately after November 13. In addition to the Prussian university clinics, which received it directly,[79] it was also used in lung sanatoria, private clinics, and by private individuals.[80] By the end of the year, at least sixty-nine scientific papers on the subject, case studies for the most part, had been published, twenty-four of them in the *Deutsche Medizinische Wochenschrift*, which was to bring together the largest number of contributions.[81] Some of them reported dramatic cures, such as the case of the eleven-year-old girl whose lupus was treated in Hamburg beginning on November 21. Despite a (mistakenly) low initial dose, the child reacted dramatically with high fever and reddening and swelling of the diseased parts of the skin. The injection of a second, tenfold higher dose was followed by the crisis and the cure.

> 23 November 12 noon, injection of 0.01 [cc]. 5:30 PM violent shivering lasting 3/4 [hour], patient in great distress, nausea. 5:45 PM, beginning of serious exudation onto lupus-affected places. 6:30 PM, temperature 40.3 C, vomiting, great fatigue, respiration 72, very frequent, barely perceptible pulse. Camphor injection.
>
> 7:30 PM, scarlatina-like exanthema over the entire body. Intense pain in the face. Strong exudation in the lupus-affected places.
>
> 24 November. After a quiet night's sleep, waning of the local and general reaction, beginning formation of a bark.
>
> The swelling has now entirely disappeared, the barks have gradually detached themselves, and the place of the lumpy lupus knots has already been taken by fresh, smooth, red scars.[82]

Dramatic, and even dangerous reactions were seen as signs of the healing process, and the lightning-quick cure described above could easily be complemented by others. Thus the *Münchener Medizinische Wochenschrift* reported on a tubercular tumor that dropped off in its entirety with amazing speed after a few injections.[83]

Anyone who had tuberculin could barely cope with the onslaught of patients. Georg Cornet had received the medication as early as October. He initially treated a few patients Koch had privately sent to him and responded to applications for treatment addressed to Koch. Although he was not affiliated with any of the Berlin clinics, "the number of patients hoping to be saved was growing every day. Because of some indiscretion my name was linked to the medication in a newspaper . . ., and now there was no stopping. Like a growing avalanche, letters and telegrams from all countries and in every language reached Koch and were passed on to me, and within a few days my correspondence also numbered in the hundreds."[84] Paeans like the following, which compares Koch's medication with the manna of the Old Testament, were obviously meant seriously. One stanza will suffice:

> Radiant dew drops now moisten the earth,
> Bright pearls of living light against the somber ground,
> And the sickroom resounds with a heartfelt sob,
> A shout of delight that pierces the sky:
> "He cures us, he cures us all!
> Now I may kiss you, my child, oh my world."[85]

Comparable praise of Koch was prevalent in the scientific literature as well, and the recipient enjoyed a string of honorary citizenships, decorations, and other honors.[86] In the provinces the "testing" of the medication could assume the character of a popular festivity. "Greifswald too finally saw the great day when the first inoculations with tuberculin were to be carried out at the Clinic for Internal Medicine. It was celebrated rather like the laying of a cornerstone or the unveiling of a monument. Against a background of laurel bushes, doctors, nurses, and patients in snowy-white garments and the chief in his black cutaway were lined up: Address by the internist, injections for a chosen group of patients, thundering hurrahs for Robert Koch."[87]

It is particularly interesting to investigate how the image of the tuberculin reaction and the assessment of its diagnostic and curative value developed after November 13. How was the medication used, and what experiences resulted from it? Judging by the published case studies, the gamut of its applications was quite wide and continued to combine diagnostic testing with therapeutic use and control experiments, such as had already been practiced in the testing stage. Reports of rapid cures were frequent. Oppenheimer of Heidelberg, for example, reported that a female patient suffering from tuberculosis of the larynx was symptom free after only three injections.[88] Helfrich of Greifswald presented several cases of tuberculosis of the skin and the joints that had rapidly healed, and concluded his report "not in the hope but in the full certainty that in the diagnostic and therapeutic area Koch's medication represents a momentous step forward."[89]

Many publications mentioned therapeutic successes in their very title, and even though control tests in healthy subjects continued to be made, Koch's views of the tuberculin reaction were considered correct and proven at this time. One author declared that by November 21 the knowledge of the general manifestations of the tuberculin reaction was so solid that one could "give it short shrift."[90] One pathologist at Virchow's institute used patients' tissue samples to confirm Koch's theory of the tuberculin effect as a curative effect.[91]

However, confusing observations occasionally did crop up, even at this time. Bouts of fever sometimes lasted longer than expected. Hermann Senator who, as we saw, was one of the first to test the medication, realized that "the intensity of the reaction, as well as its duration and the rapidity of its onset is not related to the intensity or the extent of the tubercular process."[92] Sometimes reactions appeared even in patients who did not seem to have tuberculosis at all, for instance in an eleven-year-old syphilitic boy in Berlin.[93] Such deviations could be dealt with in two ways: reactions in patients who until then were not considered to be tubercular confirmed the diagnostic value of the medication, which revealed a hitherto hidden infection.[94] But they could also be understood not as contradictory but as supplementary observations. Thus Oskar Fräntzel, a zealous defender of tuberculin therapy, remarked that "again and again one encountered hitherto unknown phenomena."[95]

And indeed, the fact that tuberculin was to be used for both diagnostic and therapeutic purposes made for a great variety of applications and also constituted an extremely flexible framework for interpretation. In conjunction with the prevalent liberal attitude toward therapeutic experiments of the time, this explains why the fatalities that did occur were initially not blamed on the medication but attributed to the known poor condition of the patients—and experiments on moribund patients were considered ethically and scientifically unimpeachable. Heuristic experiments on moribund patients were also conducted. These were injections given for the purpose of studying the effect of the tuberculin reaction on the patients' pathological anatomy after death. At the meeting of the Verein für innere Medizin of December 15, 1890, for instance, two cases of "severe phthisis" from the clinic of Ernst von Leyden were presented.[96] Both these patients had severe pulmonary tuberculosis with complications and were clearly close to death. The first patient had been given four doses, increased from 2 to 5 milligrams, between November 26 and December 2 and had shown a rather weak reaction. Only the last injection had produced a temperature rise to the usual level above 39 degrees C.[97] The second patient, a "much deteriorated"[98] woman, received eight injections between November 21 and December 8, and her dosage was increased from 0.1 to 4 milligrams. With a maximum temperature of 38.5 degrees C, she too had a weak reaction; she died on December 12, 1890. The pathologist seized on these two cases not so much because they involved almost completely destroyed lung tissue but because of the varying effects tuberculin had on additional tubercular sites in both patients' bodies. Thus the diseased tissue in the esophagus exhibited typical signs of the reaction, while recognizably tubercular structures in the intestine hardly showed any change. Ernst von Leyden made it clear that in treating these two patients, a cure was not a possibility, and that this was a matter of obtaining interesting material for pathology studies: "As soon as we gained access to the new medication, these two cases were brought into treatment, albeit not precisely with a prospect of therapeutic success but rather with the intention of gathering observations and knowledge. A therapeutic success was not intended here; indeed it was quite out of the question . . . I also note that the two patients suffered no harm from the use of the medication, in other words , there can be no thought that their exitus was hastened in any way."[99]

Another case of interest was reported by the pediatrician Eduard Henoch to a meeting of the Charité physicians on December 4, 1890. This physician had studied the effect of tuberculin injections on children aged two to eleven who suffered from rare forms of tuberculosis. In addition to control experiments in healthy children, he also injected some who suffered from tubercular meningitis or peritonitis, not to cure them but because "to my knowledge the effect of injections on such things is not yet known."[100] Barbara Elkeles has drawn attention to the contradiction between Henoch's recognizably cautious attitude toward the use of tuberculin[101] and his willingness to give an injection to a deathly ill and comatose two-and-a-half-year-old suffering from tubercular meningitis.

In fact, the human experiment seemed justified to him precisely because of the patient's desperate condition: "I freely confess that I had some reservations about using Koch's injection here, the reason being that a medication capable . . . of causing major swelling and hyperaemia can become dangerous if used in tubercular swellings within the cranium . . . In this case, which came to us in the last stages of the disease, death was a certainty anyway, and so I decided to test this example."[102] Although the child's condition hardly changed between the beginning of the testing with four injections and her death, Henoch considered his experiment a success, since the comatose child had regularly shown a feverish reaction following each of the injections. After the child's death Henoch decided not to use tuberculin in such cases again, regardless of the stage the disease had reached. Two months later, he justified his experiment by calling it a failed attempt at a cure.[103] But then, in December, the autopsy results convinced him that—notwithstanding the fever reaction—the injections had caused no change in the diseased tissue and thus had been therapeutically ineffective. He also realized that his experiment was ethically questionable. What concerned him was not, as a modern observer would expect, the problem raised by nontherapeutic experimentation on humans but the fear that his experiment had constituted an act of euthanasia: "For I do believe that as a physician one must never use a medication whose effect of increasing the intracranial pressure unquestionably opens the prospect of shortening the patient's suffering, even though this prospect might be quite desirable for the unfortunate child and his or her parents."[104]

All in all, these first weeks when tuberculin came into use leave the observer with the impression of a sustained euphoria, a state in which most people essentially ignored the problems raised by this medication. Aside from warnings that "caution in choosing the initial dose is always called for,"[105] and general reminders that doses should be increased gradually, skeptical, much less critical, observations were practically never made. Only the Prague hygienist Ferdinand Hueppe—whose quarrel with Koch was of long standing—warned against the use of the medication, although he does not seem to have had access to it.[106] One reflection with which Ernst von Leyden opened a lecture at the Charité as early as November 27, was very unusual indeed:

> I cannot deny that the clinician finds himself in a peculiar position. We have received from the hands of a scientist of genius a medication for our use, yet we have no information about its nature; save for some vague conjectures, it is

shrouded in deep mystery . . . At its very first application, every observer will immediately realize that this is a very special juice. When injected, the smallest dose of 1–2 mg. of this substance provokes strong and violent phenomena, and a powerful, albeit temporary upheaval in the organism is bound to cause us the utmost astonishment.[107]

Implied rather than stated in von Leyden's text are two elements that were to assume major importance in subsequent months. To begin with, von Leyden was one practitioner who did not trust the promise of a bacteriologically based tuberculosis treatment that Koch had attributed to his medication. It is understandable that such reservations were expressed by an internist, for Koch's innovation only made sense in competition with internal medicine, its diagnostics and its therapies. Moreover, the allusion to Goethe's *Faust*—"*Blood is a very special juice*"—indicates that the observer was haunted by the thought that a strong reaction was not necessarily related to a strong, much less a curative effect.

### IV.2.3. Clinic and Laboratory

One of the most striking characteristics of the tuberculin treatment was the dramatic reaction seen in the patients and the fact that it could be observed most clearly in the external forms of the disease. In an age that had little experience with highly effective pharmaceuticals, the very intensity of the reaction was seen as proof of the curative effect and the innovative character of Koch's medication. In October and November 1890, this reaction had been thought to be both consistent and harmless, but by the end of the year it began to appear erratic. Variations in its strength, course, and effects came to light, so that its interpretation became increasingly difficult. Enthusiasm was replaced, as one observer put it, "by the many questions that are raised by experiments with this strange medication."[108] Whereas hitherto the effects of the injections had been analyzed in terms of their diagnostic and therapeutic value, questions were now raised about the reaction itself. The most important trigger for this was a series of deaths that were unlikely to have been caused by the poor health of the patients. In Berlin, for instance, a small child died within twenty-four hours following a single injection of 0.5 milligram.[109] Another case, in which Koch's associate Libbertz and a member of the Prussian House of Representatives, Dr. Graf of Elberfeld, were the treating physicians, attracted the attention of the Prussian Ministry of Culture, especially since here the diagnosis alone brought disaster. That is why we have a detailed account of the incident.[110]

In early December Graf had approached Libbertz and asked him to give one of his patient's a diagnostic injection. Although the state of this patient's health was bad, Graf felt that he "could assume that the tuberculosis had run its course, and in particular that fresh phthisis was not present, but that he nonetheless wished to confirm this in the manner outlined." The 2 milligrams administered were, according to Libbertz, "a very small dose."[111] However, the patient's reaction was dramatic: "Only three hours after the injection, high fever set in, up to 40°C, weak pulse that could barely be felt (about 150 per minute), accompanied by vomiting and unquenchable thirst. Death ensued twelve hours after the injection with manifestations of heart failure, but no signs of the reaction could be seen in the externally visible remnants of the tubercular processes." Although the likelihood of a causal relationship between the injection and the death was strong, the family's refusal to allow an autopsy of the deceased made it possible for Libbertz to declare any connection to the injection unproven and to postulate a latent tuberculosis as the cause of death instead.

Such incidents were not isolated occurrences, and even in cases where death did not ensue, the treatments sometimes caused such severe symptoms that it seemed questionable whether the patients would live long enough to be cured.[112] The generally accepted rule that the medication should be administered in gradually increased doses could especially lead to problems in the form of sudden and violent reactions. Recognized today as states of shock, these were considered random and dangerous phenomena at the time.[113] Ottomar Rosenbach, a clinician in Breslau who eventually became an astute critic of tuberculin, pointed out as early as December 1890 that "in the interest of the patient it is best to begin with a minimal dose, i.e. with 0.01 milligram, and to increase it slowly rather than quickly if there is no reaction."[114] Ernst von Bergmann—as we saw above, one of the most eager proponents of the medication in November—declared in December that, given his experiences since then, he "considered the use of Koch's medication in children under the age of 10 extremely dangerous."[115] Initially the intensity of the reaction had been seen as proof of its effectiveness, but by year's end it was perceived as potentially dangerous, and in several fatalities a connection appeared to be undeniable.[116] Not infrequently, the treatment was ended prematurely, either because individual physicians felt that it would be irresponsible to continue or because the patient demanded that it be discontinued.[117] However, statistics with which to assess the danger were not yet available in autumn 1890.[118]

But the emerging awareness of its dangerousness was not the only reason why questions were being raised about tuberculin. By year's end it also became clear that the reaction was far less predictable and uniform than had been assumed until then. Vinzenz Czerny of Heidelberg, who had used the medication since November 24, observed worsening of conditions that might well pose a danger to patients. The reaction, Czerny pointed out, "did not correspond to the quantity of the injected fluid . . . Some of the patients simply have very violent reactions to ½ mg, whereas others can tolerate 10 mg without a problem."[119] He felt that the effect of the medication was uncertain, and its usefulness questionable. "We saw violent local irriation without fever but also fever without a local reaction. For the time being a cause for these different effects cannot be provided."[120] Other physicians as well observed a multitude of irregular phenomena: "considerable deviations from the pattern . . . indeed fundamental differences apt to influence the attitude of the practicing physician in many useful ways."[121] Such reservations led to new questions, even among those who had a positive attitude toward tuberculin. Entire series of deviant fever reactions were found, and the most baffling observation was that healthy patients, or patients who were considered cured, could exhibit extremely violent reactions.[122]

In January 1891, finally, a number of critical assessments contributed decisively to discrediting both tuberculin and its inventor. Ottomar Rosenbach, known to posterity as an uncompromising critic of laboratory medicine,[123] attempted to classify the different fever reactions.[124] His analysis brought out the dangers of tuberculin use, but even more important, the types of reaction he observed appeared to be incompatible with the diagnostic and therapeutic claims made for the medication. Rosenbach's observations allowed for the conclusion that so far the process itself had not been properly understood. Habituation and disposition, he believed, provided a more convincing explanation than dosage for the nature of the reaction. It became conceivable that fever was a side effect rather than the main manifestation, and Rosenbach therefore wondered whether it would not be appropriate "to avoid the feverish reaction, so that one would not fall prey to the opposite error, that of the supporters of antipyresis at any price."[125]

Moreover, beginning in January 1891, increasing numbers of relapses were observed in those forms of tuberculosis that had solidified the reputation of tuberculin. Thus, Dr. Schwimmer reported on the success of his therapies for tubercular skin diseases, which he had "carried out precisely according to Koch's rules and instructions."[126] After observing the usual course of events, he had to realize that the improvement was no more than temporary and superficial. Dr.

Schwimmer gained "the impression that . . . the point-shaped lupus nodules located deeper in the tissue were untouched by the healing process."[127] While this did not raise fundamental questions about the curative effect of tuberculin, it did add a very different temporal dimension—days to a cure now became months.

The critiques of Schwimmer, Rosenbach, and other observers left open the possibility of an improved application. This was not the case for the criticism that Rudolf Virchow expressed on January 7, 1891, in a widely noted lecture before the Berlin Medical Society.[128] With the help of pathology specimens, Virchow was able to demonstrate that fresh tubercles had appeared at the edge of the tissue necrotized by the effect of tuberculin. His decisive argument was not the death of patients treated with tuberculin but rather the pathological finding: it was these fresh tubercles that raised doubts about Koch's theory of the nature of the tuberculin effect. As Virchow demonstrated in one particular case, the older pathological structures in the peritoneum of a deceased subject proved to be less affected by tuberculin than should have been expected, while at the same time small morbid nodules, so-called miliar tubercles, were found throughout the body. Theoretically it was possible that these had been present all along, but according to Virchow this was unlikely. In the course of his lecture he suggested that the injections were not only ineffective but in some cases positively dangerous. Rather than impeding the pathological process, they might in fact provoke its spread! Shortly thereafter, a study by one of Virchow's associates confirmed this suspicion. Its author was "compelled to state that under certain circumstances Koch's procedure can cause the outbreak of a miliar tuberculosis."[129]

By January 1891, then, barely two months after the medication had reached the market, potential critics had marshaled basic arguments that allowed them to explain irritation phenomena. Until then, criticizing tuberculin had seemed to be a daring move that was mostly taken by outsiders like the naturopath Heinrich Lahmann.[130] Now the mystery, which had originally been part of the sensation, proved detrimental to the substance itself. First voiced by Czerny and von Leyden, the demand that Koch publish the composition of his substance became very loud. This was also true for the request—no doubt prompted by Virchow's attack—that Koch divulge the details of his animal experiments. Pathology specimens from the guinea pigs he had cured with tuberculin would undoubtedly have greatly strengthened Koch's position. But he never produced them, and subsequently it proved difficult to replicate Koch's healing processes experimentally.[131] It was this part of Koch's work that best justified the expression "tuberculin swindle " coined by Virchow's student Johannes Orth.[132] As for the composition of the medication, on January 15 Koch did publish a rather general description of the substance, which turned out to be an extract of pure cultures of tuberculosis bacteria in a glycerin base.[133] Certain observers had already guessed at such a composition.[134] The self-confidence that Koch had displayed in autumn now made an unfortunate contrast with the questionable substance on which it was based. It became clear that in keeping the composition secret, the researcher had also sought to hide his own lack of knowledge, for Koch himself did not know exactly which substances were in tuberculin, let alone which of them had an effect.[135]

Thus the relation between Koch's view of the effect of tuberculin and the analyses of such clinicians as Rosenbach, Czerny, and Fürbringer can only be described as incompatible. The clinicians examined the effect on living patients. Their description of the tuberculin reaction made use of clinical parameters, such as fever, pulse, external symptoms, and so forth. Koch, by contrast, continuing the practices of his animal experiments, relied largely on autopsy reports. A collection of case histories he had saved for his own use in the winter of 1890–91 exclusively concerned deceased patients and usually contained detailed notes only on the autopsy report.[136] In light of this, Virchow's criticism took a different approach, which made it particularly dangerous. It dealt with the same subject, namely the pathological anatomy of the diseased tissue, and directly challenged Koch's proofs, for example, in the case of the cured—but vanished—guinea pigs.

It appears that since mid-December 1890, officials in the Prussian Ministry of Culture also took a more critical view of the medication, and not only because of the financial ambitions of its inventor.[137] As it became increasingly clear that tuberculin was not only a highly effective but also a potentially dangerous substance, they considered restricting its sale to pharmacies.[138] Among other considerations, this would have been an objectively justifiable reason to take the marketing away from Koch, and especially his associate Libbertz, who was in charge of the sales. Unfortunately this was not feasible because of the trust the ministry had originally placed in Koch. According to an internal memorandum, tuberculin was without question a secret nostrum, which meant that the controls provided by restricting it to pharmacy sales could not be implemented.[139] Juridically, this was a contradictory situation. The author of the memorandum pointed out that Koch's medication was either a secret nostrum or a registered medication that "could only be offered or sold in pharmacies." This clearly was not the case, and so it was indeed a secret nostrum, which under existing law should not have been marketed at all without a police permit, and whose sale without such a permit was subject to a fine. But since in November the Prussian minister of culture in person had assumed the responsibility for keeping the composition secret, the ministry's hands were tied as long as Koch did not divulge the composition of his nostrum. Faced with this quandary, the ministerial officials had no choice but to invoke the problematic status of secret nostrums in turning down Libbertz's request for an official approval of the use of tuberculin.[140]

Comparing the medical community's attitude toward tuberculin in November 1890 with that of January 1891, one sees an unmistakable change. The enthusiastic chorus that contained few critical voices had given way to a confused mixture of supporters and critics. As the discussion increasingly leaned toward criticism, the scope of the questions also widened. Whereas initially the critics wanted to find out about the therapeutic and diagnostic effect of tuberculin, by now the enigma was the medication itself, for its composition, effect, and safety were unclear.

This undecided situation is reflected in the reports that university clinics and pathological institutes submitted in response to an official request of the Prussian government dated December 1, 1890.[141] Usually dated December 31 and probably written in January,[142] these reports convey the impression of a certain perplexity: while most of the authors expressed a guarded optimism, they avoided clear-cut conclusions, referring to the short period of observation, the subjective nature of the symptoms in cases of internal tuberculosis, and so forth. Very few of the reports depicted tuberculin treatment as effective and safe. Among them were those of the two directors of the Moabit Hospital in Berlin, Sonnenburg and Guttman. But in this institution Koch had carried out the first tuberculin treatments, and moreover this municipal clinic was by far the largest center for tuberculin treatment in all of Prussia.[143] For Paul Guttmann, the head of the Department of Internal Medicine, tuberculin was still what it had been in the beginning, "an excellent reagent for tuberculosis,"[144] whose effects were clearly related to the dosage chosen. Guttmann even considered the reactions of patients whose life, in the view of other physicians, "might be in jeopardy," stable and safe.[145] He felt that the local reaction to the medication, which was known to provide a diagnosis, was "positively surprising in its precision."[146] It should be noted that the treatments at Moabit were not only the most numerous but also the most aggressive. Relating the number of patients to the amount of tuberculin used and the number of injections given, one finds that von Leyden at the Charité clinic had given 131 patients a total of 747 injections and used 7.375 grams of tuberculin. Guttmann, by contrast, with 50 percent more patients (196), gave almost 3½ times more injections (2,728) and used a well-nigh incredible amount of tuberculin: probably more than 100 grams.[147] But even in Guttmann's report a certain shift was discernible. Whereas initially all stages of phthisis were treated at Moabit, Guttman now emphasized the particularly good results obtained in concentrating on cases of fresh tuberculosis. Of 54 such patients he had treated, 42 were discharged as improved or even cured: "that is an improvement rate of 81 percent, which is extremely favorable."[148] In the same vein, Sonnenburg, the head of the Department of Surgery at the Moabit Hospital, recommended the so-called combination therapy.[149] In his view, there were two advantages to combining the removal of mortified tissue with tuberculin therapy: it prevented irregularities in the course of the fever reaction due to further injections and improved the prognosis by removing mortified tissue that might still contain bacteria. Sonnenburg was convinced that tuberculin treatment enhanced the possibilities of surgical therapy.[150]

The positions taken in these official reports were generally positive, although actual optimism was the exception. The usefulness of Koch's substance was seen as plausible but not proven. Moreover, the judgment varied from case to case. Not many physicians doubted the diagnostic potential of the substance, but many more doubted its therapeutic value. In external forms of the disease and in incipient pulmonary tuberculosis, the prognosis appeared to be much more favorable than in other manifestations. This is also reflected in the statistics reported:[151] of 1,769 patients treated, 28 were considered cured; 9 out of 10 of the patients cured of pulmonary tuberculosis had suffered from "incipient phthisis," and 14 of the 15 patients cured of external tuberculosis had had lupus or simple tuberculosis of the bones or joints. Deducting the 55 patients who had died from the total number of persons treated, one finds that for 28 cured patients there were no fewer than 1,686 who were facing an uncertain prognosis.[152]

In other ways as well, the official reports and the published case studies provide interesting insights into the testing of tuberculin and in particular the role of the patients. Several physicians reported an enormous onslaught of patients, which made a rational evaluation rather difficult. In January 1891, Dr. Leichtenstern reported such a situation from Cologne. He had to abandon his plan of "taking his time to study the effects of Koch's cure thoroughly and in all its ramifications"[153] and found himself constrained, "contre coeur, to expand the testing more than he would have liked to do."[154] On the other hand, it was reported fairly frequently that both physicians and patients were shocked by the violence of the reaction, and that sometimes the treatment was discontinued at the patient's request.[155] Control injections to healthy subjects or patients suffering from other diseases were usual, yet the patient's permission was obtained in only one case.[156] Although at the height of the euphoria such permission would not doubt have been easily obtained, this practice is surprising in the case of control injections. It is worth noting that those who received control injections were frequently physicians or medical students. This indicates a certain awareness of the problem and illustrates the contemporary practice of legitimizing risky therapies through heroic self-experiments. It is also remarkable that the publication almost always stated the name, age, and occupation of the patient but preserved the physicians' anonymity by using initials only.[157]

The skepticism that Koch's nostrum began to encounter by the end of 1890 did not mean, however, that the experimental use of tuberculin was now curtailed. No decrease in control injections can be observed, for instance. On the contrary, only now that the substance had become questionable was it more frequently used for frankly experimental purposes. Thus Julius Schreiber of Königsberg injected newborns and toddlers with tuberculin in order to study questions of heredity in tuberculosis. Finding appropriate subjects for testing turned out to be difficult for Schreiber, who made his experiments public on January 19 and continued them even thereafter.[158] A special problem was finding older children, who had to be tuberculosis-free: "It has remained difficult to obtain such children, which is why I have so far injected only one such boy, incidentally as punishment for some misbehavior at home. The boy was from a working-class family, where the mother had contracted pulmonary tuberculosis. At first the parents did not want to allow the injection, but then, since, as I said, the boy had done something bad, the father said, "All right, now you too will get an injection; this one will be able to take it, he is healthy."[159] By February 6, 1891, Schreiber had injected no fewer than forty newborns and had reached the remarkable conclusion that newborns showed no reaction whatsoever.[160]

Another interesting series of experiments was conducted in December and January by Erich Peiper in Greifswald. Making use of control injections, he focused on the tuberculin reaction of healthy subjects and nontubercular patients.[161] Since all twenty-two of the injected subjects showed strong reactions, this series of experiments was apt to raise doubts about the diagnostic value of tuberculin, as the author himself suggested.[162] Faced with the alternative of doubting either his own findings or Koch's substance, Peiper actually chose the former option. Three of his patients had shown very prompt and strong reactions. The author described his dilemma in one of the cases: "Of particular interest is case #22... The physical yielded no indication for the existence of a tubercular affection. And yet the patient reacted promptly to 10 injections (of up to 0.02). Repeated examinations for tuberculosis have so far remained negative. The only possible assumption is that there is a completely latent tuberculosis, or that the pat[ient] is particularly susceptible to this substance."[163]

The changed assessments of Koch's medication that came to the fore in early 1891 were thus not necessarily a matter of altogether rejecting the remedy. Rather, the quantity of secure knowledge had shrunk and the area of questionable assumptions had been considerably expanded. As yet the multiplicity of confusing observations did not affect the faith in the remedy.[164] Only a very few observers, such as Virchow or Rosenbach, had already asked the questions that eventually led to the discrediting of the remedy.

Its inventor also intended tuberculin to mark a breakthrough to the era of a bacteriologically oriented clinic. For quite some time, it would appear, most physicians agreed with this goal. This is indicated not only by the comparatively mild criticism Koch encountered, but also by the fact that bacteriological diagnostics, whose scant use Koch had deplored in October, was now considered standard procedure and apparently practiced as such. Medical criticism of Koch's secrecy as well was barely heard at this point, and a voice like that of Vinzenz Czerny of Heidelberg, who cautiously expressed some doubt in December 1890, was still decidedly rare: "It would be desirable to have the secret in which the remedy is shrouded lifted soon. I do not want to speak of the unease that creeps up on every physician when he has to operate with a secret nostrum. But the enthusiasm with which even the most sober researchers have greeted Koch's discovery is to a large extent motivated by the hope that Koch will open fruitful avenues to the curing of infectious diseases once he has revealed to us the nature of the product and the paths that have led him to its discovery."[165] After Koch's disclosure of the remedy's composition and the criticism voiced by Virchow, Rosenbach, and others in January, hopes like those expressed by Czerny were difficult to maintain. The shift in opinion that now began to emerge was not uniform, however, but reflected the experiences of the different disciplines. There was "a certain contrast between the surgeons' enthusiastic reports about cures in cases of external tuberculosis and the objective and recitent statements about the value of *Kochin* (tuberculin) in pulmonary tuberculosis by clinicians in internal medicine."[166]

In this sense, the criticism that the internist Ernst von Leyden included in his official report marked a turning point. He not only criticized the secrecy and the aspects of human experimentation connected with the procedure but also contested Koch's claims to the founding of a bacteriology-based clinical practice, which he, von Leyden, considered as a competition to his own discipline, internal medicine. "If, on the basis of my experiences so far, I am to summarize what I know about the effect of Koch's secret remedy on humans, I can do so only with the utmost reserve. For to begin with, the period of observation must be called very brief. Secondly, it is difficult to judge the effect of a remedy about which one really does not know anything concrete . . . Moreover, we physicians were obliged to experiment directly with sick patients."[167] This statement led to a real outburst of anger. Von Leyden's choice of words shows that he felt that his competence was under direct attack from the claims made for tuberculin and that it was necessary to defend medicine as the art of healing against the presence of laboratory medicine at the sickbed: "I would like to add that I consider it a disparagement of our art and science to think that the treatment of tubercular patients can be based on nothing more than subcutaneous injections . . . Simply giving a subcutaneous injection does not demand a physician; any valet can do it. Anyone who sees the art of medicine as nothing more than a schematic mechanism should stay away from the sickbed. The consequences of such a barbaric proceeding cannot fail to appear, and indeed have appeared already."[168]

## IV.2.4. "The Medical World Is Deeply Divided"

In his criticism in the winter of 1890–91, Ernst von Leyden was the first to make an explicit issue of human experimentation in the use of tuberculin. Whereas in October 1890 Koch had called on the physicians to "experiment with the remedy,"[169] von Leyden criticized that "physicians were obliged to experiment directly with sick patients."[170] Considering that for contemporaries, one of the ethical criteria for judging a physician's actions was the distinction between (justified) therapeutic and (problematic) nontherapeutic human experimentation,[171] it becomes clear that initially, in autumn 1890, tuberculin was indeed seen as a therapeutic, given its character as a miracle drug. Moreover, even its purely experimental use appeared to be legitimized by the harmlessness of the injections and the presumed moribund state of the experimental subjects. Once doubts arose about the value of the therapy and its harmlessness, the above-mentioned distinction became relevant to those judging tuberculin therapy: in the face of failure, they were able to perceive what they had earlier considered therapy as human experimentation.

Not that reports of successes with tuberculin were entirely lacking after the winter of 1890–91. But now this remedy became one medication among many others—and given the idiosyncratic nature of the arsenal for tuberculosis, this was not necessarily a disadvantage. Climate therapy in the sanatorium—the usual treatment for the dominant form of the disease, pulmonary tuberculosis—was easy to combine with tuberculin treatment, to the extent it was considered useful at all. Surgical intervention was also seen as compatible with the administration of tuberculin, as we have seen. But then, around 1890, surgical therapy was still limited to the external forms of the disease, and the surgical treatment of pulmonary tuberculosis (pneumothorax) came into its own only in the twentieth century. Tuberculin was thus in true competition only with drug therapies, whose effectiveness was usually difficult to evaluate as well.[172] Hence Koch's remedy was used as a therapeutic until far into the twentieth century.[173] In 1891, however, tuberculin became an object of scandal, for its use acquired an odor of human experimentation. Dr. Wolff, the director of the Gröbersdorf Tuberculosis Clinic, who had originally welcomed the remedy, expressed this insight when he called the tuberculin euphoria a large-scale experiment. He pointed out that while it could not be justified by therapeutic successes, it had yielded incomparable information about the remedy itself.[174] The already mentioned Breslau clinician Ottomar Rosenbach went considerably further in a study he published in March 1891. Among the many reasons he enumerated for the outbreak of the tuberculin euphoria, one touched on the experimental character of its testing. Rosenbach seems to have been the first to criticize "the astonishing fact that the remedy was also a diagnostic tool."[175] This claim had opened an excessively large and—from today's perspective—tautologically structured area of discussion where, for instance, in cases of "incipient phthisis" (apparent) diagnoses and (apparent) cures mutually reinforced each other. "This fostered feverish speculation and the obsessive use of seemingly exact information for formulating real scientific hypotheses."[176] The result was that such matters as the transferability of findings from animal experiments were largely ignored, and initially the enormous fever reaction to tuberculin was simply accepted as a healing reaction. "Physicians, who in good faith believed the remedy capable of producing the strongest effects, also followed the old principle *post hoc ergo propter hoc*, the source of so many errors. They therefore made the unforgivable error of considering everything that occurred while or after they had injected the patient as an effect of the injection."[177]

The reference to the fever reaction is enlightening indeed, for no other symptom of tuberculin use was observed quite as intensively and underwent as fundamental a change in valuation.[178] Whereas initially observers believed that its onset was proof of the effect of tuberculin, and that it could be controlled, by the spring of 1891 their experience had been broadened by several cases of death, which were sometimes accompanied by high or prolonged fever. At that point both critics and supporters of tuberculin attempted to bring about the healing reaction without a fever reaction. Even two admitted defenders of tuberculin therapy now recommended a lowering of the initial dose from 1 milligram to one-tenth that amount, since "the fever reaction following it was . . . sometimes stronger and more prolonged than desired."[179] The already mentioned director of the Gröbendorf Tuberculosis Clinic, Wolff, who took a very critical view of tuberculin in the spring of 1891, felt that avoiding the fever reaction altogether was the best way to minimize the dangers of tuberculin treatment. "The thought . . . that it might be possible to get along without fever, with very small doses and fewer injections and yet achieve success is so consistent with my own experiences that I heartily endorse it."[180]

Only a few years later the concept of the fever paroxysm as a healing reaction came to be seen as a hopeless anachronism.[181] And once tuberculin was seen as potentially dangerous, earlier observations appeared in a different light. One might ask, for instance, whether the presumptively moribund patients who had been subjected to experiments in the earliest phase had actually been in that state.[182] To be sure, there were also therapeutists, such as Georg Lichtheim of Königsberg, who thought that contraindications were "thought up at a desk rather than gathered from practical experience."[183] But many physicians now admitted that the cases they had treated had not had a favorable outcome. The pediatrician Henoch reported in early February 1891 that of the twenty children he had treated, "many had gotten worse,"[184] and that even the one child in whom he had initially observed an improvement was now worse than before the treatment.

The sharpest critique of tuberculin to that point came in late February from the clinician Friedrich Schultze of Bonn.[185] Referring to Peiper's experiments and modifying his own position in the official reports, he now denied that tuberculin had any diagnostic value at all. He pointed out that fever spikes had appeared in patients who could reasonably be considered nontubercular, while such fevers sometimes failed to appear in obviously tubercular subjects. "My experience," Schultze wrote, "makes it impossible, at least for me and at this time, to make a firm diagnosis of latent tuberculosis whenever the reaction presumed to be characteristic for tuberculosis occurs."[186] Therapeutically, matters were even worse: ten of twenty-three patients had died, and a connection with the injections was probable: "Quite often it became clear that the day when we started the injection marked the beginning of a continuing deterioration."[187] Schultze felt that it was wrong to have treated severely ill patients at all, and he laid the blame on Koch, who had not disclosed the potential dangers of tuberculin use. As for the light cases touted by the supporters of tuberculin in the spring of 1891, Schultze—along with von Leyden and Rosenbach—pointed out that these had not been called incurable before tuberculin either. Indeed, before the introduction of tuberculin, none of these patients would probably have been hospitalized, so that there were no standards for comparison. The only thing that had actually been proved was the harmfulness of the remedy: "Here is how I see the matter: if the remedy can be harmful in advanced cases of tuberculosis and especially if it might be capable of spreading the tubercle bacilli, then there is no reason why this cannot also happen in cases of new infections."[188] Schultze made it clear that he had stopped administering tuberculin in February, except that he still used it at a patient's express request.[189]

Given Koch's unwillingness to disclose the results of his animal experiments, several researchers tried in February 1891 to replicate them. Neither Gravitz, who worked with monkeys, nor Baumgarten, who used rabbits, were able to reproduce the healing processes observed by Koch. Baumgarten's attempts to cure tubercular rabbits actually pointed to the dangers associated with tuberculin use. For the most part the treated animals died more quickly than the control group, and the autopsy findings were the same for both groups.[190]

Koch's eventual disclosure of his remedy's composition also gave the Prussian Ministry of Culture the desired chance to bring the sale of tuberculin under government control. Whereas in December 1890 the ministry, owing to its own collusion in the secrecy, was forced to postpone such plans, Koch's publication of January 1891 made it possible to implement them.[191] No doubt the restriction of sales to pharmacies had a certain significance in the financial aspect of the tuberculin affair. On the other hand, the increasingly skeptical medical discussion, in which the presence of bacteria in tuberculin—they eventually turned out to be dead—caused a great deal of trouble, motivated Koch to disclose its composition. It was only then that restriction of sales to pharmacies could be imposed. An internal memorandum of March 21, 1891, on "Koch's remedy against tuberculosis (tuberculin)," by Friedrich Althoff clearly shows that officials in the ministry already considered this medication a failure. Now the man of the hour was no longer Koch, it was Virchow. Althoff saw him as the one "whose critical stance was particularly meritorious because it saved German science the embarrassment of being led to a critical and questioning attitude by an impulse from abroad."[192]

The high point of the medical criticism of tuberculin came at the Tenth Congress for Internal Medicine in Wiesbaden, where on April 7 the discussion focused on physicians' experiences with Koch's remedy.[193] To be sure, this audience still heard reports of cures;[194] but since internists such as von Leyden, Rosenbach, and Schultze had recently been among Koch's sharpest critics, critical assessments were prevalent. The speakers clearly felt the need to distance themselves from the euphoria of the late autumn, which the chair of the session, Bernhard Naunyn, branded as an "orgasm" in his opening statement.[195] The only aspect to be considered more or less established was the product's diagnostic value, but otherwise reports of severe side effects, discontinued treatments, and fatalities abounded. Paul Fürbringer, who had lost 15 percent of his patients, vividly described his disillusionment as a personal conflict between his attitudes as an experimental scientist and as a physician: "The present speaker allowed Koch's discovery to affect him with the full enthusiasm that he was bound to feel, believing that it was the most beneficial deed of experimental pathology for medicine, for the art of healing. But with the passing of time his self-awareness as a physician increasingly came into conflict with that enthusiasm, and his zeal for the treatment receded step by step, so that today he no longer dares to use the remedy when treating patients with pulmonary tuberculosis."[196]

In a similar manner Friedrich Schultze also declared that administering tuberculin was incompatible with his self-understanding as a physician. He said that "since mid-February he had not dared touch the syringe, having absolutely no criteria for judging how the patient would fare once the unknown agent had entered his body."[197] The chair of the session, Bernhard Naunyn, radicalized the argument advanced by several speakers such as Dettweiler, Fürbringer, and Schultze, to the effect that the tuberculin craze had changed the composition of the cohort of patients so much that it had become impossible to estimate the usefulness of the remedy. In his summary of the discussion, Naunyn expressed the suspicion that the cures involved cases that earlier would not have been treated at all.[198]

Another culminating point in the discussion was the session of May 8, 1891, in the Prussian House of Representatives, which began with a budget debate for the continuing construction of the Institute for Infectious Diseases. On this occasion certain critical questions were raised—with reference to the Wiesbaden Congress[199]—about the boundary between therapy and human experimentation. Referring to the mission statement of the institute, which was to include a "research department for experimental work,"[200] Delegate Goldschmidt made the following statement:

> You will admit, gentlemen, that the expression "experimental work" is not exactly well chosen in this context. I am here treading on ground where I am not at home; but as far as I know, there has never been a direct connection between an experimental department and a hospital. But if we have here a medical institution that is directly connected with an institution that conducts experiments, it is only natural that many people will believe that the results of experimental procedures and scientific research could immediately be applied and tested on the patients' bodies.[201]

Thereupon Dr. Graf—who, as we saw earlier was working with Arnold Libbertz—defended the project with the pat assertion that "experiences had to be gathered," even though some of them were "rather sad," but that this did not diminish the high scientific value of the product. In his reply to Graf, Delegate Broemel then reminded the audience that the testing of pharmaceuticals had to obey not only the rules of "strict and objective science" but also adhere to "the inviolable principles of humanity." Critical in particular of the secret nature of the remedy's composition, Broemel developed the argument that the scandal surrounding tuberculin had uncovered problems in the development and testing of pharmaceuticals that in the past had not been seen very clearly: "Gentlemen, I am convinced that the simple remark of the previous speaker, namely that the high scientific value [of the planned institution] should be the main concern of this House, must not cut short the discussion of this question, which also has a very different side—to put it succinctly, a generally human side. An investigation of this human side seems particularly necessary in light of the events of the last few months."[202]

Now that the rights and duties of physicians and patients were in jeopardy,

Broemel continued, they had become visible. The medical profession, he said, was not "a closed guild, where, as in the master singers' guild, only the expressly appointed 'watchers' could point out that a mistake had been made." Instead, "the physician's art, like any other practical science, ... has to make its case before the court of public opinion, before the public, which is most directly concerned in this question, since its health is involved."[203] He demanded that in the future, "the lay public will also have to protest against the unscrupulous and reckless use of a highly toxic secret nostrum for therapeutic purposes."[204] Looking at Broemel and Goldschmidt's choice of words, it is striking that in the spring of 1891 these critics of tuberculin use could not formulate their objections by referring to therapeutic experiments, since that concept did not exist at the time. Instead they criticized the administration of tuberculin as human experimentation. Also invoking general ethics, that is, the humanitarian principles cited by Broemel, they ended by calling on the physicians to level with public opinion. This indicates that in the wake of the tuberculin scandal, the concept of mandatory consent for therapeutic trials—which until then had been applied mostly in connection with surgical therapies—now became relevant to the area of drug therapy.[205]

By the summer of 1891, then, the testing of tuberculin had in a sense come to an end. After a period of enthusiasm and disappointment, interest in this matter was generally on the wane. "The medical world is deeply divided," one observer commented in speaking astutely of the very few who still defended tuberculin. "Only relatively few still adhere firmly to unconditional acceptance, but for many of those who continue to have complete faith in the infallibility of the remedy and the method, this is more a matter of the undoubted authority of the discoverer than of incontrovertible observation."[206] Even in Robert Koch's own notes of this period, a slightly sarcastic tone can be detected, which suggests doubts about his own work more strongly than the discoverer of tuberculin was normally willing to admit to. In the private collection of case studies referred to above, Koch added a comment to the report on a patient who had died in May 1891 after having been treated since October 1890: "(*Pride and joy* of Fräntzel's department, had gained 20 pounds)."[207] For G. Siegmund, the author of the above-mentioned report, the "modern" therapeutic agent tuberculin actually signaled "an abandonment of the modern concept of pharmacology."[208] He pointed out that the belief in its effectiveness was not based on positive knowledge and that the secret nature of its composition ruled out a rational examination. In his opinion, the use of this remedy was "haphazard" and "did not meet the standards of modern science; and its introduction into the sick human body was irresponsible."[209] Over the years, more and more facts had turned out to be untrue. In the end Siegmund saw the curative effect of tuberculin just as we see it more than a century later: as an act of faith.

## IV.2.5. Experimental Therapy

Looking beyond the summer of 1891, the testing of tuberculin signifies a remarkable start to the era of modern chemotherapy. Tuberculin, which reached the market almost simultaneously with the first patented drug, antipyrin, in 1890,[210] was beholden to tradition, since it adhered to such concepts as the curative effect of fever. It also did not represent a scientific advance, considering that—except for innumerable attempts to create an improved tuberculin—Koch's concept of bacterial therapy was never successfully adopted by anyone else. Nonetheless, the product laid claim to being a highly effective specific, and it was seen as a scientific advance even by those who rejected it as a remedy. This is confirmed by the fact that the Hoechst company (Farbwerke Hoechst) took advantage of the commercial decline of tuberculin to bring down the price it paid for the know-how needed to set up a bacteriological research department. Clearly, tuberculin was thought to have potential even after it had failed.[211] Once a contract was signed transferring production from Berlin to Hoechst, Koch's associate Libbertz went to work for the company and soon became head of the bacteriological laboratory.[212] The failure of tuberculin also provided some lessons; it is hard to imagine that Behring and Kitasato would have tested their diphtheria serum as extensively as they did, and that the medical community would have been as hesitant to accept that product without this experience.[213]

The discrediting of tuberculin as a medication thus did not spell its end as a research object. Not only did it remain a therapeutic among many others,[214] it also continued to be studied. A prominent example of this involves the psychiatrist Julius Wagner von Jauregg, who from the winter of 1890 until the 1920s examined its therapeutic effect on psychoses, syphilitic paralysis, and other mostly psychiatric disease phenomena. In the context of fever therapy, the tuberculin reaction was an attractive option that after the turn of the century, according to Wagner von Jauregg, yielded to the more effective artificial infection with such diseases as malaria.[215]

By the summer of 1891, even the German medical community had a negative, often critical attitude toward tuberculin, whereas abroad the discussion had been critical for some time.[216] It was to be expected that under these circumstances physicians increasingly shied away from further experimental therapies, but this was related to a changed attitude toward tuberculin rather than toward experimental therapies. Those who continued to believe in tuberculin felt free to work with it. As for Koch, he remained convinced of the therapeutic value of his remedy to the end of his days.[217] Moreover, the hospital section of the newly founded Institute for Infectious Diseases afforded him better working conditions than he had had before. He was no longer dependent on local clinics for patients with whom to experiment. It is therefore hardly surprising that in October and November 1892 the new institute became the site of a series of human experiments with tuberculin that, unlike the earlier testing of the product, could be conducted in total secrecy.

In October 1892 Koch had been asked to give an opinion on a preparation named tuberculocidin that Edwin Klebs had developed in Bern.[218] Klebs claimed that he had improved tuberculin to the point where it was free of side effects, and in particular no longer produced fever. Unlike Koch's remedy, which was supposed to work on the diseased tissue, tuberculocidin was meant to attack the bacteria directly.[219] Klebs had tested his preparation in Bern on both humans and animals. Now Koch checked it, in agreement with Klebs, and sometimes in his presence. They tested for the absence of a fever reaction, for the curative effect, and for the effective bactericidal mechanism postulated by Klebs. Since in the hospital wing of the Institute for Infectious Diseases, tuberculin therapy seems to have been a continuing practice, there was no lack of appropriate human subjects for experimentation. As for the fever reaction, "some patients who reacted strongly to very small quantities of tuberculin were injected with tuberculocidin alternately with tuberculin." In that case there was no fever reaction, and on this point Klebs's claims were clearly justified. In order to test the curative properties and the bactericidal effect in a second phase, it was decided to "administer the tuberculocidin treatment to six patients who did not have very high hectic fever and in whom the disease process was not too far advanced." Two of them, who had only slight temperatures of up to 38 degrees C, "experienced a decided improvement after several weeks"; the fever ceased, and both were discharged "at their own request" as "considerably improved." The remaining four patients, whose temperatures had initially been higher, "did not show a clear change, even though two of them were treated for weeks—at the recommendation of Prof. Klebs, who saw these patients when he visited the Institute—with what he considered the highest permissible dose."[220]

Koch interpreted these findings in analogy to his tuberculin, which was also effective in slight cases but had little chance of improving advanced tuberculosis. As for testing the bactericidal effect of tuberculocidin, the team now decided to conduct an experiment that became possible because of the presence of an unusual patient: "In the hospital wing of the Institute we had a patient who had earlier been treated with tuberculin and was thereby much improved. The disease process had been arrested, but he still had in his right lung a fairly large caverna, and his expectorate contained a quantity of tubercle bacilli that had remained unchanged for a long time." The constant bacteria count in the patient's expectorate made it possible to examine his sputum in conjunction with regular injections of Klebs's medication. Since the patient tolerated these injections very well, it was possible to look for even very small effects. To this end "we eventually went to doses that were so high . . . that they could probably never be used in normal practice, given the costliness of such a course. The fact is that eventually we gave as much as 5000 mg. a day." We can assume that the injections were given subcutaneously, as was tuberculin. To this extent, the experiment must have been quite painful, at least for the patient, and if it had involved Koch's tuberculin, it might well have endangered his life. With Klebs's medication, however, the result was negative only in the sense that practically nothing happened—simply a temporary decline in the number of bacteria, which at times exhibited a "brittle look." This alteration had frequently been observed with tuberculin use as well; it therefore could hardly be seen as a specific effect of Klebs's medication, even if it had been permanent and more pronounced.[221]

The report does not mention that any of the patients involved were asked to give permission for treatment. At least in the cited case of a potentially very dangerous medication for a nonmoribund patient, this does seem surprising. Even though many physicians had learned from the tuberculin experience to perceive new drug therapies as human experimentation potentially in need of legitimation, this was not the case with the bacteriological hygienists at the Institute for Infectious Diseases. These men held fast to the attitude with which they had first approached the matter, aided by the fact that they now had access to their own hospital ward. Here doctors could conduct clinical research in their own way, without contact with the public or potential competitors.

As for the medical and general public, it does not seem as if the tuberculin affair had prompted anyone to engage in durable reflections about the boundaries between therapeutic and nontherapeutic human experimentation. Memories seemed to fade, and analyses of the kind that Rosenbach or Siegmund had produced remained the exception. Nonetheless one change can be noted. It was not so much that new standards had been established as that there were lingering doubts about the scientific rigor and the humaneness in the testing of Koch's remedy. After a south German satirical magazine had suggested to Koch, back in 1891, that he should develop a remedy against the "fraud bacillus,"[222] Leo Lewin harshly judged the tuberculin episode as "shot through with unscientific shenanigans."[223] He produced a list of the various side effects and illustrated his charge of unscientific testing by pointing out that those who favored the remedy still relied on the obsolete concept of the curative effect of fever. "Even after the failures of this treatment had become obvious, this concept of fever continued to be seen as a scientific achievement," he wrote.[224] He also pointed out that in an age that had worked hard to understand fever as a quantifiable symptom, a qualitative understanding of fever as healing was redolent of vitalism.[225]

Equally as harsh as Lewin's criticism from the pharmacological perspective was Albert Moll's assessment of the tuberculin affair in terms of medical ethics. The very first page of his *Ärztliche Ethik* (Medical ethics) cited Koch's concealment of the composition of tuberculin as a prime example of ethically impermissible behavior:

> If a physician believes that he can recommend a remedy of whose composition and production he knows nothing, but which he has seen to be successful, this is considered impermissible because it would involve a secret remedy. But when a great authority recommends a remedy against tuberculosis, declaring *expressis verbis* that he is as yet unable (i.e., unwilling) to provide any information as to its provenance and its production, we witness a general race to obtain this secret remedy whose hazards are not yet known and whose usefulness is by no means proven, and indeed is later altogether denied by many practitioners.[226]

What was remarkable to Moll in this affair was not only Koch's secrecy and the experiments with children and the moribund, but the fact that such things did not really strike his contemporaries as problematic. As we have seen, this judgment was correct only for the very first weeks of testing, which were characterized by an atmosphere of intoxication. Nonetheless, we can share Moll's astonishment about the fact that the kind of criticism that was leveled after 1899 against Albert Neisser's syphilis experiments or that Paul Ehrlich encountered when testing salvarsan, did not materialize in the case of tuberculin: "I am by no means inclined to defend Neisser's experiments, but to single out one author and to attack him every day in the press while keeping silent about analogous proceedings is unfair. Once again I wish to bring up the tuberculin inoculations.

What right did hospital physicians have to treat children who had been hospitalized for quite different reasons with tuberculin?"[227]

A decade after the event, the tuberculin affair provided a very different set of insights. It seems justified to see it as the first manifestation of disillusionment in the history of modern chemotherapy. To begin with, as has been mentioned, the introduction of tuberculin coincided with the rise of the chemical-pharmaceutical industry: in 1890 antipyrin was patented almost simultaneously with the introduction of Koch's medicine.[228] Even though the legislation did not provide for patent protection for drugs or the method of producing them, this form of safeguarding material interests eventually overcame the customary secrecy surrounding some or all of the ingredients, particularly in the case of the few industrially produced drugs, the so-called specialties.[229] Whereas Koch continued the older method, his associates Behring and Ehrlich preferred to go the route of patent law.[230] The clinical testing situation also changed in the last decade of the nineteenth century. The transfer of bacteriological knowledge from the laboratory to the clinic brought about—as has been widely noted—an enormous expansion of the practice of human experimentation. In the euphoric mood surrounding the testing of tuberculin in the year 1890, the way from the laboratory to the clinic had been very short, and what contemporaries believed to be treatment looks like a large-scale therapeutic experiment to the modern observer. The boundaries between therapy and human experimentation proved to be unclear and well worth discussing, and this recognition was fostered above all by the disillusionment about the testing of tuberculin. When Koch's associates Behring and Ehrlich introduced new medications, they had to reckon with the skepticism of their contemporaries, and they did so by means of extended testing phases.[231] And this time, the intoxication of the first weeks that had been so typical for tuberculin did not occur or, in the case of salvarsan, assumed more modest proportions.[232]

The conflict that erupted after 1899 over the syphilis experiments that Neisser conducted with prostitutes—and which eventually resulted in the Prussian government decree of 1900—took place in a changed situation.[233] Questions about physicians' activities were now increasingly asked by outsiders, among them an odd coalition of liberal politicians, naturopaths, antivaccination activists, and antisemites that has been analyzed by Barbara Elkeles. In the wake of the scandal and the Prussian decree, which was the first to provide for the patients' consent for nontherapeutic interventions,[234] the practices of experimental medicine now had to be justified. But this does not mean that the medical researchers of the time tried to find justifications that would satisfy us today. On the contrary, modern authors have pointed out that the decree probably had few practical consequences,[235] and that Albert Moll's book, which criticized the lack of differentiation between human and animal experimentation, expressed an isolated opinion.[236] For the moment, scientists could adopt a purely defensive position, as did the bacteriologist Wilhelm Kolle, who noted the incompetence of laypersons to judge experiments and justified the endangerment of a few individuals with the beneficent consequences for the many. As Kolle wrote in March 1900, "It is a peculiar feeling for professionals when laymen do not understand medical experiments and depict them in a biased way."[237] Kolle referred to anesthesia as well as smallpox and rabies sera, none of which could have been developed without human experimentation. However, he did stress that "it is an important principle of modern medicine that in testing new remedies in humans, the harmlessness of the substance studied must be established through animal experiment . . . Koch's tuberculin and Behring's diphtheria serum, by which thousands have been saved from certain death, were brought into science and medical practice in this manner." Kolle was unwilling to go beyond the physician's self-imposed obligation of *nihil nocere* or to even discuss obligations toward the patient and the public: "And while the physicians who, sometimes struggling with severe material difficulties, are doing their best for their fellow humans . . . they have to watch as long-winded debates about their internal issues, their ethics, and their science are conducted by laymen to the detriment of the medical profession."

By the turn of the century, the self-confidence and the optimism with which experimental medical researchers had approached the testing of tuberculin had thus given way to a rather tense defensive posture. The researchers defended the autonomy of the physician's action, but the standards for such action had begun to shift. Paul Ehrlich, for example, aware of the problem of transferability from animal to human experimentation, decided in 1909 to test his salvarsan extensively before bringing it on the market. His contemporaries considered this proceeding unusual, indeed unethical, since the scientist was holding back a valuable medication.[238] At the same time there were critical public discussions about the possible side effects and dangers of salvarsan therapy.[239]

## IV.3. Experimental Therapy for Sleeping Sickness, 1906–1908

### IV.3.1. Sleeping Sickness in Berlin

A summary analysis of the relation between clinical and experimental medicine after 1900 is provided by a remarkable therapeutic experience that took place in Berlin in the years 1906 to 1908. A group of scientists, most of them connected with the Institute for Infectious Diseases, found themselves confronted with an extremely dangerous infection of a laboratory attendant. In their attempts to cure the patient, they had to find a balance between the interests of the patient, physicians, research, and therapy. Examining the strategies they chose illustrates the standards they adopted and the mentality of experimental physicians that came to light in the tuberculin scandal.

Looking at the history of chemotherapy, one realizes that in 1906 intense research focused on two diseases that researchers assumed to be related because of the morphological similarity of their pathogens, namely, sleeping sickness and syphilis.[240] While Paul Ehrlich conducted experiments on this topic in Germany, Koch traveled in East Africa to test therapeutic agents against sleeping sickness.[241] Sleeping sickness was of interest to medical research and of major importance for Germany's colonial policy after the Uganda epidemic devastated parts of East Africa and the Congo shortly after the turn of the century.[242] But of course it was very difficult to conduct clinical studies for it in Europe. A researcher who was not satisfied with studying only the pathogens of the disease, the trypanosoma, had access to very few sick persons. When Koch was preparing for his 1906–7 expedition to East Africa, for example, the discussion was limited to a single acute case of the disease, which was being treated in Hamburg's Institute for Tropical Medicine.[243]

Against this background, the trypanosoma infection that the laboratory attendant Berthold Schmidt had contracted in August 1906 in Koch's laboratory in Berlin was a dramatic occurrence in several respects. First of all, for the patient sleeping sickness was believed to be extremely dangerous and probably fatal.[244] On the other hand, this was a rare opportunity to expand the information about the disease beyond the experimental into the clinical realm without having to travel to Africa. The scientists involved also seem to have anticipated a possible dramatic public reaction and therefore tried—successfully it seems—"to avoid public knowledge of this regrettable case."[245] At the time the experimental treatment therefore remained unpublished and was only mentioned in passing in Bernhard Möllers's Koch biography of 1950.

According to Möllers, Schmidt had contracted the disease when he was bitten by a laboratory rat (or mouse) infected with trypanosoma.[246] His treatment lasted two years and was directed by Koch, first by correspondence and then in person.[247] In the first eighteen months the patient was treated with atoxyl, without conclusive success. A possible reason is that, according to Möllers' description, the treatment did not follow the protocol recommended by Koch.[248] In the spring of 1907 Koch wrote to Wilhelm Dönitz, the head of the treatment center at the Institute for Infectious Diseases, "That poor trypanosoma-infected institute attendant Schmidt does not seem to be doing well, according to the latest news to reach us. I would be extremely sorry if he were to become a victim of Science."[249] In the end, a mercury inunction cured the patient, who apparently was still alive decades later.[250]

This case raises a number of questions that cannot be answered satisfactorily on the basis of Möllers's information, which gives us little more than what was reported above.

> Why was the matter kept secret, who made this decision, and what were the reasons given?
>
> Möllers mentions Wilhelm Dönitz as the treating physician. Were there other treating physicians in addition to the director of the treatment center at the Institute for Infectious Diseases, and how did they coordinate their activities?
>
> Given the length of the treatment, one must ask about the strategy. In addition to atoxyl and mercury, were other therapeutic agents available, and if so, when were they deployed?
>
> What about the patient and his rights? Was he informed, and did he have any input—of whatever kind—into the course of his treatment? Was the proceeding discussed in the context of the above-mentioned Prussian guidelines concerning therapeutic experimentation on humans of 1900?
>
> And finally, it is important to ask whether and in what way Schmidt's therapy was considered to be experimental research. Even without impugning the motives of the physicians involved, one has to wonder whether

they saw the illness of the laboratory attendant primarily as a therapeutic or a scientific challenge, and how did they reconcile these two interests?

Fortunately, at the time, the Ministry of Culture considered this case so important that it charged the director of the Institute for Infectious Diseases, Georg Gaffky, with producing a written report. I was able to consult Gaffky's detailed report, complete with attached notes from the treating physician, Dönitz. Together with the personnel file of Berthold Schmidt, who had worked as an attendant in the Institute for Infectious Diseases since 1901, they make it possible to describe a remarkable patient history in the early days of chemotherapy.[251]

The very first report, which Gaffky submitted on August 4, contains fairly precise information about the distribution of responsibilities. In Koch's absence, his laboratory was under the exclusive direction of his subsequent biographer Möllers. Schmidt had contacted Möllers on the morning of August 3, after he had fallen ill with shivering, fever, headaches, and diarrhea the night before. Möllers found a temperature of 39.4 degrees C and initially thought that the illness was caused by a relapsing fever, which seems to have been rather common in his laboratory. However, the laboratory test of the blood said otherwise. "The microscopic test for spirochetes of relapsing fever was negative. But in the afternoon one trypanosome was found in each of two microscopic culture cover slip preparations." This raised the suspicion of sleeping sickness, which was confirmed on the next day. "The newly obtained blood samples were again stained and made into cover slip preparations, and again one tryposome was found in each of two preparations. Thus we have now found altogether four typical parasites. They completely correspond to *Trypanosoma gambiense*, which is pathogenic to humans, so that unfortunately there can no longer be any doubt about the diagnosis."

Schmidt had apparently become infected by one of the mice kept in Möllers' laboratory. These mice kept the trypanosoma cultures alive (in animal passages) while producing the trypanosoma extract that Koch needed in Africa. After Schmidt had been bitten by one of the white mice—apparently fourteen days before the outbreak of the disease—he had "contented himself at the time with thoroughly washing the small wound, but had not reported the incident to the chief physician Möllers." Even though Gaffky was skeptical about an infection by way of the mouse, he considered it his duty to structure the work in the labo-

Berthold Schmidt in the group of staff members of the Institute for Infectious Diseases, 1912. Schmidt, in white smock, standing in second row, far right. Picture of Robert Koch in the background. RKI/Koch papers.

ratory differently. Since in the same year two other attendants had fallen ill with relapsing fever, "this new infection with trypanosoma in the same laboratory makes it mandatory to institute new precautonary measures." Gaffky did not feel, however, that halting the testing was warranted, pointing out that the dangers incurred were inherent in the nature of his institute. But he did order the infected animals to be isolated and directed "that attendants are not to handle them at all and that the work be done exclusively by physicians, who are to be informed of the danger of infection."

As for the patient himself, it was decided to isolate him as well and to send him to the municipal hospital at Moabit, where the physicians of the Institute for Infectious Diseases could conduct the treatment themselves.[252] Infection experiments conducted meanwhile made this step appear necessary, as one can gather from a remark in Gaffky's report: "By now the patient has been isolated in the Moabit Hospital. The diagnosis has been further confirmed by the fact that the animals inoculated with his blood, as mentioned on p. 2 of the report, have since developed trypanosomiasis."[253]

Although the official in charge at the Ministry of Culture did not feel that any blame attached to the institute or to Möllers, the Institute for Infectious Diseases decided to keep the case secret—so secret, in fact, that even the patient was not given the information about his case. "He and his family believe that it was a case of relapsing fever. Of course we have not destroyed this belief," Gaffky reported to Paul Ehrlich.[254] Such infections were not rare at the institute, as the reports indicate. On the other hand, as Gaffky admitted in an internal memorandum, "this illness could not be attributed to simple bad luck," so Schmidt's wages should continue to be paid.[255]

### IV.3.2. Atoxyl

Before beginning the treatment, Gaffky contacted the two prime researchers in this field: Koch, who was in East Africa investigating therapies for sleeping sickness, and Paul Ehrlich in Frankfurt, who was experimentally studying the chemotherapy of the disease. But since Koch's reply from Africa did not arrive until mid-September, Gaffky began by following Ehrlich's advice. Both Ehrlich and Koch recommended the use of atoxyl, an arsenic derivate, with Ehrlich recommending a decidedly lower initial dose than Koch. Starting on August 5, the treatment Berthold Schmidt was subjected to for months on end in consultation with Ehrlich was risky:[256] hitherto the effectiveness of this remedy against trypanosoma had been tested only in animals, and its effect on humans was difficult to predict: "The first dose was kept very low (0.04 gr.) because reports about this kind of treatment for humans are not yet available and because one could anticipate that an extensive destruction of the blood parasites caused by the remedy would unduly increase the already present fever."[257] Fortunately, not only was the dose tolerated well by the patient, it also led to a "decided decrease in the temperature" so that it could be gradually raised until it reached 0.1 grams and the fever disappeared altogether—albeit only temporarily, to the doctors' dismay: "On 8 August the fever had abated and the patient was fever-free for three whole days. On the two last days he again received the same dose of atoxyl he had received the time before (0.1 gr.). Nonetheless, the next day his temperature rose so much that one had to conclude that this was a new attack."

As time went on, trypanosoma were isolated again and again, and fever spikes occurred repeatedly, so that Dönitz found himself compelled to experiment with varied dosages (quantity per day, size of single doses). His therapy experiments took place at the same time as Koch's testing of atoxyl on the Sese Islands in Lake Victoria. It is worth noting that in Berlin, diagnosis and therapy were more attentive to the patient's interest than was the case in East Africa. In making the diagnosis, the Berlin team thus did not perform the very painful punc-

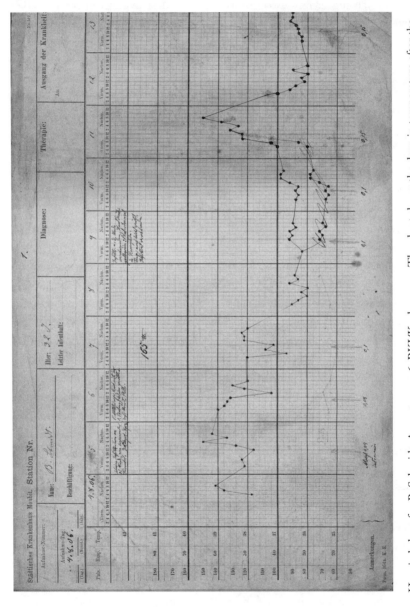

Hospital chart for B. Schmidt, August 4, 1906. RKI/Koch papers. The chart shows the drop in temperature after the first injection as well as a renewed spike on August 8.

ture of glands that Koch practiced in East Africa and recommended by letter.[258] Instead they made do with the technically much more demanding identification of the parasites from swabs. Similarly, the treatment protocol was modified according to the patient's wishes.[259] Clearly the advice concerning the therapy that Koch willingly provided from Sese was not followed. For instance, the Berlin team chose a decidedly lower dosage than Koch had recommended and also experimented with the dosages.[260] In fact, Dönitz's treatment did not adhere to any firm protocol; instead, the medication was administered intravenously in changing dosages. According to Dönitz, his therapy took place in difficult, indeed experimental conditions. "After Schmidt's most unfortunate infection with Tryp. gambiense, Gaffky, following a remark of Koch's, ordered a halt to the work with this trypanosoma; but I, being in charge of the hospital section, had no choice but to deal with the trypanosoma patient. I therefore had to try out everything on this man, because I could not conduct any parallel animal experiments, and so, of course, my progress was very slow."[261] He began with injections during the attacks, and then stopped them in order to observe the effect. This effect consisted of recurring bouts lasting two to three days, separated by fever-free periods lasting three to five days. Dönitz then decided to increase the daily dose, not so much because parasites continued to be isolated and the patient's glands remained swollen, but because he had heard of experimental work with animals that suggested a habituation effect, which he hoped to circumvent by means of higher dosages and the resulting shortening of the treatment.[262]

But this modification of the treatment was not continued when in mid-September August von Wasserman filled in for Dönitz, who went on vacation. Wassermann substituted intramuscular for intravenous injections in order to "place the patient under the continuous effect of the remedy," as he put it.[263] Although the first bout of fever thereafter was relatively mild with a maximum of 37.6 degrees C, another violent attack soon followed. Wassermann modified his schema in the same manner as Dönitz had done, namely by increasing the dosage (from 0.05 to 0.1 gr.), and combining it with intravenous administrations, so that in one twenty-four-hour period the patient received 0.75 grams of the medication. Here too decisive success did not materialize, even though the fever-free interval between attacks seems to have increased to ten days. Nonetheless, Dönitz resumed the intravenous injections after he returned from vacation, now administering "fairly large doses" of atoxyl "with intervals on three consecutive days."[264] The result was a decrease of the fever-free period to an average of seven days. Dönitz therefore adopted Wassermann's treatment protocol.

Consequently, Dönitz reported on December 13 that the trypanosoma could no longer be found between attacks, and that the attacks not only occurred more rarely but were weaker. Since at this point the trypanosoma were difficult to find, judging the success of the treatment was now a matter of interpreting the bouts of fever. One can gather from Dönitz's convoluted formulation in his report of December 13, 1906, that this was by no means an easy task. Still, the patient was obviously on the road to recovery: "After that attack, the fever-free period now [on 13 Dec. 1906] lasts as long as 20 days, not counting one temporary rise to 37.60 C, which is legitimate, since in an attack the body temperature rises as high as 39.5 C and usually remains quite high for some time. But even interpreting that 37.6 C as an attack, there have been 13 fever-free days since then. All in all, one has the impression that the patient is on the road to recovery."

The improvement in Schmidt's condition that began toward the end of the year mirrored Koch's experiences on Sese, where he became convinced in November that atoxyl was a remedy for sleeping sickness.[265] Schmidt was discharged from the hospital on December 22, and it should be added that the certainty of a successful treatment was partly motivated by the fact that the patient would not have withstood the cure much longer. On January 11 Gaffky reported to Ehrlich that the patient had been "discharged as cured," adding, however, that "shortly before his discharge from the hospital, certain symptoms had pointed to an incipient arsenic poisoning; they very quickly disappeared once the medication was discontinued. I presume that Herr Wassermann has already confidentially told you about the way atoxyl was administered. For the time being I hold with *'Quieta non movere'* [Don't move quiet things]; we will simply wait and see."[266] Consequently, by the end of the year, it even seemed conceivable that the work with pathogenic trypanosoma in humans—which for a time seems to have been discontinued in Koch's laboratory—might be resumed. After Gaffky had considered halting this work in August, Koch gave the order in September.[267] Then, in December, he consented to its resumption, "as long as I am not held responsible."[268]

## IV.3.3. Dye Therapy

Berthold Schmidt's treatment in the winter of 1907 gave rise to the very problems that had put an end to the positive mood on Sese Island: shortly after the patient's discharge from Moabit Hospital, he suffered a renewed fever attack. Schmidt was readmitted, this time to the Infectious Disease Unit of the newly founded Virchow Hospital. He was intravenously given high doses of atoxyl, since the continual administration of small doses had come to be considered problematic: "The idea was that in the earlier atoxyl treatment we had started with small doses, only gradually moving to larger ones. Perhaps one can attribute the very poor success of the occasional larger doses to the fact that—in analogy to the observations made by Prof. Dr. Ehrlich—the parasites had gradually become habituated to the remedy, in other words, become atoxyl resistant."[269] This approach failed, not only because a new fever attack followed but also because the patient showed alarming side effects: "At this point the patient began to complain about feeling cold and about dull sensations in the feet, slight vertigo, and impaired vision. An examination of the eyes revealed a cloudy corpus luteum, presumably caused by a penetration of trypanosoma. The other complaints suggested that the parasites are now also present in the spinal canal."

Side effects had already appeared in late autumn 1906, and impaired vision in the wake of atoxyl therapies was a known phenomenon at the time,[270] so that the note of surprise Gaffky sounds in his report is not altogether credible. And although the shift to lower dosages had brought a decrease in side effects, the therapy had reached a turning point. This was particularly true for the patient, on whom it now dawned that he was not suffering from relapsing fever as he had been led to believe but from sleeping sickness. "The mood of Schmidt, who seems to suspect that his illness was caused by an infection with trypanosoma, is very depressed," Gaffky reported on March 9. Moreover, given the obvious failure of atoxyl, the medical staff had to decide whether this course should be continued purely as a matter of treating the symptoms and in hopes that the effect would come in the long run. This, of course, was unlikely. By this time Ehrlich had completed his studies on the atoxyl resistance of certain trypanosoma strains, and Schmidt's trypanosoma proved resistant in this sense.[271] The alternative was to expand the already given experimental character of the proceedings and begin testing entirely new chemotherapies. For Gaffky the decision was clear. Given the latest research of Ehrlich, with whom he was in close contact, "it seemed time to try one of the modern medications that have been used in animal experiments against trypanosoma diseases, in particular the *pararosanilin basis* described by Prof. Dr. Ehrlich. Thanks to his good offices, Dr. Weinberg of the Casella Company in Frankfurt am Main provided us with this preparation, which is not commercially available."

More important than ever for this cooperation was the person of Dr. Dönitz, who as a former colleague of Ehrlich's in Frankfurt was in a position to speak personally to his "dear friend." In his letter to Ehrlich of March 1907, in which he noted that the closing of Koch's laboratory had practically forced him to experiment with humans, he also stressed the importance of the insights gained from the treatment of the laboratory attendant. Having regularly reported to Koch, he said he now "knew that intermittent and also heavy applications of atoxyl will fail, but that by saturating the blood with atoxyl we have eliminated acute bouts of fever, without, however, preventing the reappearance of the trypanosoma in the circulating blood."[272] He now intended "to make use of color therapy," and this had brought him into conflict with Klaus Schilling, who also worked on trypanosoma at the Institute for Infectious Diseases and also needed dyes.[273]

Ehrlich now provided help with the application of dye therapy, although he had originally advised against it. When Dönitz informed him about the incident, he replied in August 1906: "My dear friend, . . . as you probably know, I have continued to work on trypanosoma therapy, and I am indeed going after a number of effective substances, but unfortunately these were tests with mice, who tolerate medications better than other animals. I therefore have no idea whether what I have in hand can already be tried in humans."[274] In February 1907, however, he supported this procedure and provided Dönitz with recommendations on how to use saponified olein of rosaniline, a fuchsin: "As for the dosage of fuchsin, it is very difficult to determine. But I believe that one can safely give 1 gr. of fuchsin base spread throughout the day. If this were tolerated well, you could quickly increase the dosage further. It might be advisable to test [the medication] simultaneously in some healthy patients in the hospital ward in order to find the *dosis bene tolerata* of the oleic acidy paranin at the same time."[275]

Starting on March 1, 1907, Schmidt was treated accordingly. The result was, as Gaffky reported on March 9, that "the symptoms mentioned above have almost entirely disappeared and that the patient himself is convinced that the medication has agreed with him well." Thereupon an attempt was made to combine atoxyl and rosaniline,[276] but by mid-March—after a renewed bout of fever had lifted the last doubts about the atoxyl resistance of Schmidt's parasites—only the dye therapy was employed.[277] As for its effect, the patient's blood was now free of parasites,[278] but on the other hand, renewed (albeit rare) bouts of fever showed that the trypanosoma had disappeared only from the peripheral blood. Moreover, the course of the treatment brought severe side effects that could not be eliminated, even by lowering the dosage. Indeed, lower doses brought back stronger symptoms of the sleeping sickness and on February 9 led to another bout of fever. Thereupon the patient received the rosaniline in a new, water-soluble preparation, which he tolerated well only initially. By the end of April he had to be taken off the medication "because it irritated the stomach and came up in vomit shortly after ingestion."[279] Such setbacks fed beginning doubts about the effectiveness of dye therapy, since they almost always led to new attacks. The medical team tried to counteract the side effects that again and again caused the patient to vomit up the different rosaniline preparations by constantly changing treatment protocols. In this manner they were at least able to produce somewhat longer fever-free phases: after one bout of fever in late April, new attacks did not occur until May 11 to 12 and June 19. But the price to be paid for this was that, aside from the subjective problems of the patient, a measurable danger to his health emerged in the form of incipient kidney failure. Under these circumstances it was appropriate the interrupt the therapy, "partly in order to provide respite for the stressed urinary tract and partly in order to see how long the fever-free period would last under these conditions." The answer was provided in short order by a renewed fever attack. As Gaffky summarized in his report of July 11, the continuation of the therapy depended on the control of the side effects: "Looking back at the dye treatment administered over approximately three months, this preparation appears to have had a beneficial effect on the patient to the extent that the fever-free periods have become considerably longer. Unfortunately, the preparation has a disagreeable effect on the kidneys and the bladder. The next task will therefore be to find a mode of administering the dye that will serve the therapeutic purpose without producing harmful side effects."

The fact that there were few indications that the therapy treated more than the symptoms did not seem to bother him. The treatment was continued throughout the summer of 1907, with acidic rosaniline replacing the hydrochloric acid form. Gaffky had to report further attacks of fever, and though they were fairly weak—maximally 38.9 degrees C—the therapeutic advantage was questionable. The only remarkable finding was the body's rapid elimination of the medication, which, judging from the red color of the urine, was completed after twenty-four hours. Thereupon "the administration of the remedy was modified to the extent that Schmidt took 3x0.1 gr. of the dye every other day."[280] These numerous treatment modifications suggest that they were seen as interesting research alternatives arising in the transition from animal to human experimentation.[281] Although the report does not indicate that this medication had any effect on the patient's state of health, the decision to administer it three times daily instead of once did bring a change: the patient, whose "urine had a permanent red coloration" forced an end to his isolation. "Following his repeated and urgent requests, he was granted a leave of absence from the hospital, where, given the more than a year-long duration of his illness, an effective isolation was not feasible anyway."

At this point, the state of Schmidt's health was actually rather good, so he was now "occupied with light work in the Institute." There it was possible to observe him and to continue the medication, possibly in a reduced form, although this is unfortunately not specified in Gaffky's report of September 16. At any rate, after a lengthy pause, another severe attack with a 39.9 degrees C fever occurred in early September, together with the reappearance of trypanosoma in the patient's blood.[282] This caused the medical staff to think about alternatives to dye therapy. They considered—and even practiced from October 7 to 25—an arsenic acid therapy. That acid had been found to be effective against trypanosoma cultured from Schmidt's blood in 1907; these cultures were still atoxyl-resistant after six months.

### IV.3.4. A Demonstration

The dye-chemical phase of Schmidt's career as a patient ended with Robert Koch's return from Africa in November 1907. In Africa Koch had become convinced that rosaniline was ineffective against sleeping sickness. He also took the view that the way the Berlin physicians had proceeded brought about the atoxyl resistance of Schmidt's trypanosoma: "In observing the many patients that passed through our hands, many of whom had undergone lengthy treatments, we never saw anything that would have pointed to this kind of atoxyl resistance of the trypanosoma. If something of this kind developed in the Institute attendant Schmidt, it was probably caused by the frequent small doses. We never treated our patients in this manner."[283] In May, Koch's guess was confirmed when he read Ehrlich's lecture "Chemotherapeutische Trypanosomenstudien,"[284] which described atoxyl resistance. Immediately after his return in November, Koch seems to have taken over the management of the treatment. The dye therapy was stopped in early November, and the patient received—following Koch's practice in East Africa—injections of large doses of atoxyl. Each time, two days of injections were followed by one-week pauses. Given the rather good general state of the patient's health, the side effects he had already experienced, and especially the notorious atoxyl resistance of his trypanosoma, it is hard to see a reason for this decision. It seems to have been legitimized more by the authority of Koch—to whom Gaffky had immediately handed over the case[285]—than by sound arguments. As one would expect, it yielded no positive results. The trypanosoma continued to be demonstrable. In December and January, the patient suffered renewed and rather severe—39 degrees C—attacks of fever.[286]

Robert Koch's stay in Berlin was brief. By March 1908 he again left the city and traveled more in a private capacity.[287] Nonetheless, in the winter of 1907–8 he made one last and quite surprising change in Schmidt's treatment: having taken the patient off the chemotherapy in November, he now also stopped his own atoxyl treatment. Beginning on January 22, the patient received a new treatment, a short-term mercury inunction, which was a traditional remedy for syphilis.[288] "On three successive days 4 gr is rubbed into the skin, and on the fourth day a warm bath is administered. This new treatment agrees very well with Schmidt, and despite the strong action of the mercury, it has not, so far, produced any detrimental effects."[289] Obviously spawned by the closeness between syphilis research and sleeping sickness research, this new method of treatment never gained any significance in the subsequent therapy for sleeping sickness.[290] Judging from Gaffky's report, however, this course of treatment was successful, meaning that Schmidt was fever-free until at least late March,[291] and that the parasite had permanently disappeared from his blood. In several of his letters Koch asked Möllers to repeat the treatment.[292] In early September 1908—after a third course of mercury rubs, and more than two years after Schmidt had fallen ill—Gaffky declared him cured. But Gaffky's final report decidedly betrays a certain distance toward the mercury rubs, which he considered a follow-up treatment: "Since Schmidt had never been free of fever for so long a time, one may consider this remarkable improvement a result of the follow-up treatment with mercury."[293]

## IV.3.5. Therapy and Experiment

From today's perspective, it is well-nigh impossible to decide from which disease the laboratory attendant Berthold Schmidt suffered. It is true that the animal experiments conducted with his blood allow for the possibility of a trypanosoma infection, but on the other hand, the path of the infection via a mouse bite is difficult to imagine. The symptoms and the presumable incubation period of two weeks hardly permit an unequivocal attribution, considering that the patient was almost constantly medicated, which makes an interpretation particularly difficult. And finally, modern science would consider his surviving a case of *Trypanosomiasis gambiense* lasting several years unthinkable.[294]

Thus the medications tried on this patient over two years were not necessarily tested in a person suffering from sleeping sickness. However, this objection to the proceeding of the Berlin physicians—obvious though it is to an observer in the twenty-first century—would not have registered with the protagonists who, with the exception of the deceived laboratory attendant himself, remained convinced of the presence of sleeping sickness.

It seems appropriate to divide this story into three quite different phases of therapeutic-experimental proceedings. Initially, the physicians operated with atoxyl, which they knew from their animal experiments. This continued until, in the winter of 1906–7, the atoxyl resistance of the parasites began to emerge, first from Ehrlich's research and then in the patient's therapy. Then followed a strongly experimental phase. The patient received a series of different fuchsin derivates, and it should be noted that these too were substances that Ehrlich was testing at this time in animal experiments, as he indicated in his Berlin lectures of February 1907. Dönitz's connection to Ehrlich and Ehrlich's interest in experimental chemotherapy were thus dominant factors in the treatment until late summer 1907, and it is not surprising that the suggestions Koch sent from Africa were by and large ignored.

Koch's return from Africa then opened the third phase, which one might characterize as the dogmatic turning away from the pursuit of a strongly heuristic proceeding: aside from Koch's authority it is hard to see any reason for the resumption of atoxyl therapy in autumn 1907. After all, all the physicians involved were convinced of the atoxyl resistance of Schmidt's trypanosoma. The decision to employ external mercury applications also reflects the preponderance of Koch's authority. Significantly, the reports after December 1907 no longer contain evidence of experimental treatments. Indeed, the intensity of the reporting declined considerably in that period: whereas between August 1906 and December 1907 Gaffky had sent the Ministry of Culture ten reports—some of them very long and usually supplemented by detailed accounts from Dönitz's pen—he sent only three short notes until September 1908. After Koch had taken over the treatment, Dönitz was not heard from at all.

It is rather strange that in examining this example of early "high-tech" medicine, one cannot discern an actual strategy for treatment or experimentation— except for a general escalating tendency. It looks as if various projects were carried out, or dropped, as opportunities presented themselves. Wassermann did not hesitate to modify Dönitz's treatment plan as soon as the opportunity arose; Loeffler had suggested his medication long before it was used—namely as soon as the dye therapy was abandoned. The use of atoxyl was discontinued when the patient's eyesight seemed to be in jeopardy. These proceedings lost their experimental character when Dönitz lost control over his patient following Koch's return from Africa.

In comparing the attitude of the treating physicians in this case with that prevailing in the testing of tuberculin, one finds few changes. Gaffky's semiofficial reports may not have been the ideal place for such reflections, but the absence of any discussion of the boundaries between animal and human experimentation is still striking. Gaffky raised no questions at all about the transferability—and the legitimacy—of procedures developed in animal experiments to humans. Dönitz simply stated that the only thing he could do after the cessation of the animal experiments was to experiment on the patient himself. This remark at least reveals some awareness of the problem, but in view of the long drawn-out experimental treatment of a patient who was by no means moribund and suffered many side effects, it has the unfortunate character of self-justification. It should be added that this phase of the treatment eventually came to an end through the action of third parties, the patient himself and the returning Koch. As had been the case with tuberculin, the only option open to the patient was to refuse the continuation of the treatment. Considering that the diagnosis was hidden from Schmidt until he suspected it and that he was never asked for his consent to be treated, it is clear that here again neither the Prussian decree on this topic nor Albert Moll's disquisitions about the physician's duties toward his patient had any impact.

On the other hand, Berthold Schmidt's role was not as simple as it might seem. There were obvious attempts to buy his consent. Dönitz modified the treatment protocol to suit Schmidt, and the continued wage payments throughout the entire period—even including a raise—are informative in this respect. It also seems remarkable that over the years Schmidt, who had started life as a dragoon, rose from simple laboratory attendant to head Präparator, a title he held when he finally retired on March 31, 1949. For many years, Schmidt was able to make use of his illness in a manner that would have been impossible without the ambivalence between his clinical symptoms and the laboratory diagnosis on the one hand and the secrecy surrounding various incidents on the other. Decades after the disappearance of the trypanosoma from his blood, he still experienced vague symptoms, such as weakness and fatigue, sweating, and buzzing in the head. "The slightest exertion would cause him to see black before his eyes," he told a public health physician who had been called in by the institute's management in an effort to put a stop to the rest cures that Schmidt was taking at the expense of the Robert Koch Institute.[295]

There was, however, one characteristic difference between the tuberculin testing of 1890 to 1891 and the sleeping sickness experiments of the years 1906 to 1908, and that was their secrecy. It was so strict that, as I pointed out above, the public learned of these events only forty years later, from Möller's biography of Koch.[296] No one had worried about secrecy at the time of the tuberculin testing; on the contrary, even the experiments of Henoch or Schreiber, which many contemporaries considered objectionable, were published in toto as soon as possible. By comparison, the conduct of the physicians who treated the laboratory attendant Schmidt shows the same defensive attitude that is evident in the cited article by Kolle, who worked at the Institute for Infectious Diseases between 1893 and 1906 and was a personal friend of Koch's.[297]

As medical bacteriology made its way from the laboratory to the clinic, it triggered an extraordinary expansion of human experimentation. By the turn of the century, the lay public and a few physicians began to see ethical problems in these practices, but the medical bacteriologists initially reacted with stubborn insistence on their absolute control of their patients and refused to acknowledge the problem raised by the transition from animal to human experiments. Against the background of the alleged incompetence of the lay public, the secrecy surrounding the treatment experiments—or their export to the colonies, which ultimately served the same purpose[298]—apparently seemed perfectly justified.

In the years around 1900, distinguishing between human experimentation and therapy was a decisive criterion for judging the ethics of medical conduct. As I have shown above, only failed treatment efforts were branded as experiments. Given that in 1908 the physicians involved in Berthold Schmidt's case were convinced that they had cured him of sleeping sickness, there is another irony in this story: aside from being kept secret, the therapeutic success—measured by the microscopic findings—was the result of a questionable diagnosis. At the end of the day, it was this fact that spared Dönitz and his colleagues from being accused of using Berthold Schmidt for human experimentation.

CHAPTER V

# Traveling

Robert Koch's Research Expeditions as Private and Scientific Undertakings

## V.1. Traveling Bacteriologists, Bacteriological Travels

In the preceding chapters the history of medical bacteriology has been examined from three perspectives: that of the early history of the medical-bacteriological laboratory and its objects; that of a research program and its dynamic; and that of the relations between medical bacteriology and the clinical medicine of that period. These chapters also examine three phases in the history of the discipline. The first chapter dealt with establishing the boundaries and the methods of the subject, which reached a first completion with Koch's animal-experimental studies of traumatic infections. The second chapter examined the dynamic of the fully developed bacteriology of the 1880s, using the example of tuberculosis research. The third chapter looked into connections that appeared in the wake of the transfer of bacteriology into clinical practice around 1890.

In this manner the history of medical bacteriology became visible in a series of snapshots. By way of a conclusion to this study, I will now set out to examine the special dynamic of this process that extended over more than thirty years. In doing so I will pay close attention to the period from 1870 to after the turn of the century, in which medical bacteriology developed from a niche area for botanists and experimental pathologists to one of the major disciplines of scientifically oriented medicine.

It stands to reason that over a span of more than thirty years a researcher would change as much as the science he practices. The perspective chosen here therefore makes it possible to raise a biographical question and to investigate the relation between the development of a science and a biography in a period that practically encompasses Robert Koch's entire professional life.[1] The bacteriologist himself considered this a decidedly difficult relationship, as can be seen from several statements he made after the turn of the century. Not surprisingly for a person beyond middle age, they have a retrospective character. Rather more surprising, though, is their melancholic, if not surly tone. Koch, a greatly honored and respected scientist, no longer saw much that was positive in the bacteriology of his time and was nostalgic for the 1880s. As he saw it, the terrain of the discipline had changed tremendously. Instead of abundant research topics, there was now (around 1900) a surfeit of researchers who made life difficult for a founding father of the discipline. "Your letter so vividly brings the old days back to my mind," Koch emotes in a letter of 1904 to his old friend of the Breslau days, Carl Weigert. At the time, Koch recalls, one could "work quite without worries and disturbances." This, he said, had now fundamentally changed: "Nowadays, whatever I touch or undertake, immediately a crowd of jealous and envious people comes along and tackles the same subject, trying to contest my work or, if unable to do that, spoil my pleasure in it."[2]

Knowing that he had such problems in dealing with critics, competitors, and increased specialization can be a first step toward understanding Koch's enthusiasm for scientific travel, the subject of this chapter. By 1900 working in Berlin had become a burden to him. But far from there, and especially in the tropics, working conditions were clearly reminiscent of the happy decade around 1880. In October he wrote to Georg Gaffky from Rhodesia, where he was working on epizootics: "At home [in Berlin] everything has been gone over so thoroughly that it really is no longer worthwhile to do research there. But out here the streets are still paved with the gold of science. How many new things did I see and learn when I first came to Africa!"[3] When studying Koch, his travel practices virtually cry out to be examined. Particularly after 1896 he traveled abroad numerous times, mostly on expeditions to tropical areas, for the declared purpose of furthering his etiological, epidemiological, and therapeutic research. Its subjects were principally parasitic tropical diseases such as malaria or sleeping sickness, as well as epizootics such as rinderpest or the so-called east coast fever.[4]

The sponsors and financial backers were usually governmental organizations, whether it was the Imperial Ministry of the Interior in Berlin or the Foreign Office in London. Whereas these expeditions sometimes kept the researcher and his associates away from Berlin for years on end, the only earlier research trip had been an expedition to study cholera that took Koch to India by way of Egypt in 1883–84.

There are two sets of questions that must be raised when investigating Koch's scientific travels. First, how did this passion originate and develop; which circumstances of Koch's life gave rise to it; what influence did it have on his work, for instance, on the choice of certain research topics; and how did it shape the way others perceived this work? And second, does an investigation of this subject lead to more than simply biographical information? In other words, could travel as a mode of scientific work be significant not just for Koch but also for bacteriological hygiene and its history?

As for the first point, I reported earlier that Koch came from a family in which emigration was by no means rare. Only two of nine brothers permanently remained in Germany![5] As a young man he had dreamed of becoming a world-traveling naturalist. Later, in the first materially difficult years of his professional life, "Robert's adventuresome plans for going abroad" had been a constant source of worry for his parents and his first wife.[6] The expedition of 1883–84 for the study of cholera was thus also the fulfillment of a youthful dream.[7] The close examination of this expedition will be one of the two topics of this chapter. What special conditions and possibilities for scientific work did it afford, and how did these shape Koch's experiences and his attitude toward travel? What did it do for his career; and what special characteristics of bacteriological hygiene become visible as we examine this subject?

Starting in the mid-1890s, Koch spent the majority of his time on such expeditions. It is appropriate, of course, to link his scientific career as a whole to the city of Berlin; yet for the period after 1895 it would be misleading to call him a Berlin scientist in the narrow sense. He was indeed the director of the Berlin Institute for Infectious Diseases, but he frequently turned over its day-to-day business to others so that even lengthy absences did not interfere with its functioning. Even his retirement in 1904 by no means marked the end of research travel. On the contrary, it was the starting point of a long expedition—his last—undertaken to study the African sleeping sickness. This expedition took Koch and his associates to East Africa for more than a year. It will be the second topic of this chapter, which will include a comparison between this and the cholera expedition. Accordingly, an effort will be made to investigate the special conditions of such work and Koch's experiences with it. Beyond that, one can ask what an examination of this expedition can teach us about the development of medical bacteriology after the turn of the century and about the relations between the pioneer Koch and that discipline.

By way of further refining the questions broached above, it seems appropriate to treat two important aspects of travel, namely, its implications for Koch's biography and for the history of science, in this introductory section of chapter V. To begin with, one has to ask how Koch's attitude to travel changed between 1885 and the turn of the century and whether that change was related to the development of medicine and tropical hygiene around 1900. Then too one has to discuss to what extent travel was a form of working specific to experimental bacteriology and hygiene as a whole, and whether looking into the travels of the hygienist Koch can yield insights into the development of the discipline.

As for Koch's increasing passion for travel, it is worthwhile to begin by exploring the mutual relationship between his biography and the development of the discipline from about 1885 on. At that time Koch could look back on a period of successful scientific work: within three years his methodology for bacteriological investigations had made it possible to explain the etiologies of the two most important infectious diseases of the time, tuberculosis and cholera. It had made its discoverer a famous man who, at the age of barely forty, became deputy director of the Imperial Health Office (IHO) and member of the Prussian Staatsrat. His self-confidence as a prominent representative of the hygienist community, who no longer had to worry about being compared to major figures in the field such as Pettenkofer or Pasteur, is revealed by his demeanor at the two cholera conferences in 1884–85.[8] It was also the reason why he was no longer satisfied with his assigned activities at the IHO. To be sure, the independent imperial institute he envisioned at that time remained a dream, but in 1885 he did become director of the new Institute for Hygiene at Berlin University, which took its place beside the two existing ones in Munich and Göttingen.[9]

The founding of this institute symbolizes a number of changes that became significant for Koch's later travels. As courses began to be taught at the IHO, and especially as a training program at the Institute for Hygiene at Berlin University began to experience rapid growth, bacteriological hygiene was transformed from an esoteric science to a field of teaching that was in great demand, not only from Berlin students but also from other scientists, public administrators, and so forth.[10] The simultaneous founding of the *Zeitschrift für Hygiene*, edited by Koch and Flügge, and the publication of the first textbooks for the young field also indicated that the discipline was changing.[11] The pioneering period was definitely drawing to a close. In 1882 one might have called he tubercle bacterium a Berlin bacterium, for its identification demanded the use of the methods employed in Koch's laboratory at the IHO, and at the time this was the only place to learn them. In the years after the Berlin Institute for Hygiene had begun offering bacteriological training courses, the Paris microbiologists sent spies to take these courses so that they could report to Paris on the methods used in Berlin.[12] To be sure, when these courses became more numerous and when the first textbooks were published,[13] these methods soon lost their exclusive character. As they could be learned more easily, they increasingly became common knowledge.[14]

As a result of these developments, a small group of researchers grew into a many-voiced chorus of colleagues and critics.[15] For Koch this meant that he had to consider competing views, and this was something he did not like to do. As the letter to Weigert cited above shows, it was a step in which he could find very little positive value. One example of the changes with which he had to contend is the field of tuberculosis research we have examined above. Here, after the tuberculin scandal of 1890–91, the question of the heritability of the disease, which he felt he had solved once and for all, surfaced once again—this time in the up-to-date guise of individual constitutions.[16] A few years later he had the disconcerting experience of standing by as his assertion of the nonidentity of the pathogens of human and bovine tuberculosis foundered in a storm of criticism. In the letter to Weigert that was quoted above, he concluded ill-humoredly: "Besides, I believe that in my work I have encountered particularly unfortunate circumstances and that I have run into more opposition—completely unjustified opposition—than anyone else."[17]

By the 1890s, moreover, the field of medical bacteriology as a whole became more differentiated: immunology, serology, parasitology, and tropical medicine appeared as separate fields of research. For Koch this development also presented problems. The stance of the pioneer that he had so successfully cultivated became increasingly difficult to uphold. What had been the uncharted terrain of microbial life in the 1880s had now become a popular area where many sought to stake their claims and where there was no lack of competition. Nor did the enormous growth of the field, which was accompanied by a democratization of its methods, the proliferation of subjects, and a concomitant specialization of its researchers, strike Koch as a positive development. The reply to his students' congratulations on his sixtieth birthday in 1904—published in the *Deutsche Medizinische Wochenschrift*—was decidedly grumpy:

> Long gone are the great days when the few bacteriologists could be counted on one's fingers, and when each one of them could go through large areas of research undisturbed . . . So [now] in making the most modest and most careful delineation of a research area, one is bound to step on one person's toes, knock over another, and come too close to the area of a third one, so that, before one knows it, one is surrounded by enemies on all sides . . . In this respect I must complain about particular misfortune, for I . . . always run into passionate opposition, and that, alas, from people who have little or no understanding of these matters and who are least qualified to judge them.[18]

The crowd of resentful opponents included, as Koch had to acknowledge to his regret, not only competing bacteriologists of other schools,[19] physicians and naturopaths,[20] but also former associates and students such as Hueppe,[21] Behring,[22] or Nocht.[23] This lack of respect greatly saddened him—regardless of whether to us his perception appears correct or not.[24] In 1899, when Richard Pfeiffer left the Institute for Infectious Diseases for a professorship in Königsberg, Koch shared his misgivings on this point with him : "I hope that you will faithfully remember the Institute and will not become estranged as quickly as most of the others who have come from it."[25] It looks as if to Koch, his position as director of the Institute for Infectious Diseases was not so much a source of power as a threat. In a letter from this period, he wrote in frustration to Wilhelm Dönitz that "all the intrigues hatched against the Institute were directed more against my person than against the Institute."[26] In 1904 he took his retirement as early as possible, right after his sixtieth birthday, and indeed had applied for permission to do so even earlier.[27]

After 1885, Koch's work did not advance very well. He no longer achieved any notable breakthroughs in the area of etiological research and technical innovation, the origin of his reputation. The discussion of tuberculin, as we saw, had taken a critical turn. As for the studies of the subsequent years, it is difficult to find a common denominator for them. However, a few elements can be pointed out. Over time Koch apparently redefined his area of research; various indications for this can be found after 1892. The first striking fact is that after combating cholera in Hamburg in 1892, Koch increasingly turned toward epidemiological questions and on this basis developed preventive measures designed to control epidemics in entire populations. While these studies continued to focus on pathogens, these were now treated not so much in etiological as in epidemiological contexts. This is well documented, for example, by Koch's epidemiological studies of typhus in the region around Trier, which in 1901–2 led him to formulate the concept of the asymptomatic carrier of diseases.[28] It is therefore not surprising to learn that quite a few of his studies concerned epizootics, since here such carrier models were fairly easy to examine. And it is worth noting that he worked on typhus research in Germany and on epizootics such as rinderpest, horse fever, and east coast fever in East Africa at about the same time.[29] Another and related shift in focus is visible in the area of parasitology and vector-dependent diseases, for Koch worked not only on epizootics but with equal intensity on malaria and sleeping sickness.[30]

That such research required travel to faraway countries was not, in Koch's eyes, a necessary evil but indeed the actual attraction. It was a way to combine several objectives, for it enabled Koch to work far from his Berlin colleagues and competitors and to bring his etiological interest to bear on objects that, as he enthusiastically reported, were "complex, and at times positively tricky."[31] In a sense this reorientation matches the contemporary development of science, which in the wake of the intensifying colonization of the tropics led to the creation of the discipline of tropical medicine.[32] And while Koch showed little genuine interest in some of these areas, such as serum research or immunology, he found that the parasitical, vector-dependent diseases of tropical countries provided him with a chance to reorient his traditional research strategies, with their focus on the pathogen, toward complex new problems. He attempted, for instance "to solve the great enigma of the etiology of malaria."[33]

In addition to the scientific attraction of the tropics, Koch also appreciated their exotic nature and warm climate. Thus he wrote in some disappointment from Egypt, where he was vacationing in February 1908, "The weather, though, is less tropical than I had expected to find it, . . . that is why in a few days we will go further south."[34] About New Guinea, from which German settlers fled in droves at the turn of the century—unless they had died of malaria—he remarked in all seriousness that it was "one of the most beautiful and interesting countries in the tropics."[35] Well aware of the health risks of such places, he did not seem to care. "If I could, I would stay here for years,"[36] he wrote in 1903 from Rhodesia, a country in which the danger of tropical epidemics was small, but which afforded fascinating possibilities for research. Among the pluses of the sleeping sickness expedition of 1906–7 was the fact that it took him to warm countries. Suffering from a cold in the spring of 1906, Koch was "really longing for a southern climate, where it surely will clear up soon."[37] After arriving in East Africa, the researcher, who at home frequently complained about his poor health, amazed his entourage by showing great physical endurance.[38] In a letter to Wilhelm Kolle, Koch described the Sese Islands in Lake Victoria, where in the early twentieth century tens of thousands of people had probably died of sleeping sickness—and where he himself was laid up for weeks because of an infection in his feet caused by insect bites—as a place of "wonderful landscapes and climate. Truly a perpetual mild summer reigns here."[39] Returned from there to Berlin, he wrote to Paul Ehrlich in November 1907, "But I would not like to stay in Berlin for too long. The climate does not agree with me at all, and I have promptly come down with the first catarrh."[40] A few days later he confessed in a letter to Flügge, "I am already impatient to leave again. But it is not anxiety that drives me, nor a new work project, it is just the longing for sunshine."[41]

In terms of the history of Koch's life, his desire to travel was quite plausible. The professional aspiration to become a (world) traveling naturalist and the dreams of emigration he had nurtured as a young man are the biographical background to the development that by about 1895 turned a Berlin bacteriologist into a traveling researcher on tropical medicine. It also seems plausible that the cholera-expedition of the years 1883–84 played an important role in refashioning his youthful enthusiasm into a style of scientific work. Being his first research expedition, it molded his style as an experimental scientist in decisive ways. As we shall see in examining the sleeping sickness expedition of 1906–7, this style remained remarkably stable. It should be noted that in this comparison, biographical relevance and scientific productivity are but two different aspects of the story. In choosing his subjects and his methods, Koch recognizably combined themes of current interest, such as parasitology and vector-dependent diseases, with the goal of perpetuating the research style of the heroic days. In other words, he wanted to continue his pathogen-centered work and to maintain exclusive control over the objects of such research. This chapter does not intend to evaluate Koch's work after 1895;[42] rather, the goal is to use the example of the sleeping sickness expedition to find out how the microbe hunter of the 1880s coped with the changed environment of the years after the turn of the century.

Beyond the biographical framework outlined so far, we must now focus on the research expeditions in the wider context of the history of science. This is to raise the second question about the role of travel in the history of bacteriological hygiene. Whatever the personal motivation for Koch's travels, they also modified a traditional form of scientific work and in some sense were typical for the evolution of bacteriological hygiene as a whole. The special importance of traveling establishes this discipline as a specific example of the "experimentalization of life" that recent work in the history of science and medicine has identified as an important characteristic of the period. On the one hand bacteriological hygiene was one of the new laboratory-based disciplines that characterized the medicine of that period. On the other hand, however, it differed in one important respect from physiology, the prime example of the "laboratory revolution in medicine" for Koch's contemporaries and the subsequent historiography.[43] Unlike the physiologists, the bacteriological hygienists had to reproduce the findings elaborated in the laboratory beyond its confines. Whereas the physiologists were able to calculate such matters as the nerve conduction speed in the laboratory, the hygienists were always obliged to prove—in areas like the purification of drinking water, disinfection, or clinical diagnostics—that the relationships they had discovered in the laboratory could also be reproduced in the outside world. Turning the (outside) world into a laboratory was the objective, and bacteriological hygiene became a science practiced by traveling experts.[44]

The laboratory of bacteriological hygiene was shaped by its relations with the outside world in another respect as well. When scientists attempted to work experimentally far away from the infrastructures of the industrialized world, the conditions required for such work sometimes became visible precisely when they were lacking or difficult to establish. To give one example: in Berlin regular access to patients was a matter of course for therapeutic research. But in Africa, where sick persons were not socialized as patients by hospitals, insurance plans, and so forth, access to them had to be created—if need be by force of arms[45]—and this identifies them to the historian as part of the experimental setup.

A cursory look at the travel activities of hygienists and bacteriologists thus makes it clear that one should not dismiss them as a basically obsolete form of scientific work that had formed the basis of the historical-geographical pathology of the years before 1850.[46] Indeed, in looking at the work of microbiology-centered hygienists, one realizes that theirs was an important field of scientific research that assumed a variety of forms. Some of these were simply projects that applied the tenets of bacteriological hygiene to a specific object—such as inspections of drinking water facilities.[47] Others were research expeditions for the purpose of making contact with research objects that were nonexistent or difficult to find at home, such as malaria or sleeping sickness.[48] Some travels, finally, were undertaken in pursuit of therapeutic experiments that for one reason or another were more easily carried out on the spot.[49]

However, I should point out that despite the extensive, varied, and conspicuous traveling practice of the hygienists and bacteriologists—and here the history of public health or that of tropical medicine provides the most telling examples—these practices have not so far become the object of specific studies in the history of medicine or science. To be sure, it might be said with some reason that in spatial terms, the pathogens of many diseases have been discovered outside as well as inside the laboratory, but the special conditions under which knowledge is acquired while traveling have not so far been studied in their own right. As far as the era and the subject of bacteriology are concerned, historiographical reflections on expeditions as a mode of scientific work or the travel account as a genre of scientific literature are very rare and not even comparable to cutting-edge studies of the travels of Darwin, Humboldt, and the traveling science of earlier centuries.[50]

In placing this investigation into a general context, one must pay attention to more than the specific scientific purposes of traveling. Travel frequently created privileged access to the public at home and could afford a laboratory scientist social prominence as a world traveler. The example of Alexander von Humboldt was well known to Koch and his contemporaries.[51] To be sure, times had changed since Humboldt's great voyages, yet it was often still easier to gain public attention from the other side of the world—and also to guide it, given the absence of competition and the press.[52] And finally, travel also changed the scientist's way of life, sometimes beyond the temporary loss of his accustomed environment of clinic, university, and laboratory. I am referring here to the voluntary or involuntary adaptation to customs in foreign lands, the diseases encountered there, changed social contacts, and the challenge posed to sober science by a spell-struck local population.[53]

And there is one more thing: traveling as such was a traditional form of scientific work, but it increasingly changed in the nineteenth century. The place of the versatile discoverers, many of whom still traveled the world as amateurs in the eighteenth century, was gradually taken by specialized researchers who expanded their expert knowledge through their travels.[54] This development reflected the progressive specialization in medicine and the sciences, and became possible only in the industrialized environment. The fact is that David Livingstone's "Dark Africa," though perceived as more and more exotic in the contemporary imagination,[55] was being colonized. By the turn of the twentieth century, a researcher could expect, if he was lucky, to find railroads, colonial administrations, post offices, and telegraph facilities. At the same time, science and medicine had become indispensable and widely used resources for colonization.[56] These changes continued apace all through Koch's professional life, and the two diseases he investigated in 1883–84 and 1906–7 were discussed in ways characteristic of their respective eras: cholera was the Great Scourge from the moment it appeared, and combating it became the driving force behind the building of a public health system in nineteenth-century Europe.[57] Then, around 1900, sleeping sickness, along with malaria, was the epidemic that provided an argument for the medical-scientific foundation of colonialization.[58]

## V.2. The Cholera Expedition of 1883–1884
### V.2.1. Preconditions and Preparations

To understand the significance of the cholera expedition in the context of this study, we must take a look at its director's professional situation in 1883. On August 16, 1883, the participants to the expedition set out for Egypt. In addition to Koch, they were the physicians Georg Gaffky, Bernhard Fischer, and the chemist and laboratory dissector (Präparator) Treskow, all of them employees of the IHO. At the time, Koch had been working at this institution for about three years. His career was steeply rising, along with the reputation of his institution. Recognized as an outstanding figure in the world of science at least since the identification of the tubercle bacillus in March 1882,[59] he had subsequently directed a working group engaged in several projects of public importance. This increased his fame as much as it enhanced the reputation of the IHO as a modern center of large-scale research.[60] Thus the Berlin Hygiene Exhibition of 1882–83 provided the first opportunity to introduce the public to the IHO's newly developed methods, from solid culture media to disinfection apparatuses. The pavilion of the IHO was, as a French observer wrote, "the high point of the exhibition from the scientific point of view."[61] Koch reported in a letter to his daughter that his bacteria were now beginning to have a public life in addition to their existence in the laboratory: "In our pavilion we have had a lot of visitors, who wanted us to explain the dangerous and non-dangerous bacteria we are showing, as well as the photographs and the many apparatuses."[62]

In a sense, Koch's controversy with Pasteur also belongs in this context. Objectively, it concerned questions about "correct" microbiological techniques, matters of priority, and the phenomenon of virulence in bacteria in connection with Pasteur's anthrax vaccine. But when the controversy was fought out at the international Congress of Hygiene at Geneva in 1882[63], it became a public spectacle from which nationalist rhetoric was by no means absent. This was one of the ways in which bacteriological research found its way from the professional literature to the daily newspapers.[64]

This is the environment in which we must also place the cholera expedition. The outbreak of the epidemic, which was reported from Egypt in June 1886, provided the opportunity to test a new set of bacteriological laboratory techniques, disinfectants and such, on one of the most dangerous infectious diseases of that era.[65] At the same time the organization and execution of the expedition would also be proof of the capabilities of the still young Imperial Health Office.[66]

Keep in mind that cholera was not just any infectious disease; it was *the* epidemic of the era of industrialization and urbanization in Central Europe. Ever since its first appearance in Europe around 1830, it had remained a constant challenge to medicine and public health systems. The sanitation of cities and the rise of hygiene as an academic discipline would be hard to explain without the repeated appearance of the scourge from Asia, which was of vital interest to public health.[67] Demonstrating the presence of a bacterial pathogen and on that basis providing means to combat and prevent an epidemic would certainly be proof of the effectiveness of bacteriological hygiene in the most important area of public health care. But the successful identification of such a pathogen would constitute a challenge to the hitherto dominating cholera theory of the Munich hygienist Max von Pettenkofer. Pettenkofer had postulated a complex web of causes for the etiology of this disease, in which the soil and its composition played a decisive role. Accordingly, a biological pathogen—which Pettenkofer also accepted, but did not identify—could cause disease only under special local conditions, such as soil composition or ground water levels, and in combination with the constitution of local individuals.[68] Koch, by contrast, espoused a reductionist model, which identified the bacterium as the necessary cause of the disease. Demonstrating such a bacterium was thus a way to refute Pettenkofer's entire theory.

At this point Koch had already acquired professional experience with the disease. In 1866 he was working as junior physician at the Hamburg General Hospital.[69] There the ongoing epidemic gave him the opportunity to conduct microscopic pathology studies on cholera cadavers. However, the significance of these investigations for his subsequent cholera studies was probably slight. After all, seventeen years had passed since then, and Koch himself stated that by 1883 his earlier work had "largely disappeared from [his] mind."[70] Nonetheless, Koch's papers contain drawings of microscopic preparations that can be dated to 1866 and show microorganisms taken from cholera-infected intestines. But whether, as his biographer Heymann assumes, the young physician actually encountered the cholera bacterium at this early date is a matter of definition. The drawings are rather imprecise and do not point to etiological reflections of any kind.[71] It had been known for some time that cholera-infected intestines were full of microorganisms, and to that extent there was nothing new in Koch's observations.[72] Instead, the drawings point to a familiarity with the work of the microbiologist Hallier.[73]

The 1883 epidemic in Egypt provided the occasion for studying a disease to which Europeans felt they had to travel, lest they meet up with it at their own doorstep. The IHO was certainly interested. Back in 1881 its leadership had made plans to investigate this disease on the basis of pathology samples brought in from Brazil.[74] It is not known, however, whether these plans were carried through. In any case, this material, preserved in alcohol, would have allowed for no more than work on the pathological anatomy of the disease. Culturing experiments, which were a necessary part of Koch's biological research, would have been impossible due to the effect of the conservation. Contamination of the material by putrefactive bacteria was also a significant problem. Traveling was therefore necessary, and for several reasons the epidemic of the summer of 1883 provided a good opportunity: it took place in Egypt, which was reasonably close to Germany. Moreover, Egypt was a country marked by the existence of roads and a few hospitals where medicine of the European type was practiced. And finally, it was a major advantage that France, at the suggestion of Pasteur (who did not participate in person), had also sent an expedition. This added the political legitimation of nation-state competition to the scientific justification of the expedition.[75] On August 4, 1883, the Imperial Ministry of the Interior, "having been informed that several governments have sent medical expeditions to Egypt for the purpose of gathering scientific observations about cholera," contacted the IHO. Holding out the prospect of financing by the Imperial Chancellery, it also suggested that Koch direct the expedition, "unless this choice were to encounter serious reservations."[76] This was not the case, and so on August 9, Koch was officially appointed to lead the expedition.[77] The three-week period of preparation and the traveling time of the expedition was about the same as what the French expedition had required.[78]

### V.2.2. In Alexandria

On August 23, the four-man German research team stepped onto Egyptian soil. Considering that planning had started only a few weeks earlier, this was fast; yet given the rapid course of an epidemic like cholera, it was not fast enough. As Koch wrote to his "dear Emmy" only two days after his arrival, the epidemic was already waning.[79] Hence it had become more difficult to come close to cholera victims or to obtain fresh cholera cadavers.[80] In the end, the group had no more than ten such cadavers. The work started with autopsies and microscopic pathology studies of cadavers that "could be autopsied immediately after death or at least a few hours later."[81] Koch clearly followed the strategy he had successfully used in the past when he first looked for pathological and anatomical signs of infection; he sought a bacterium that could be linked to the symptoms, and then performed culturing experiments and animal testing. The first step could be carried out with nine of the ten cadavers at his disposal. In the tenth case, death had ensued some weeks after the patient had survived the cholera. The autopsies yielded the picture of a clear correlation between the distribution of bacteria and the symptoms of the disease: the microorganisms were found exclusively in the digestive tract, particularly in the small intestine, and never in other organs.[82] But then, this was initially nothing more and nothing less than the confirmation of observations about the pathological anatomy of Asian cholera that had already been made on the basis of cadaver samples at the IHO and by other researchers. As Koch remarked, they did not allow for statements about a bacterial etiology.[83]

In Egypt it subsequently proved impossible to experiment with this material. The material from the cadavers would not infect either the fifty mice brought from Berlin or the monkeys, dogs, and chickens purchased locally. This was particularly regrettable because Koch was rather keen on the infection experiments carried out by Carl Thiersch back in 1856 and their recent confirmation by the British bacteriologist Burden-Sanderson.[84] Animal experiments failed to produce infections, and it was also impossible to confirm the specificity of the identified bacterium by means of culturing. The reasons for these failures probably lay in technical difficulties related to the heat prevailing in North Africa in August. Thus Koch's famous gelatin-based solid culture media lost their special characteristic in higher temperatures, liquefied, and became useless for the production of pure cultures. Not even the recourse to a specially designed refrigerator solved the problem.[85]

In view of the unsatisfactory results and the deteriorating working conditions as the epidemic waned, it became clear by mid-September that "the ongoing investigations can keep us occupied for approximately two more weeks."[86] Apparently the plan had been to follow the epidemic from Egypt to Syria, but since no epidemic broke out in Syria, this no longer made sense. Koch hoped to continue to Upper Egypt, where a sufficient number of cases could apparently still be expected, but the expedition was officially advised against it. Autopsies of cadavers probably could not have been performed there against the religious opposition of the local population.[87] Nonetheless the travelers remained in Egypt for two more months, until November 13, 1883. Their activities in the remaining time were manifold and not restricted to bacteriological-hygienic matters. Thus the researchers performed autopsies on typhus or dysentery cadavers and carried out some bacteriological testing of water and air.

"When I departed for Egypt I would not have dreamed that the expedition would last this long, but I am not at all displeased with this development," Koch wrote to his brother Hugo shortly before continuing to India.[88] This venture became, as can be seen especially in Koch's private correspondence, a formative experience that offered "a plethora of unforgettable impressions, experiences, and enlightening episodes," from which he expected "to profit for the rest of [his] life." In this spirit the members of the expedition took a trip to the town of Damiette. Since the epidemic had first appeared there in June 1883, it was worth a closer look. Similarly, the scientists later visited a camp for Mecca pilgrims by the Red Sea, where Koch's cultural fascination obviously outweighed his concern for the hygienic conditions in the camp: "We saw towns and came into contact with people who are completely untouched by European culture and exhibit oriental ways in pure form. The high point of our experience was a camel ride into the desert, where we were led by a Bedouin sheik . . . A troop of his Bedouins, some on foot, some on camels, but all of them in picturesque attire, accompanied us and sought to entertain us with their war chants and their warlike games."[89]

Koch greatly enjoyed the fulfillment of his long-standing desire for travel and proudly presented himself as a world traveler in letters to his relatives. To his daughter, Gertrud, he not only sent the stamps "you so avidly desire," he also managed to combine this present with a scientific curiosity: held down by a

The members of the cholera expedition of 1883-84. Left to right: Gaffky, Treskow, Koch, Fischer. RKI/Koch papers.

stamp hinge, he sent her "a genuine mosquito, one of the most vicious kind, by which one is so often bitten here."[90] His brother Hugo, a mine owner in Mexico, he regaled with geological rarities for his collection, such as petrified specimens and a supposedly genuine "piece from the tip of the Cheops pyramid."[91]

However fascinating traveling in Egypt may have been for Koch as a private person, from a scientific point of view it was not particularly productive. There were few tangible results; after all, the pathological anatomy of the cholera-infected small intestine and particularly the massive presence of microorganisms in this site were not unknown to contemporary science. As long as Koch and his associates were unable to identify the microorganism as a bacterial species and to define its etiological significance in animal experiments, they really were not offering any new observations. The French team, by contrast, had succeeded in finding a cholera pathogen in the blood, thereby confirming Pasteur's suppositions on the subject.[92]

Nonetheless, the expedition made the front pages of German newspapers as early as September 1883, thanks to circumstances that had little to do with cholera. Involved here was the competition with other research teams, particularly the four-man French one, which had reached Egypt a few days before the German team.[93] As an extremely nasty article in the *Berliner Tageblatt* of September 26 indicates,[94] the press was eager to interpret the relation between the two research teams according to the pattern of Franco-German competition with which the readers were familiar from the days of the Hygiene Congress at Geneva. Whereas the German team sought scientific truth, the article said, the French were there only to show off: "Within short order, rumor had it throughout town [Alexandria] that the German doctors had already achieved very good results. The French gentlemen pricked up their ears and of course, faithful to their national character, tried to wreck the Germans' parade with a giant cholera bacillus they claimed to have discovered. But with a few words of refutation from Geheimrath Koch they were sent back into their own narrow scientific backyard." It is interesting that Koch and the scientific institutions at whose behest he was traveling organized their own publicity: all along the way the researcher regularly sent reports to von Boetticher, state secretary in the Imperial Ministry of the Interior. These were—with one exception—slightly edited and published in different places, such as the *Reichsanzeiger* or the *Deutsche Medizinische Wochenschrift*. They purported to give the public authentic insights into Koch's scientific work of the kind that could not have been carried out in a laboratory at home.[95]

These reports are interesting hybrids, part scientific article, part travel literature. Quite personal in tone since they were written in the first person singular, they made use of the presumptive authority of the travel account to report on scientific work.[96] In reading descriptions of such things as the hygienic conditions in Damiette or learning about the findings of an autopsy, the readers felt that they were participating in an adventure that touched on both the macro- and the microcosm. Such narratives could easily accommodate information that had not even been gained on the expedition. This was the case, for instance, on September 17, 1883, when Koch wove a short survey of the animal-experimental cholera studies by other scientists into the description of his own infection experiments.

These travel accounts also accomplished something else. They depicted Egypt—no doubt accurately—as a country "very rich in parasitical and infectious diseases."[97] In his reports Koch made considerable use of this *genius loci*, partly by way of emphasizing the dangerous character of the disease, and partly to stylize the scientific work directed against the epidemic as a heroic struggle. Instead of working, as they had done before, in modern and hygienic laboratories in Berlin, scientists now went after epidemics in exotic and dangerous places and worked there under most hazardous conditions. And although the reports were not necessarily emotional in tone, they did cast the "Father of the Bacilli" Koch in the role of the crusader who, fighting the enemy in faraway lands, confronts him at the front lines.

Reflecting the deteriorating working conditions, the first published report of September still deals largely with animal experiments and work in pathological anatomy, while the second one, written in November, went in for medical geography and epidemiology. Koch never gave a forthright account of the very unfavorable working conditions, except in his very first and unpublished report of August 25, 1883. Here he simply stated that "in Alexandria cholera is rapidly receding and there is not a moment to lose if any material for the investigation is to be found."[98] Yet in September he depicted the stay in Egypt as part of a larger travel plan that only provided for preparatory investigations in Egypt and called for following the epidemic from there. When no outbreak of cholera occurred in Syria, this intention proved impossible to carry out, and it was decided to pursue the epidemic to its presumed place of origin in India. The expedition sailed from Suez to that subcontinent on November 13.

## V.2.3. In India

This second stage of the expedition featured a team reduced from four to three persons, for the chemist and dissector Treskow had returned from Egypt to Berlin. About four weeks after sailing from Suez, Fischer, Gaffky, and Koch reached their destination, Calcutta.[99] The work of the following four months would yield the results that won Koch the reputation of having discovered the cholera pathogen, known as the comma bacillus because of its curved shape.

Calcutta did offer far better working conditions than Alexandria. Here cholera was endemic, albeit with certain seasonal fluctuations, and both the British colonial administration and the head of the Medical College Hospital, Dr. J. M. Cunningham—incidentally a vehement proponent of the soil theory of cholera—supported the work of the German commission, which before long was installed in the hospital building and could count on a regular supply of cholera cadavers.[100] Above all, it was expected that the culturing experiments that had failed in Alexandria could be resumed with better prospects of success in the cool climate of Calcutta's winter months.[101]

Immediately upon his arrival, Koch wrote a report outlining a research program for the coming months. Spelling them out in detail, he listed the problems with which the expedition had made little headway in Egypt. These were the microscopic and pathological examination of cadavers, the production of pure cultures, the description of the biological characteristics of the cholera bacteria, attempts at disinfection, investigation of the viability of the bacteria outside the human body, and finally—in case of a repeated failure of the bacteriological experiments—epidemiological studies in India.[102]

Aided by the cool weather, Koch was able to report a first breakthrough in January—pure cultures. He stressed that these were a fruit of the methods developed at the IHO. "With the help of the methods developed at the Imperial Health Office, methods that worked extremely well here too, we were able to isolate the bacilli from the intestinal content of the purest cholera cases and to grow them in pure culture."[103] On this basis it became possible not only to provide an exact description of the biological characteristics of the bacteria and their colonies, but also to check and refine the pathological anatomy of the disease against the findings obtained in Egypt. The cholera bacteria could now be distinguished from other bacteria found in the intestine, and so their specific presence in cholera victims could be established. Clearly in excellent spirits, Koch also reported that "recently some of the experiments conducted on animals have yielded results that raise hopes for further success."[104] Feeling that he was headed straight for the goal, he ended his report of January 7 with a critical review of competing research. Without naming the main proponent of the soil theory, Max von Pettenkofer, he cited statistics for Calcutta that established a connection between an increased supply of drinking water and cholera morbidity since 1870. This argued against soil composition as an etiological factor and for drinking water as a source of infection, especially since "according to the almost unanimous opinion of the local physicians, the regression of cholera is exclusively attributable to the introduction of a drinking water pipe."[105] Koch dismissed the French commission's earlier report on a cholera bacterium found in the bloodstream by pointing out not only that this was not a cholera bacterium, but that it was actually not a microorganism at all, and that "the French commission was caught in the same error to which other researchers had fallen prey before, by mistaking the blood platelets for specific organisms."[106]

By February 2, 1884, Koch followed up his optimistic initial assessment of January by reporting that "the question I had left open in my last report of 7 January, namely whether the bacilli found in the cholera-infected intestine are parasites elusively related to cholera, . . . can now be considered solved."[107] However, the demonstration of a bacterial etiology for cholera that was implied here—and acknowledged by contemporaries[108]—did show a few peculiarities. It is worthwhile to examine these more closely, since some of them are directly related to the specific working conditions of a research expedition.

The first surprise is the fact that Koch had not made this announcement as early as January. Factually, the February report contained nothing new. Rather, it elaborated on matters that had already been communicated earlier. For instance, it gave a detailed description of the characteristics of the cholera bacterium and its behavior in culture media. Added to this was a survey of the pathological findings obtained thus far. At the center of the report was the discussion of the knowledge gained, which Koch interpreted to mean that a bacterial etiology of cholera was now an established fact. As scholarship has pointed out,[109] this demonstration was incomplete by the standards of Koch's own methodology, which—briefly put—called not only for the identification of a pathogen but also its cultivation and use in animal experiments. The observation on which Koch based his claim was that the bacterium could always be demonstrated in the digestive tract, particularly in the small intestine, of cholera victims and at the same time could never be found in other parts of the body or in other diseases. To Koch this meant that the bacteria could be only one of two things: opportunistic parasites of cholera or its pathogens. In the former case they would also have to be present in healthy humans. Since this was not the case and "since the vegetation of these bacteria in the intestine cannot be caused by cholera, we are left with the assumption that they are the cause of the cholera."[110] In backing up this argument *ex negativo* with a description of the pathological anatomy of the disease, Koch was particularly careful to show that the pathological process corresponded to the appearance, the quantity, the distribution, and the behavior of the bacteria.[111] Moreover, the description reflected the invasion model that Koch and others favored at the time. That model presupposed a healthy body completely free of pathogenic germs. The appearance of such germs marked the beginning of the disease. Koch's description of the typical circumstances of a cholera outbreak in Egypt followed this model: "Thus the cholera bacilli behave exactly like all other pathogenic bacteria. They are found exclusively in the disease to which they are related; their first appearance coincides with the outbreak of the disease, they increase in number as the disease process intensifies and disappear again when the disease has run its course. Their location as well corresponds to the expansion of the disease process, and their quantity at the height of disease is so great that it accounts for their pernicious effect on the intestinal mucus."[112]

The final element of this argumentation, the successful animal experiment that Koch had promised in January, was still missing. While he did admit this, he now—in stark contrast to his only recently completed studies on the etiology of tuberculosis, where he had insisted on the absolute necessity of such a demonstration—depicted the animal pathology experiment as merely desirable, yet impossible in the case of cholera. "One is bound to wonder whether it [the animal experiment] will ever succeed, for there are indications that animals are not susceptible to cholera. If any animal could contract cholera, someone would surely have reliably observed it in Bengal, where all year long and throughout the country infectious cholera material is present."[113] The scientist also pointed out that in some other infectious diseases whose bacterial etiology was undisputed, such as leprosy and typhoid fever, reproduction in animal experiments had never succeeded. Nonetheless, in terms of Koch's own standards, such as the variant of the postulates he had used in the tuberculosis studies, the demonstration was incomplete.[114] He had refined a pathological anatomy of cholera whose main tenets were already known, albeit somewhat controversial, and he had identified a bacterium that probably was the pathogen. This finding was supported by epidemiological observations. But as long as it was impossible to produce the disease in animal experiments and through the use of pure cultures, Koch's argument remained circular and postulated in general what it sought to prove in particular. It was thus much more than "desirable"[115] to carry out these experiments, and Koch was no doubt aware of this. The last report, sent to Bötticher from India on March 4, dealt exclusively with this problem and presented his solution to it. The fact is that while Koch was working in the laboratory, "chance produced [nearby] a fortuitous experiment with humans that in this case made up for the lack of animal experimentation."[116] He was referring to locally circumscribed epidemics that could be observed in areas surrounding so-called tanks (ponds or reservoirs), which served, among other things, as supplies of drinking water. Since it was possible to demonstrate the presence of the bacilli in the tanks, in clothing, in the excreta of the residents, and so forth, the path of the infection could be documented at every step: "Considering that hitherto researchers have in vain searched for cholera bacilli in samples of tank water, sewage, river water and all other kinds of potentially polluted water, and that these have now been found with all their characteristic qualities in a tank surrounded by a cholera epidemic, this result has to be seen as extremely important."[117]

Ultimately, one is bound to wonder why Koch did not continue the experimental work begun in India and instead decided in February to content himself with a state of knowledge that was unsatisfactory by his own standards. One of the reasons was that he had indeed achieved significant results, namely, informative epidemiological investigations, a very precise pathological anatomy of the disease, as well as a bacterium grown as a pure culture, whose relation to the disease was probably of an etiological nature. On the other hand, however, it was not even possible to continue the work, since "further investigations on this subject had to be halted because of the hot weather that came very early this year."[118] Koch had worked in a limited window of time and space, defined by the presence of cholera on the one hand, and by the temperature necessary for proper work in the laboratory on the other. Such conditions were very rare in most parts of the world and prevailed in Calcutta only from December to February. By the end of January it was already getting warmer; the researcher urged his team to make haste, aware that the end of proper working conditions was in sight. To his daughter he wrote on February 12, "We won't be able to go on like this for long, for the weather is already heating up mightily, and soon everyone who can possibly leave the city will flee northward, to the Himalayas."[119] A few days later the culture media liquefied for good. The researchers realized that they had to be satisfied with what they had achieved if they did not want to let ten months go by before resuming their work. Koch, Gaffky, and Fischer left Calcutta for Darjeeling, the summer quarters of their British hosts, enjoyed a few days of vacation, and on March 14 sailed for Europe.[120]

All in all, the reports from India convey the impression that this was a time of intense work, for in addition to the laboratory work discussed here, it brought epidemiological studies, disinfection experiments, and other research opportunities. This might give us the idea that in Egypt Koch's cultural activities had been merely the result of insufficient scientific activity. But this was by no means the case. To be sure, there were phases of intense work in India, particularly around the New Year, when Koch complained in a letter to his spouse that "because of this constant scientific activity [he] had not yet had a chance to take a closer look at the city and its inhabitants."[121] On a later occasion he was so busy that he even failed to remember her birthday in time.[122] On the other hand, as his letter suggests, he considered the work load to be unusually heavy. He normally was quite willing to take time out to resume the program of cultural enrichment. India had much to offer in this respect, and Koch was clearly determined not to miss this opportunity. Even while traveling to India he took advantage of a scheduled stopover in Ceylon "to make an excursion to the interior of this fairytale island, which still has virgin forests and wild elephants."[123]

Once arrived in Calcutta, the scientist did work hard, but not to the extent suggested by the official reports. In a letter to his daughter on February 12, Koch describes the course of a typical day: "I get up at seven in the morning, go or rather am driven to my laboratory, where I work until two; then we eat, and in the afternoon I write, read, visit the fair, or perhaps just go for a ride."[124]

In other words, there were regular working hours, and there was also room for cultural activities. In Koch's case, visits to the fair were a priority. This was a major exhibition that was held in Calcutta at this time. Koch particularly liked to visit the so-called Indian Court to see Indian arts and crafts. He enjoyed the exotic character of the exhibits there, and just as he had sent presents from Egypt, so he was pleased that here too "almost everything shown is also for sale."[125] As he enthusiastically reported to his wife, "Whenever I go to the fair, which is quite often, I practically limit myself to visiting this so-called Indian Court and enjoying the sight of these magnificent things, such as metal items, jewelry, rugs, shawls, carvings, and so forth." To his daughter he also sent a colorful description: "When I am at the fair, I often regret that you cannot see all these wonderful things that are gathered from all over India. But I have bought a few little things for you, so that you can get at least an idea of India's treasures. It will be marvelous when I open my suitcase and they come out one by one."[126] Whether "a few little things" were really all he bought is questionable, for it appears that in Calcutta Koch succumbed to a regular collecting frenzy, which was not so much focused on aesthetic or other standards but rather aimed to acquire curiosities. These, after all, would allow the traveler to demonstrate his own status and the exotic nature of the places visited upon his return home. Only when there was nothing to buy did he search on his own. In the same letter to his daughter he wrote: "Where there was nothing to buy, I have at least collected for you a few shells, a flower or pretty leaves. Besides, I have bought a lot of photographs, in which you can see how beautiful it is in Egypt and in the tropical countries of Asia; that is where we will travel in our minds, and then we will go walking in the jungles of Ceylon and in the streets of Calcutta."[127]

Aside from the curiosities, Koch had little interest in the countries he visited. For all the enthusiasm he expressed for the Indian Court, jungles, and elephants, great contempt marked his comments about the city and its inhabitants, these "lazy and clumsy people."[128] His interests hardly went beyond the tourist's souvenir hunting. At the Indian Court he finally fell into a regular shopping frenzy: "I have already acquired many beautiful things there and would like to buy more, except that my funds are exhausted. I would therefore appreciate it if you sent me all the money you can spare. The more the better for you and the family, for almost everything I bring back will be for you. I don't know how much you will be able to send, but I do know that a thousand marks would not be too much for me. It's just that I only buy good things, and these can't be gotten cheaply."[129]

## V.2.4. Return to Berlin

The return voyage to Berlin took almost two months. Koch's health was poor and there were also opportunities for additional epidemiological studies, for instance in Bombay, where the members of the expedition looked into the special conditions of that city by the sea and their effect on the appearance of cholera.[130] On May 2, 1884, the travelers finally reached Berlin. Their arrival had all the marks of a triumphal entry. The themes that had informed the public perception of the expedition from the start now appeared with even greater clarity.

"Welcome, Victors!" read the headline of the *Berliner Tageblatt* on May 3,[131] greeting the returning scientists in the "proudly armed New Germany." The expedition headed by Koch was presented as a campaign of German science. Had not the scientists under their "commander" Koch been dispatched to "discover the secret paths of one of the horrible pestilences that have plagued humanity for the last fifty years or more?" And had they not pursued "this enemy all the way to his East-Indian lairs?"[132] This seemed to provide a firm foundation for the glory of "German Learning" in a new area, that of the natural sciences, in addition to all the others:[133] "Germany is not only the most outstanding workshop for the study of Antiquity; no indeed: it is also the most industrious laboratory for the recently opened areas of the experimental natural sciences."[134] On the occasion of a banquet honoring the returning scientists, the expedition was

Title page of the program for the banquet honoring the returning cholera expedition, 1884. RKI/Koch papers.

literally categorized as a victorious military action that could easily stand up to comparisons with the Franco-Prussian War: "Just as thirteen years ago the German people celebrated a glorious victory against the hereditary enemy of our nation, so does German Science today celebrate a brilliant triumph over one of humanity's most menacing enemies, one of the most dreaded and murderous epidemics of modern times: Cholera."[135]

The title page of the program for this banquet showed humorous stylizations of the most important elements of the traveling bacteriological laboratory and its public presentation. Special emphasis is placed on the exotic character of the countries visited. Thus nineteenth-century Egypt is naively portrayed as the land of the pharaohs; yet laboratory needs such as microscopes and experimental animals can also be found there. Included as well are classic elements in the allegorical depiction of epidemics, such as the sick, grim reapers, and the angels of mercy. The latter are flanked by the members of the expedition in antique garb.

The public resonance the expedition enjoyed was not altogether spontaneous, but rather, as I pointed out above, promoted to some extent by the efforts of the Imperial Ministry of the Interior. It is thus not surprising that officials there were thinking about staging an appropriate conclusion at the time of the return. By the end of April they had already prepared a report for the emperor, in which they proposed that the members of the expedition be given decorations. For Koch they envisaged the Imperial Crown Medal II Class—normally a military decoration—and not without reason, according to Karl Köhler, the ministerial councilor and subsequent director of the Imperial Health Office who wrote the relevant memorandum: "If there were a special reward for such deeds in the fight against the natural enemies of human life, analogous to the Iron Cross that is given for fighting the enemies of the Fatherland, I would suggest that in the present case that distinction be included as well."[136] These decorations were bestowed on Koch and his companions in addition to an audience with the emperor and financial rewards specifically authorized by the Reichstag. For Koch this sum was 100,000 marks; his companions received 35,000 in all.[137] Around that time Koch was also appointed director of the IHO and given a seat on the Prussian Council of State.[138] When looking back on it, Koch himself particularly appreciated receiving the military decoration: "My favorite decoration is my Imperial Crown Medal, which our old Emperor personally bestowed on me after we returned from the cholera campaign. It is to be worn on a black-and-white ribbon, like a military decoration. And that is exactly what it was."[139] William Coleman has shown that this enthusiastic reception had its counterpart in the lack of controversy within the scientific community.[140] It is interesting to note that here too the public relations work of the IHO and the Imperial Ministry of the Interior had yielded specific results. After all, the publication of scientific studies—in the *Reichsanzeiger* and in well-written reports to the Imperial Ministry of the Interior—had lent the results a semiofficial character, which could only enhance their credibility. Upon the return of the expedition, the full value of such exclusive publishing venues became clear when competing scientists were denied access to them.

Recall that Koch's reports from India and Egypt had done more than simply document the work in the field. They had actually made a point of advancing the drinking water theory of cholera, which postulated the spread of the epidemic by this means and favored control measures such as sand filtration of the water. At the same time the reports also contained more or less explicit appraisals of competing research. Although in the area of microbiology, this criticism was primarily directed against the French competitors, with respect to epidemiological theory, they were aimed in particular against Max von Pettenkofer's soil theory of cholera. Although Pettenkofer did not deny the existence of a possibly specific bacterium, he did not accept Koch's claim of its etiological importance as the necessary cause of disease. For his part, he maintained that the maturing of an infectious agent under special climatic and local conditions was necessary.[141]

Koch's travels and the semiofficial publication of his reports had given the Berlin bacteriologist an advantage over Pettenkofer, as the following marginal episode illustrates rather well. Visiting Pettenkofer when passing through Munich, Koch had insisted on the independent character of his findings, which he said did not need the framework of Pettenkofer's cholera theory.[142] This must be seen against the background of a discussion of Koch's findings published in March-April 1884, in which the Munich hygienist endeavored to incorporate the Berlin bacterium into his soil theory. The text came out in a daily newspaper, the *Münchener Neueste Nachrichten.*[143] Pettenkofer had of course tried to have it published also in the *Reichsanzeiger*, but this was turned down by the Imperial Interior Ministry upon consultation with the IHO. As Dr. Struck, the director of this institution, emphasized, this was a matter of not giving Pettenkofer the opportunity to express "premature" criticism of Koch's findings.[144] And indeed, Koch's reports from India acquired their final cachet only upon his triumphal return and because of the quasi-governmental publication of his findings. A critique by Pettenkofer published in the same place would not have been beneficial to the semiofficial aura and reputation of the enterprise.

Surveying the cholera expedition as a whole, one finds that the circumstances of this venture profoundly marked the proceedings, the results, and the public perception of the undertaking. The following elements appear to be most important: the travel reports emphasize the race against the French researchers, which was a way to create a new and presumably more widespread interest in bacteriology in the public at large. Koch became, as the (women's magazine) *Gartenlaube* put it in the spring of 1884, "a man whose name is now on everyone's lips."[145]

The excessive rhetoric that touted bacteriological hygiene as a war against epidemics was particularly effective against the background of the exotic and dangerous places that were now presented as the home of disease. It was again the *Gartenlaube* that, a few weeks later, showed its readers the "much discussed and much feared comma bacillus in a faithful rendering."[146] Largely using Koch's own language, the same magazine also provided a description of the Bengal jungle as the place of origin of the epidemic, leaving no doubt as to the dangerous character of the disease: "Lush vegetation and a rich fauna have developed in this area of the world, which, inaccessible to humans not only because of frequent flooding and the presence of many tigers, is above all avoided because of the malignant fevers . . . It will be easy to understand that here microorganisms have a better chance to develop than almost anywhere else on earth . . . Under peculiar conditions a very peculiar fauna and flora is bound to develop, and in all probability the cholera bacillus is part of it."[147]

Here it becomes clear that the most important result of the popular reception was to hide the unfinished and incomplete aspect of the picture, which surely characterized the subject and indeed was still perceptible in Koch's reports. Thus a complex research process became—and this is not unusual in popularized science—a simple discovery.[148] Beyond that, the success of the cholera expedition greatly enhanced the renown and the credibility of the set of methods and institutions that had made it possible to identify bacteria as functional representatives of disease.[149] At first glance the idea that a few black lines in the image of a microscopic preparation should embody the cholera was unusual, but the *Gartenlaube* did its best to make it accessible to its readers: "The lay reader who sees these illustrations for the first time will be hard put to believe that these bent lines grouped in a jumbled cluster in the first image are organized creatures endowed with their own peculiar way of moving. And yet they are pure comma bacilli."[150]

The reification of the disease in the bacterium also seemed to make its control imminent. In a further article about the "Cholera Peril," the *Gartenlaube* enlightened the reader about the connection between the search for causes and the possibilities of intervention.[151] Until now "there was a lack of knowledge about the causes of these epidemics, which is why all actions against them lacked the necessary clarity." Koch's merit, it continued, was that he "demonstrated that the earlier supposition was true and that cholera was indeed caused by the intrusion of one of these tiny fungi into the human body."[152] In the end, success not only legitimized special research methods but also enhanced the reputation of the persons and institutions involved: "He [Koch] was the right man and—to add this right away—in the right place. The Imperial government has provided the Health Office with the most perfect means, not only unstintingly but also with thorough understanding of the situation."[153]

The expedition was decidedly shaped by the changing scientific culture that was gradually emerging in the industrialized world. The place of the heroic discoverer was now taken by teams of specialized researchers who, working in large research institutions like the Physikalisch-Technische Reichsanstalt or Koch's IHO, where division of labor was the norm, increasingly competed with the universities.[154] Although the expedition with its four and then three members had actually been rather small, it had relied on the technical and financial resources of the IHO, and hence of the Imperial government. Preparing for an undertaking of this scope within a few days would hardly have been possible with the means of a university. The sizable equipment needed was either on hand at the IHO or could be procured in short order. The expenses that, not counting personnel costs, amounted to roughly 33,000 marks, were paid by the government, that is, covered by the respective German consulates general in Egypt and India.[155] And finally, the expedition knew how to use the possibilities offered around the globe in this era of colonialism. Both Alexandria and Calcutta had hospitals run by western standards, which must have made experimental work considerably easier and was a reliable source of cholera cadavers. The telegraph provided constant communication with the homeland, and in India the travelers even had access to electricity and gas. At the same time the work was that of specialists who, beyond the field of their medical interest, moved around the country like tourists.

The scientific work of the expedition was also shaped by travel. To begin with, researching the etiology of cholera in Central Europe would have been a gamble, given the sporadic appearance of the epidemics. It therefore seemed reasonable to confront the cholera in the only region where it was known to be endemic. Moreover, the conditions obtaining in the field also shaped the modus operandi. Thus the microbiological identification of the pathogen was feasible only in the special climatic condition of the Indian winter; at the same time the epidemiological situation in Bengal offered the opportunity to observe unique phenomena such as the tank epidemics. For the modern observer it is also interesting to see that the temporary absence of certain external conditions highlights their importance for the laboratory's functioning. One example is the locals' religious attitude toward the handling of cadavers, which in Upper Egypt would have made work in pathological anatomy impossible.[156] Finally, an undertaking originally projected to be short term turned into an eight-month voyage.

Its scientific yield is most clearly shown by the fact that as a result of the expedition, further travels for the purpose of etiological studies had become unnecessary.[157] The pure cultures first produced in the field made it possible to identify cholera bacteria also in Europe, for instance in Toulon, where Koch traveled in the summer of 1884. Moreover, the control of cholera was now closely tied to the methods developed in Berlin, and hence was practically the property of the IHO and the head of the department of bacteriology. Indeed, the yield of the expedition was more than the precise description and identification of a hitherto not altogether unknown bacterium as the pathogen for *Cholera asiatica*, which was thereby definitively separated from *Cholera nostras* because the disease was now an object tied to Berlin. Identifying the bacteria and interpreting the findings of pathological anatomy could only be done by using the method of the working group in the IHO. Only a few months after the identification, critics were simply told that they had "embarked on their difficult and sensitive task without sufficient prior knowledge and preparation."[158] And in the sessions of the German cholera commission held in 1884 and 1885, Koch's findings—now supplemented with a number of somewhat problematic animal experiments[159]—still carried the day. Even Rudolf Virchow, normally not a supporter of bacteriology, rallied to Koch's position in the deliberations.

In the end, the successful conclusion of the expedition also changed the Berlin bacteriologist's perception of himself and others. Koch had fulfilled a dream of his youth. As a scientist he now became a public figure. The decoration he received, the grant authorized by the Reichstag, his membership on the Prussian Council of State, the banquet in his honor given by the Berlin physicians, and finally the promotion to deputy director of the Imperial Health Office—all of this gave him prestige.[160] Adding the personally enriching experience the expedition had been for Koch, one can gauge how much his personality and the position he occupied had changed since he had joined the IHO in July 1880. As a result of the work done by him and his associates, a set of highly promising methods had led to a bacteriological theory of infectious diseases that was beginning to exert a dominant influence on contemporary hygiene and medicine. Medical bacteriology had changed from a marginal area of knowledge to a phenomenon which, as Virchow grudgingly remarked, "dominated not only the thinking but also the dreams of many older and almost all young physicians."[161] It is true that Koch later came to complain about his overwhelming administrative obligations, writing in 1884 to his friend and colleague Carl Flügge that he could "barely catch a moment for [himself]."[162] Yet until 1885 he had above all enjoyed the pleasant sides of glory and learned about traveling as an agreeable and profitable way of doing scientific work—an experience to which he subsequently would return again and again.

## V.3. The Sleeping Sickness Expedition of 1906–1907
### V.3.1. Protozoa Research

Even if everything in the cholera expedition did not go exactly as Koch might have hoped, that undertaking was marked by its purposeful and efficient proceeding. This trip is part of what is called the microbe-hunting phase in the development of medical bacteriology.[163] In the early 1880s this field became one of the leading disciplines of that time, and the identification of pathogens for major infectious diseases played a decisive role in this development. Such undertakings as the cholera expedition received much public attention.

The sleeping sickness expedition, which took place more than twenty years later, between April 1906 and August 1907, was quite different, even in its outward characteristics. It appears as an extremely protracted undertaking marked by extensive and difficult preparations, a jumble of personnel and institutional actors, imprecise and changing goals, and scientifically problematic results.[164]

What is striking is not so much the long duration—although it was eighteen months—as the extremely involved preparation. What had been taken care of in a few weeks in the case of the cholera expedition took about three years for the sleeping sickness expedition.[165] The reason was not so much that the subject itself commanded little interest. On the contrary: around 1900 this disease, whose name had been known since the Middle Ages, was, along with malaria, one of those tropical diseases around which the discipline of tropical hygiene began to develop. Its endemic presence in the Congo Basin and its spread to East Africa during the Uganda epidemic after 1900 ensured that it was perceived as *the* African epidemic. As a result, the disease and the need to fight it were increasingly seen as necessary elements in the colonization of Africa.[166] But while researchers from other nations mounted expeditions to pursue this subject, and while in 1902 a British research team succeeded in identifying *Trypanosoma gambiense* as the pathogen,[167] the involvement of German science was slight. Whereas Britons and Belgians traveled in the Congo, for instance, German activities were of modest dimensions, and the two cases of sleeping sickness that were treated in the Hamburg Institut für Schiffs-und Tropenhygiene (Institute for Maritime and Tropical Hygiene) were considered important sources of information.[168]

This late start had to do with several factors. From the researchers' perspective it was not necessarily an advantage that Germany had now become a colonial power.[169] The expedition of 1883–84 had been financed from the budget of the Interior Ministry by way of the IHO at a time when the German Reich still endeavored to become a colonial power. After the turn of the century, the Reich did have colonies. Thus the financing of research expeditions became the responsibility of the Foreign Office (Auswärtiges Amt, or AA) and its colonial department. But the under-financing of this department was notorious, which is why it attempted to keep the cost of the colonial administration as low as it possibly could. Hence, when the Institute for Infectious Diseases submitted a plan for a sleeping sickness expedition to the Foreign Office in 1903, the answer was a flat No.[170] Nor was it possible to obtain the support of the Imperial Health Office in the form of financing from the budget of the Imperial Medical Administration, for that office also considered such an expedition unnecessary.[171] The main reason for this refusal was information to the effect that the epidemic raging in British and Belgian colonial areas had not affected the German possessions in East Africa.[172] In the summer of 1903 the Foreign Office took the position that the few cases found had been brought in from the British Mandate. No infections had taken place in the German territory, so an expedition was not necessary: "Perhaps it may prove useful later to follow the British example and send out a commission consisting of a pathologist and a bacteriologist. Given the no doubt very considerable expense to be expected, the colonial administration feels that under the existing circumstances the need for this undertaking is not urgent at this time."[173] And yet the plans made in 1903–4 were of modest dimensions. At that point no one was thinking of the expedition that would take place in 1906. Rather, the plan was to send two younger specialists for tropical medicine, Fritz Martini and Ernst Fülleborn, to explore the situation in East Africa.[174] But the Foreign Office refused to consider sending even a single physician, which the local colonial administration had requested at Robert Koch's suggestion.[175] Instead it charged the medical officers of the Bukoba and Muanza districts, Feldmann and Lott—who had had some bacteriological training in Germany—with keeping an eye on the situation. For this purpose they were issued "bacteriological equipment." For the rest it was decided to wait for the publication of the British researchers.

Once the endangerment of the German Mandated Territory had been eliminated as a reason for sending an expedition, pleading for its scientific interest also proved difficult. In the winter of 1903–4 David Bruce and Aldo Castellani had identified *Trypanosoma gambiense* as the pathogen for the disease, so that the need for research no longer seemed urgent.[176] Accordingly, the foreign minister upheld his negative attitude in 1904 as well. On November 4 he stated that "I do not have funds from which a subsidy for the expedition could be given."[177] In an appended statement the colonial office took the precaution of refusing to be put in charge of matters of research. Moreover, it was felt that the gains to be expected from an expedition would be insignificant: "Concerning the proposed commission for the study of sleeping sickness, it would no doubt be useful to science, but it is not clear that the expected practical advantage would be an improvement over a constant supervision of the threatened districts by the relevant colonial medical officers. Therefore, since what is involved here is mainly the competition between German science and that of other civilized states, the colonial office should not have to bear the brunt of the cost of such an undertaking."[178] However, the grounds for this refusal left open certain possibilities. For one thing, the local medical officers were now charged with observing the situation in the field, and this gave hope for new information. For another, the refusal stated that the German colonial territories were not in danger. If this were to change, an expedition would obviously become absolutely necessary.

Koch, whose research on epizootics and various tropical diseases of humans had occasionally taken him to Africa, particularly Rhodesia, had alerted the Prussian Ministry of Culture to the importance of this matter as early as 1902.[179] As we have seen, he supported the sleeping sickness expedition. But there is no indication that at this time he was thinking about joining it in person. In May 1903, after the Foreign Office—with the approval of the IHO—had turned down the application, Wilhelm Dönitz, the director of the Institute for Infectious Diseases during Koch's absence, once again stated the institute's position. He not only pointed out that the epidemic was rapidly spreading but also emphasized the need for autonomous German research. "It must be stressed that this epidemic is approaching the borders of the German colonial territory with astonishing speed . . . Hence it seems urgently necessary to authorize, as soon as possible, studies to be conducted independently of other countries on the nature and the spreading of this murderous and expanding disease."[180]

In the course of the year 1903 a number of new cases of sleeping sickness were identified with the help of the district health officers on Lake Victoria,[181] and this resulted in an altogether changed picture of the disease, for it became clear that the epidemic that had earlier been thought to be confined to the British mandate had now spread in the direction of the German colonial territory. To some extent Koch was responsible for this new vision of things. Not only had he meanwhile begun to conduct experimental studies of trypanosoma,[182] on July 30, 1904 he also submitted a formal report to the Prussian minister of culture. His hypothesis on the epidemic's path in relation to the German territory was based on the latest state of the research, the reports of the district medical officers, and conversations he had had in East Africa.[183] He argued that the epidemic had begun to move more or less along the coast of the Indian Ocean: "Since railroad traffic is sure to bring in more of the disease, and since Dr. MacDonald informs me that *Glossina palpalis* has already traveled along the railroad as far as 30 English miles from Mombasa, I consider it highly likely that sooner or later both factors necessary for the outbreak of the disease will come together in Mombasa and that the epidemic will then permanently settle along the African coast."[184]

The increase in trade and traffic had created a dangerous situation on the shores of Lake Victoria as elsewhere. Both had fostered the spreading of the tsetse fly that, according to Koch, was already present at the southern part of the lake. It was expected that as the fly spread, so would the disease.[185] Koch demonstrated the looming danger by means of a map that compellingly illustrated the advance of the epidemic from the northern side of the lake. Hence it seemed "urgently necessary to institute the appropriate measures right now" and "above all [to plan] sending a scientific mission whose task it would be to find some kinds of protective measures,"[186] which the English so far had not been able to do.

The fact that the sleeping sickness, having long plagued the other powers, was gradually becoming a threat to the German colonies caused the IHO to soften its stance. In a statement requested by the Interior Ministry in August 1904 Köhler, the director of the IHO, did mention his earlier refusal but now, citing the changed circumstances, agreed with the position of the Interior Ministry, which had favored the expedition all along: "There is no question that the spread of the disease into our Protectorate is only a matter of time." Considering that so far no medication existed, "it is certain that with respect to sleeping sickness a scientific commission would find an important field of activity."[187] At this point the IHO took charge of coordinating the planning, which in the spring of 1905 led to the decision in favor of the expedition—to be financed by the Imperial government in the year 1906. The decision was made in early April

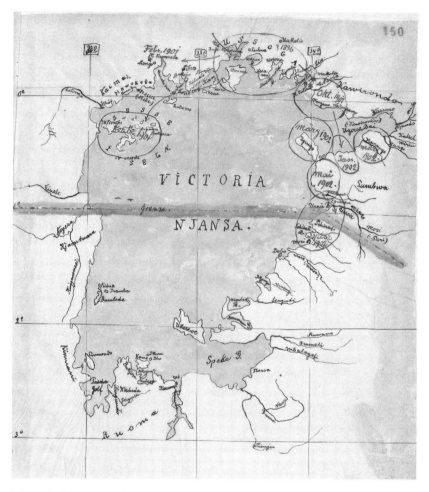

Map of Lake Victoria, from a letter of Robert Koch to the minister of cultural affairs, July 30, 1904. GStAPK, I HA, Rep. 76 VIII B, 4117. The circles and arrows, with dates, indicate the spreading of the disease.

1904 at a meeting that brought together representatives of all the institutions involved.[188] As for the epidemic, it had by then definitively affected the German territory.[189] Everyone agreed that even though at this point the actual danger was still slight, the threat was real and the colony in danger. By way of explaining his consent, the representative of the Colonial Office referred to what was called the colonial human economy and remarked that "for the further development of our colonies we are primarily dependent on the labor force of the native population," so that an epidemic would also be economically harmful. For his part, the director of the IHO spoke of the "high scientific value of the undertaking." And indeed the chairman of the meeting defined the purpose of the prospective expedition as primarily scientific: "The principal task of the expedition will be to investigate the etiology and the spreading mechanisms of sleeping sickness and to find means to combat the epidemic. The expedition will have to be limited to two or three experts at most, i.e., one physician who is quite familiar with East-African conditions and mores, one experienced expert in protozoa, and if necessary one bacteriologist."[190]

The extent to which the planning had by now come into the hands of the IHO is also indicated by the fact that a member of this institution was to lead the expedition. This was the biologist Friedrich Schaudinn, a protozoologist who in the context of his syphilis research had already done experimental work with trypanosoma, which he thought were morphologically related to the pathogen of syphilis, *Spirochaeta pallida*.[191] Together with the bacteriologist Fred Neufeld, a former assistant of Koch's who also worked at the IHO,[192] Schaudinn wrote a memorandum about the planned expedition in the summer of 1905. This document expressly named his own research, the relationship between spirochetes and the trypanosoma, as the objective of the study to be conducted. A first version of the memorandum (June 1905) had described the undertaking as a research expedition that would place the work on sleeping sickness into a wider framework of studies in parasitology and tropical medicine. The expedition was to conduct culturing and infection experiments and therapeutic trials;[193] and it would also study a number of other diseases in addition to sleeping sickness.[194] According to the second memorandum, the findings to be expected concerned Schaudinn's area of research, protozoology. The biologist considered trypanosoma to be a developmental stage of spirochetes; and the expedition was meant to clarify the biological position of the pathogen identified by Bruce in this sense.[195] At this point, Schaudinn's etiology of syphilis, which involved a spirochete, had already been formulated but not yet unequivocally proven. Thus, "for the purpose of resolving this matter, the study of other species of spirochetes that occur mainly in the tropics would be of major, perhaps decisive importance."

Koch was not involved in this decision, and there is no indication that at this time he intended to participate in or direct the expedition.[196] On the other hand, his preoccupation with the subject was quite intense. Having done experimental work on trypanosoma in Berlin in 1904,[197] he spent a good part of the year 1905 in East Africa, where he could continue his experiments, mainly at the agricultural research station at Amani.[198] As far as the trypanosoma were concerned, his interest was focused on the host cycle of parasite and fly, and here the connection with sleeping sickness did not engage his attention as much as the biology of the parasite and its development in the intermediary host.[199] Whereas for Schaudinn the biology of the pathogen was of central importance, Koch, looking back to his work on malaria, was interested in describing the developmental stages of the parasite in the tsetse fly. "Particularly close to my heart are the trypanosoma studies, which actually are the key to further investigations of sleeping sickness,"[200] he wrote to Gaffky in June 1905, shortly before his departure for Amani. "Here I am completely absorbed in the protozoa diseases," he reported to Kolle on September 1,[201] and it was only then that he began to be interested in leading the planned expedition. He saw himself as a pioneer in trypanosoma research and felt that the sleeping sickness expedition would be a means to further his goals: "Anyway, the trypanosoma studies must be diligently pursued here [in East Africa], since this is of vital interest to the colony's economy. The investigation of sleeping sickness cannot be done separately; it can only be part of the trypanosoma studies. I would therefore consider it a mistake to limit the projected expedition to the study of this disease. But if it were given the larger task, on which I have already done considerable work, I would be glad to be appointed to lead the expedition."[202]

In the summer of 1905 Koch's interest in trypanosoma research thus began to come into competition with the expedition planned by the IHO. There are several reasons why this competition turned into confrontation and why Koch, after his return to Berlin in late October, demanded not only the directorship of the expedition but also the exclusion of Schaudinn and Friedrich Fülleborn, the second prospective participant. Objectively, he considered Schaudinn's thesis that spirochetes were a developmental stage of trypanosoma to be mistaken.[203] As for Fülleborn, he came from the Hamburg Institute for Naval and Tropical Hygiene, a competitor of the Berlin Institute for Infectious Diseases.[204] Added to this were personal reasons. Rumor had it that both Koch and Schaudinn were being considered for the 1905 Nobel Prize in Medicine.[205] It is noteworthy that in December 1905, when Koch asked the new head of the IHO, Bumm, to appoint him head of the expedition, he betrayed personal animosity against Schaudinn. Bumm complied with Koch's wish and explained the situation in a secret memorandum to the minister of the interior. A draft of this document reads: "My esteemed predecessor Dr. Köhler had planned to entrust the direction of the expedition that the Reich will send for the purpose of investigating sleeping sickness to the member of the Health Office, Regierungsrat Dr. Schaudinn. Since, however, the latter is a zoologist without special medical training, our staff physician Dr. Fülleborn—temporarily assigned to the Institute of Hygiene in Hamburg—was proposed as the medical participant of the expedition."[206]

But then, as we saw, Koch had demanded not only the direction of the expedition but also the exclusion of Fülleborn—he considered him "unqualified"—probably because he wanted to take along his own student Kleine. Bumm, who had initially tried to mediate between Koch and Schaudinn, reported to the Interior Ministry that "the scientific disagreements between the two scientists are so great and their personal attitudes toward each other are such that a fruitful collaboration appears to be impossible." Even beyond the support of the Foreign Office for Koch's wishes, there was now a new argument that greatly strengthened his position and eventually would point the whole expedition in a new direction:

> On the other hand it is obvious that the aim of the expedition is above all the practical control of sleeping sickness. When a seasoned scientist with long experience in combating epidemics such as Robert Koch—to whom the Reich . . . owes a great debt of gratitude—offers his services, we must seriously consider whether we can refuse such an offer. Since Regierungsrat Schaudinn has no experience with combating epidemics and indeed is not even a physician, I feel obliged to support Geheimrath Koch's wish to be given the direction of the expedition.

In other words, the protozoologist Schaudinn was not only undesirable but above all unqualified. In the winter of 1906 the expedition therefore changed both its leadership and its objective. Instead of protozoology it now focused on the practical control of the epidemic. Koch was undoubtedly the right man to direct such an expedition, but at the same time he ran the risk of acting against his own interests, which in the years 1904–5 had also largely turned toward protozoology. But in a meeting of February 14, 1906, he fully embraced the medical objectives that had now come to the forefront.[207] He stressed that "now the most important thing was no longer to study a hitherto unknown disease . . . but above all to find ways and means to combat it." To justify the changed focus of the expedition, he referred to the studies on the etiology of the disease that British researchers had meanwhile published. However, these findings had actually been known by the summer of 1905 and to that extent probably already served as a justification for redefining the objectives of the expedition. The memorandum of summer 1905 had clearly referred to the British research but also dealt extensively with Schaudinn's program of research. Hence the memorandum was now reworked, and its final text clearly differed from the original version.[208] It still mentioned protozoa research, but in more general terms and without naming Schaudinn. By contrast, the danger posed by the epidemic and the task of developing appropriate control measures now occupied considerable space.[209]

Koch's actions can be called ruthless. Not only had he pushed aside Schaudinn and Fülleborn, two researchers connected with the Hamburg Institute for Hygiene,[210] he had also spun a successful intrigue against the Imperial Health Office and its former director, Köhler, who had initiated the plans. Köhler assessed Koch's sudden decision to lay claim to the directorship of the expedition as follows:

> Before his departure last December . . . I expressly asked Koch whether he was planning to investigate sleeping sickness in Africa. He said he did not, and it was only then that new plans were made which, although primarily concerned with sleeping sickness, also—responding to a strong wish of Schaudinn—proposed to take a look at trypanosoma and other protozoa-related diseases in general. Objectively it would of course be better if Koch and Schaudinn were to cooperate, so that they would complement each other; but, given the characters of these two gentlemen, I do not think this is possible. Schaudinn, whom I sent to see Koch several times in order to establish contact, must have been treated by him in such a way that he does not feel at ease.[211]

However, this manner of proceeding carried some risk for Koch. It was the reason why the expedition that finally left for East Africa in April 1906 did so with an undefined and in part contradictory set of goals: after all Koch's own trypanosoma studies had also had a rather biological-protozoological cast. In committing himself to medical questions of combating sleeping sickness, he thus endangered his own interests. Moreover, there was a danger that the whole undertaking would succumb to the need to prove successful in the crusade against an epidemic and that research would be pushed into the background.

The differentiation within bacteriological hygiene that had taken place since the days of the cholera expedition thus left its traces in the very planning of the sleeping sickness expedition: it was a conflict between basic and applied research, biological and medical objectives, the investigation and the control of a disease—all of which would have been unthinkable two decades earlier. Even though Koch was personally critical of this process of internal differentiation, he did know how to use the changed conditions to his advantage. In 1884 the biology of a microorganism and the identity of a disease had been well-nigh identical, but by 1906 it was possible—at least in the rhetoric—to play off the disease against its pathogen. Koch's actions in this situation were contradictory: while in his research he was interested in one subarea, namely the host cycle of the trypanosoma, at the level of science policy he questioned the value of this preference by postulating the overriding interest of combating epidemics.

### V.3.2. Feldmann's Epidemic

The composition of the small team illustrates to what extent Koch considered the expedition his private affair. The expert in tropical medicine who accompanied him was his former student Karl Friedrich Kleine, who had taken Fülleborn's place; traveling at their own expense were Arnold Libbertz, Koch's childhood friend and confidant in the tuberculin days and the director of the bacteriological research department at the Hoechst Company, and Koch's wife Hedwig. Only Sanitätsrat Max Beck of the Imperial Health Office did not belong to Koch's inner circle. His participation had evidently been requested by the IHO. Eventually it became clear that Koch did not entirely trust him.[212] After its arrival in Mombasa on May 3, 1906, the group was not particularly anxious to begin fighting the epidemic. Rather, its first destination was the research station of Amani, which offered agreeable living and working conditions. In the laboratory one could experiment with trypanosoma and *Glossina* (tsetse flies) that were brought in from the lower-lying environs without being incommoded by them, thanks to the station's elevated location. Koch was planning extensive experiments with *Glossina* and trypanosoma that were to continue the studies he had begun at Amani in 1905.[213] In addition to these, which involved such work as breeding and infection experiments, he also arranged for trials in the area concerning possible control measures, such as the clearing of the vegetation necessary for the survival of the flies.

Amani provided relaxed working conditions in the best scholarly style, so for the time being, the crusade against the epidemic that had been announced in Germany was put off to a later phase of the expedition. "There is nothing dangerous about this expedition. Aside from the scientific work, it can actually be considered a restorative and refreshing voyage," Koch had written to Libbertz even before their departure.[214] In another letter he praised the local attractions:

"We are installed in the institute's laboratory and are working steadily . . . In the pauses between the rain my wife and I have taken some walks through the jungle and to a nearby mountain, from where one has a wonderful view onto the jungle-covered mountains and the sea in the distance. We both liked it very much."[215]

Koch had originally planed to leave Amani after a few months and go to British Uganda to test some control measures there.[216] But by the end of May it became clear that his stay in the Umbara Mountains would be much shorter than planned. It appeared that the few scattered cases of sleeping sickness that had signaled a threat to the German colonial territory in 1904 and 1906 were not the whole story. As soon as he arrived in German East Africa, Koch had found alarming reports by the district medical officer of Muanza, Karl Feldmann, stating that the epidemic had now reached Lake Victoria and that thousands had already become ill. "The further news about the sleeping sickness in the Muanza district I have received allows for little doubt that the disease has broken out there," Koch reported to Germany on May 7.[217] The threat of an epidemic that had been so carefully dramatized in order to justify the expedition now in turn forced a modification of the plans. The laboratory work was shortened and the therapeutic trials in the field were given priority. At the same time the team considered canceling the stay in British Uganda altogether and carrying out the therapeutic experiments in German territory.[218]

But when the members of the expedition arrived in Muanza on Lake Victoria on June 30, they discovered an amazing fact: "In this place, which teems with tropical diseases, precisely the one on which I had pinned my hopes, sleeping sickness, is missing."[219] The researchers found only one victim of the disease, and he had come in from British Uganda.[220] After a thorough investigation of the terrain, to which Feldmann was also summoned, Koch noted in the draft for his report on the expedition, "Having thoroughly investigated the area in every direction, I feel certain that the fears harbored by Dr. Feldmann are groundless and that so far the German territory is still completely free of endemic sleeping sickness. All of the few scattered cases that have been observed until now originated in English areas."[221] The threat turned out to be an artifact of intense observation, and the epidemic that had developed in German East Africa was a paper tiger that had existed only in Berlin and was nowhere to be found in the field. Nonetheless it was irritating that the *Glossina*, while not as widespread as Feldmann had claimed, were decidedly present. Moreover, they had not appeared just recently—which would have left open the possibility of an imminent outbreak of the epidemic—but had always been there. It gradually became clear that Feldmann's errors had been caused not only by his modest microscopic skills[222] but above all by the assumption—which Koch shared until then—that the area occupied by the fly coincided more or less with that of the disease. When Feldmann had crisscrossed the lake in the summer of 1905 and assembled an impressive documentation for the impending epidemic, he had essentially been looking for the fly and extrapolated from its presence to that of the disease.[223] So, why was the fly present, but not the disease? An unusual epidemiology on the lake provided the answer. It was precisely the high incidence of the disease among the predatory tribes who lived on the islands in the lake that had so far prevented its spread along the shores. "In earlier times the Waganda, the tribe that dominated the lake through its pillaging raids, saw to it that other dwellers on the lake stayed away from the *Glossina*-infested shore, and therefore had very little contact with the flies. But lately this seems to have changed because conditions on the lake have become more peaceful, now that most of the Waganda who crisscrossed the lake have been carried off by the sleeping sickness. In the German territory the shore dwellers are beginning to go in for fishing."[224]

What had been missing, then, was not the tsetse fly but humans to be infected. The thesis that colonization had brought pacification and hence more intensive trade relations, which in turn favored the outbreak of epidemics, was not new. It had been espoused by British sleeping sickness researchers, in particular John Todd.[225] An added advantage of this interpretation, which showed the colonial powers as bringers of peace, was that this explanation of the absence or presence of the epidemic did not require questioning the existing knowledge about its etiology and epidemiology.

Eventually, however, the expedition could not help making other observations that finally forced a correction of the current views of the etiology of sleeping sickness. The examination of numerous animals (birds, mammals, fish, lizards), frequently turned up trypanosoma, particularly in the case of crocodiles. Sometimes difficult to differentiate from each other and from *Trypanosoma gambiense*, they had to be different from the pathogens that affected humans, since infections in humans were not found in these animal habitats. Koch had expected that in areas without sleeping sickness the tsetse flies would also be free of *Trypanosoma*. But this was not the case. In the draft of a report, he outlined the confusing situation: "The types found in Muanza, to be sure, have nothing to do with *Tr. Gambiense*, . . . but whether they are separate species or related to each other, and whether the type found in Entebbe might be part of *Tr. gambiense*, these are still open questions that can only be answered by further careful investigations. In any case our finding makes it clear that if trypanosoma are found in the *Glossina palpalis*, this does not ipso facto justify . . . considering them part of *Tr. Gamb.*"[226]

All in all, then, the expedition had gotten off to a bad start. The experimental work in Amani had to be broken off in order to fight an epidemic that, although it had had made the financing of the expedition possible, turned out on closer inspection to be nonexistent. Subsequent investigations made it clear that the existing assumptions concerning the etiological relationship between humans, flies, and parasites had to be questioned. In a letter to his daughter of August 3, 1906, Koch somewhat peevishly summarized his efforts so far: "I am still wandering around, searching for the sleeping sickness . . . We searched very hard, and we found the fly that belongs to the sleeping sickness, but not yet the disease. That's very nice for our colony, but our expedition must continue to wander."[227]

### V.3.3. On Sese

However interesting they were from the protozoological point of view, the results obtained in the area around Muanza were of little use for medicine. On the contrary, the microscopic diagnosis had become much more difficult; indeed it was bound to remain uncertain as long as the different types of trypanosoma could not be reliably differentiated. Only the fact that the three nonpathogenic forms were rare in areas where sleeping sickness was present made it possible to hold on to the microscopic diagnosis.

In this situation in the summer of 1906 Koch decided to return to the original plan for the expedition and travel to British Uganda, where the disease was endemic. Refusing the offer to work in well-equipped laboratories in Entebbe,[228] he decided to search for a place where the expedition "could have at its disposition not only the *Glossina palpalis* but also cases of sleeping sickness contracted on the spot."[229] Such a place was the Sese Islands in Lake Victoria. Koch accepted the British invitation to work there. These islands, located not too far from Entebbe, offered sleeping sickness in endemic form that within a few years had probably carried off the major part of the population, originally numbering about 30,000 inhabitants. Moreover, they had two mission stations, which the expedition could use as a base. The station at Bumangi, which was operated by the missionaries known as the White Fathers of the Société des Missionaires d'Afrique, allowed the expedition to study its patients, although their number had shrunk to no more than a dozen by the time it arrived. The other station, a mile from there, belonged to a British Protestant mission in Bugala. It was empty and served as the expedition's camp.

In a sense, the decision to go to an area where sleeping sickness was endemic also determined the direction the research would have to take. Such a terrain had little to offer for studying unsolved questions, such as finding distinctions among different kinds of trypanosoma.[230] This is why a now reduced team—Libbertz and later Koch's wife had gone home—began to conduct therapeutic experiments, in particular the testing of atoxyl. Koch had brought this preparation and also Ehrlich's trypan red from Germany.[231] On the island of Sese the team could assume that practically the entire population was affected by the epidemic in the sense of an actual infection.[232] Nonetheless, proper testing of the therapeutic agents required a certain control over the patients and dependable diagnostic methods. Both were difficult to attain. Outside the German colonial territory it was nearly impossible to force patients to remain in place against their will. The team therefore had to trust that the sick would stay in the huts that had been built especially for them, and that they could be reliably identified by little badges they were made to wear around the neck. As for the diagnostics, apparently swollen lymph glands were used as the main symptom for a suspicious diagnosis, which would subsequently be confirmed microscopically. In this operation the physicians tried to avoid painful measures so that they would not jeopardize the patient's cooperation: "Since we had been told again and again that according to general experience lumbar taps and tapping of the enlarged lymph nodes on the neck would meet with the decided resistance of the natives, we began with the simple testing of the patients' blood."[233] This blood test, however, proved unreliable when quite often no parasites were found in obviously sick persons. Eventually a trick developed by the British researchers was used to perform the tapping of the lymph nodes on the neck. The patients were told that this intervention was done for therapeutic purposes: "The tapping of the nodes thus gradually became an integral part of the treatment . . . In fact, patients quite often ask for it themselves,"[234] Koch reported in autumn 1906. But autopsies of deceased patients, as the team had still done in Muanza, had to be given up completely for fear that they might cause the patients to flee, since they clearly would have seen this as cannibalism.[235]

The therapeutic experiments themselves soon focused on atoxyl, whose effect on trypanosoma in the laboratory was already known, and which almost simultaneously (in December 1906) was also used for the first time in the Congo.[236] By experimenting with frequency and dosage, the team developed a standard operating procedure that called for two injections of 0.5 grams each on two subsequent days. If this procedure was followed, further taps of the nodes usually no longer showed trypanosoma, and the swelling of the nodes receded.[237] But these were not the only effects: after three to four weeks, a clear improvement in the patient's general condition would set in. While in light cases this improvement could be definitively proved only by a microscopic diagnosis, the treatment of severely affected patients yielded impressive results. Once the first therapeutic successes had been observed, the team focused on these severe cases, "in order to see with our own eyes—independently of the often unreliable statements of the patients—how well the atoxyl is working."[238] This allowed them to document such cases as that of twenty-five-year-old F. "Been ill for three years. In treatment since 25 September. She is so weak that she is unable to walk. She cannot even sit. Unresponsive. She is incontinent. Now she is able to walk well. Responds better and enuresis diminished. Mentally alert, she shows a lively personality."[239]

After a certain period of waiting and observation, Koch in early November—this was about two months after the first injections[240]—expressed the view that it had "turned out that in atoxyl we have a medication that seems to be a similar specific against sleeping sickness as quinine is against malaria."[241] Moreover, the treatment had the advantage that it "could be applied without difficulty to populations of any size."[242] After its difficult beginnings, the expedition seemed poised to become a full success. In early November Koch was therefore in a euphoric mood tempered only by slight cautiousness, as can be seen in a letter to Libbertz, who had returned to Germany: "It truly is a wonderful thing when one is able to snatch people who are inexorably promised to death from that fate. This is when one is happy to be a physician. It does seem that the means have been found to cure sleeping sickness. At least that is what it looks like so far. I don't quite trust the thing yet and am afraid that there might be some reversal that would bring the whole beautiful structure tumbling down about our ears. But if things continue as they have done until now, our task will have been solved brilliantly."[243] At this point, caution was only a private matter. In a letter to the president of the IHO, Bumm, Koch presented an unequivocal assessment:

In my report I purposely spoke rather reservedly about the curative effect of atoxyl. But it has been almost three weeks since then, and in this time the improvement in the severe cases, patients who without atoxyl would no doubt have already perished, has made such significant and visible progress that there can no longer be any doubt about the specific effectiveness of this medication. In using atoxyl to combat sleeping sickness, the main task will be to set up the therapy in such a way that mass treatment of the natives can be carried out without difficulty. In this respect I feel that the method I have followed represents a good choice.[244]

The unexpectedly clear and rapid success of the atoxyl treatment led to a further change in the expedition's plan, in which combating the epidemic now almost completely crowded out parasitological research. "Our research has gone quite differently than I originally thought. We are not dealing with the flies so much, and all the more with the sick people."[245] The expedition, which had been thankful for the few patients it had found on its arrival, began to see increasing numbers of them by the end of August.[246] In early November the camp was practically swamped by the sick. Instead of the dozen patients originally available, there were now almost a thousand, all of whom were treated in the two camps of Bumangi and Bugala.[247] In this situation, Koch declared in November 1906 that the use of atoxyl was the successfully tested standard treatment. At the same time he planned a large-scale campaign to combat the disease. He asked Berlin for more time, more money, and above all more atoxyl. Having brought with him one and a half kilograms, he already had to order five more kilograms by October and finally asked for another twelve in November.[248] He felt that this quantity was needed, since by now he not only had to supply his field hospital on Sese, but was also thinking about sending smaller expeditions to the various areas of the German colonial territory.[249] As a further consequence of the decision to consider the atoxyl therapy sufficiently tested, he now decided to publish the results obtained so far. As in the case of the cholera expedition, he had written reports that had remained unpublished. In early November he urged their publication, concerned above all about retaining his priority vis-à-vis his British hosts. "To strengthen the case for a speedy publication, I should like to mention the fact that some time ago the government of Uganda sent three of its physicians to us for the purpose of studying our equipment, the method of treatment and its success. As a result it decided to establish, as soon as possible, a number of stations where natives can be treated with atoxyl. Surely we can expect these stations to have, and to report, the same success as we do. It would indeed be undesirable if these reports reached the public before ours."[250]

Taking note of the parallel to the successful cholera expedition, the Imperial Health Office arranged for the publication of the report in the *Deutsche Medizinische Wochenschrift*. Since in nearby Entebbe the expedition had access to that means of communication of the industrial age, the telegraph, a speedy publication was possible. The first four reports, the last of which was completed on November 27, appeared before New Year's in issue 51 of the *DMW*. Here too, as in the case of the cholera expedition, the public's enthusiasm was easily aroused. Thus the *Kreuzzeitung*, an important conservative paper, announced on December 18, 1906 that "German medical science, under the proven and always victorious leadership of Robert Koch, [had] gained a great victory."[251] After his return to Germany in November 1907, the celebrations for Koch were similar to those of 1884. He now received the title Kaiserlicher Wirklicher Geheimer Rat and the right to be addressed as Excellency. Then there were public lectures and in February 1908—exactly as after the cholera expedition—a festive banquet given by the Berlin physicians.[252] Even the title page of the program for this event resembled the one of 1884. It brought together the emblems of hygiene as tools for combating epidemics, from the microscope to vials of atoxyl. What had disappeared since 1884 were the other members of the expedition, who were now replaced by Aesculapius paying homage to Koch under a tropical sun.

While the expedition was thus celebrated as a great success in Germany, its course was taking a new and unfavorable turn in Africa. In early 1907 the team had temporarily run out of atoxyl, so that it was reduced to observing the patients. To Koch's surprise, the healing process did not continue when the atoxyl injections were stopped; instead it "came to a halt, and some patients even suffered an unmistakable deterioration."[253] Yet renewed tapping of the lymph nodes no longer showed any parasites. This meant that "the trypanosoma have not been completely eliminated . . . but are still somewhere in the body." Clearly, tapping the nodes was diagnostically insufficient. But it would not do to simply replace it with the established lumbar taps because the team could use only such methods of examination "as the natives are willing to put up with." Instead the team was able to improve the technique of blood testing to the point where although "extraordinarily laborious and time-consuming," it yielded good results. Changing the diagnostic procedure brought to light another interesting fact. While the disappearance of the trypanosoma from the lymph nodes had

Title page of the program for the banquet given by the Berlin physicians in honor of Robert Koch on November 2, 1908. From Kolle, *Robert Koch, Briefe an Wilhelm Kolle*, 63.

not been related to the number of injections given, "the blood test taught us that in this respect clear differences do exist."[254] This led Koch to a conclusion that was to have dire consequences for the patients. In a letter to the president of the IHO of February 18, 1907, he stated that "the patients will have to be treated much longer and perhaps also more intensively."[255] He now increased the number of injections with doses of up to 1 gram per day, which worked well.[256] "On this point I only wish to report that after we resumed the treatment, the atoxyl did not fail us and immediately caused the trypanosoma to disappear again. I fully expect that with longer treatments and perhaps also stronger doses, they will eventually not appear at all."

To be sure, the higher doses were painful for the patients, who now began to refuse treatment or leave the camp. Moreover, cases of blindness cropped up, which Koch at first considered temporary but which proved to be long lasting.[257] Even if after the appearance of such blindness he apparently returned to the earlier smaller doses, his actions were imprudent. Cases of blindness due to atoxyl poisoning had been reported in the literature as far back as 1905.[258] In January 1907, moreover, the president of the IHO, Bumm, had expressly warned of this danger.[259]

In the spring of 1907, the expedition thus once again faced a choice. In light of the latest complications, the further publication of reports did not seem advisable. Koch himself delayed his reports: having sent four of them by November 27, 1906, he did not submit the next one until five months later, on April 24.[260] The IHO classified it as "secret" and published it only after Koch's return in autumn 1907.[261] The question also arose as to how the expedition could continue in this situation and what results it had achieved with respect to sleeping sickness. Koch's answer to this last point is surprising, for he referred to the planning of a campaign against sleeping sickness based on atoxyl treatment. In the spring of 1907, Koch had become convinced that instead of higher doses a longer treatment with the medication was appropriate,[262] and by summer he definitely considered the expedition a success: "Some questions concerning trypanosomiasis in humans, mostly of a theoretical nature, we had to leave aside. Investigating them would require many more years of strenuous work. But the essential fundamentals for combating the epidemic are now available."[263]

Under the given circumstances on Sese, however, Koch's conviction was speculative, and the questions left aside had by no means been only theoretical: the patients had never stayed long enough in the camp to allow for a definitive evaluation of the therapy. Koch admitted that "in no case" extended treatments of patients "could be carried out . . . in a completely exact manner. As soon as they [the natives] feel somewhat better, or when the cure is lengthy and they become bored, they break it off and run away."[264]

In 1909, Robert Kudicke was interviewed by the Imperial Health Board about the campaign against sleeping sickness on Sese Island. When asked whether any definitive cures had been observed after treatment with atoxyl, he replied under an oath of disclosure "that he knew of two cases in which 6 months after the end of the treatment no trypanosoma had been found—but neither could trypanosoma be discovered in these patients before treatment."[265] Kudicke therefore felt that the positive mortality figures as well were the result of the patients' brief stays in the camp rather than of therapeutic successes.[266]

But Koch's continued defense of atoxyl focused not so much on individual patients anyway. "For me the main thing is that the trypanosoma disappear, for on the basis of this fact, which is no longer in dispute, one can effectively fight the epidemic,"[267] he had written to Gaffky in December 1906. The medication was, as he put it later, "not perhaps an unfailing remedy but in any case so powerful a weapon in the fight against sleeping sickness that we must use it right now as much as possible . . . This therapy provides the means of keeping persons suffering from trypanosomiasis free of trypanosoma in their blood for at least 10 months, with the effect that they are not susceptible to infection by *Glossina*, which makes them unable to spread the disease, that is to say, harmless."[268]

On this basis Koch planned to establish serial testing facilities and anti-sleeping sickness camps, so-called concentration camps, in the German colonial territory.[269] Here patients were to be interned even against their will and treated with atoxyl. In particularly endangered regions that could not be secured by other means, such as the clearing of vegetation, he also advocated the "transfer of this population to *Glossina*-free areas." Two such camps were established while the expedition was still working there at Kisiba and Shirati on Lake Victoria, and another would follow on Lake Tanganyika.[270] The subsequent German anti-sleeping sickness campaigns in East Africa and Togo that originated here are not part of this study—but their characteristics seem to go back to Koch's camp on Sese, characteristics that include tampering with scientific evidence and the brutal treatment of the patients.[271]

Even allowing for the historical evolution of ethical standards, Koch's behavior on Sese remains problematic. And indeed, in the detailed report of the expedition that he published later, he felt the need to justify his actions. "If I am speaking here of experiments on patients, it is to be kept in mind that the latter were suffering from an absolutely deadly disease and that there was no way to save them unless a remedy was found."²⁷² The application of an (ineffective) therapy was thus used to justify therapeutic research at the expense of the patients' well-being, which by this time would no longer have been permissible in Germany. This point is proved by the secrecy with which the laboratory attendant Schmidt, who had contracted sleeping sickness, was treated in Berlin at this very time.²⁷³ It is also clear that when Koch placed the fight against epidemics above individual therapy, he bought into the ideological framework of a colonialist human economy. As he emphasized, the value of the campaign he had conceived was not that it cured individuals, but that it preserved the working capacity of the population as a whole for the colonial masters.²⁷⁴

This interpretation is strongly supported by the expedition's activities in the months before autumn 1907. In the first years after the turn of the century, the search for chemotherapeuticals against sleeping sickness was intense. Unfortunately, however, most of the substances that were effective against trypanosoma in the test tube proved to be highly toxic and hardly suitable for use in humans.²⁷⁵ Nonetheless testing the chemotherapeutica mentioned by Ehrlich in the remaining patients in Africa continued to be one of the essential tasks of the expedition in the summer of 1907. In addition to the almost classic procedures, such as the repeated mercury applications known from syphilis therapy, the team now tested a half dozen of the substances proposed by Ehrlich and made available by the German chemical-pharmaceutical industry: nucleole came from the Rosenberg Company, arsenferatin from Boehringer, trypanred from Hoechst, afridolblue and afridolviolett from Bayer, oleic acid parosanilin and parafuchsine acetate from Casella. The results of the sometimes very painful injections were negative: "The reason why these preparations have shown a decided effect in animal experiments is that animals can be given considerably larger doses than humans," a disappointed Koch concluded.²⁷⁶

### V.3.4. Hunting

Although therapeutic experiments were the focus of the expedition's work on Sese, the picture of this undertaking would be incomplete without a brief look at Koch's and his companions' other activities. Among these were extensive investigations that could be mentioned here only in passing, such as experiments with clearing vegetation for the purpose of controlling the tsetse flies, infection experiments with crocodile embryos, and so forth. Moreover, life on Sese was not all work. Aside from the monotonous and usually poorly prepared food,

Koch definitely enjoyed his stay.[277] "Aside from the many thunderstorms, the climate here is ideal. Never too hot and never too cold. Early in the morning 18–20 C, up to 25 C in the afternoon. In other words, a permanent summer. I don't think you have any idea what a horrible climate you live in," he wrote to his daughter in May 1907.[278] In the same letter he reported that microscopy had become very time consuming, since the team had been obliged to change over to blood testing. Still, there was also time for leisure activities, for instance, the training of three domestic animals: a dog to guard Koch's bed, and two parrots that had been taught to speak reasonably well. Even though Koch ironically characterized his life on Sese as that of a hermit,[279] the islands were not really that remote, but were connected by way of nearby Entebbe to the communications network of the industrial age with its railroads and postal and telegraph offices. Books and newspaper regularly arrived from Berlin. In May 1907, for example, the chess enthusiast Koch sent for no fewer than nine books about championship matches.[280]

Visitors stopped by occasionally. Koch's insistence on the prompt publication of his reports in November 1906 was in no small measure motivated by the visit of two British physicians accompanied by a colonel in the colonial administration on October 16.[281] In autumn 1906 two Reichstag deputies also stopped by, and later regular tourists came. In April 1907 two professors at the Vienna medical school who traveled around Lake Victoria paid a visit to Koch's camp. And the following June brought a certain "Duke Adolf Friedrich with his caravan."[282] In the spring of 1907 Koch must have been thankful for such diversions, since at that time, a painful infection on the soles of his feet kept him immobilized for weeks on end.[283] As he noted, these infected sand flea bites were to some extent the result of his poor eyesight, "since with my myopic eyes I am unable to see and remove these initially tiny insects in time."[284]

Recovered from this infection by late July 1907, Koch, accompanied by a medical orderly named Sacher, undertook a somewhat strange inspection tour on Lake Victoria. It lasted from July 28 until September 30, shortly before he departed for Germany. Not mentioned in the published report of the expedition, it is documented in one of Koch's diaries.[285] Although officially the purpose of the trip was to check on patients who had been treated,[286] the diary entries convey the impression of a hunting trip rather than a scientific undertaking. To be sure, hunting had been part of the expedition all along. Animals were shot and autopsied in order to identify animal hosts of the trypanosoma and learn to distinguish among different forms of trypanosoma. Koch himself was an enthusiastic though not experienced hunter; in fact he had never gone hunting until the 1890s. In September 1906 he proudly reported to his family that he had now killed his second crocodile.[287] As for the crocodiles, he considered them the most important hosts for *Trypanosoma gambiense* on Lake Victoria and therefore forcefully advocated hunting them and destroying their clutches.[288] This was not a particularly original idea but an activity recommended long ago by David Livingstone that had been practiced all along. The discoverer of the etiology of sleeping sickness, David Bruce, had agreed as well.[289] But it did serve to justify the intensive hunting in which Koch and Sergeant Sacher engaged for a good month. The two of them shot at practically everything that moved, from hippopotamuses to crocodiles, from herons to eagles. "Shot two pelicans," Koch noted on August 8, "shot birds around the valley," on September 9, and "shot a lot of other birds, among others a beautiful blue heron," on September 11. On September 17 he saw "a giant snake; we first demolished its vertebral column with a shot of pellets and then smashed its head with gunshot." Many more entries of this kind could be cited. The diary also contains colorful descriptions of strenuous stalking: "12 September . . . Sat there for almost two hours, but the crocodile didn't come. Then we took off, crept along the edge of the island, suddenly saw between rocks and bushes the tail of a crocodile. The animal stayed on the spot because its rear was paralyzed. Writhed and bucked in front and opened its jaws to catch our scent. Brought it down with a bullet to the head."

Even if some meat was needed to feed Koch's rowers, the number of animals killed was much too high for this purpose, and moreover some of them, for instance the eagle, were hardly suitable food. At the same time, references to the scientific character of the undertaking were rather sparse: on occasion the dispatched animals were photographed and autopsied, and preparations for subsequent use were produced. But it is often difficult to distinguish between the sporting and the scientific interest. Thus Koch noted on September 27, "Very soon come to a small clearing, where two crocodiles lie. One of them immediately flees, and at the same moment the other receives a shot through the vertebral column. It can't get away but snaps its jaws, hisses and growls. Is photographed when still alive, then killed and the head cut off." All told, a fixation on crocodiles in this undertaking is unmistakable. This was not only a matter of shooting the animals, but also of finding their nests, counting their clutches, and then usually destroying them. Koch also acquired two so-called crocodile fishing rods, but they did not work. He had better success with the use of poisoned bait. All in all, then, Koch and Sacher engaged in an extermination campaign against crocodiles, which, considering the number of crocodiles and the means at the hunters' disposal, had something quixotic about it. Judging by the diary entries, the undertaking gave Koch considerable pleasure, in part perhaps because its tangible results compensated for the disappointing outcome of the atoxyl treatment on Sese.

"Under the beautiful cover sleeps doom,"[290] the bacteriologist wrote in a letter from Muanza. The hunting campaign now provides the occasion to offer some reflections on Koch's perception of nature and his understanding of disease, complementing what was said earlier in this chapter about his fascination with Africa. The scientist's enthusiasm for African nature obviously went hand in hand with an aggressive attitude and to this extent is reminiscent of the contemporary clichés that depicted "Black Africa" as a fascinating and dangerous place. As a physician, Koch could add specificity to this image. The characteristics of the Dark Continent that others viewed through the sights of a gun were also visible, albeit in a different form, through the lens of a microscope. In Koch's case, these were trivial images of hidden and menacing things. For all the beauty of nature, the bacteriologist uncovered innumerable dangers lurking under the surface: "Malaria, dysentery, relapsing fever, all of this is amply present here as almost nowhere else. The place is swarming with anopheles; even on my mountain we have them almost in pure culture,"[291] he wrote from Muanza, providing a fine example of his perception of nature. An examination of his own rowers, "52 strong young men," yielded the results he had expected: "And by the way, of these 52 seemingly healthy young people, 47 had *Filaria perstans*, 26 had malaria parasites, and 2 recurrent fever spirilla in their blood. This is what the supposedly still healthy local population looks like."[292]

That nature is a source of dirt, danger, and above all disease was a fundamental conviction of Koch's. This attitude was already revealed in his naturalist reflections on the origin of cholera in the jungles of Bengal. One can assume that for him Africa was the apotheosis of all this—a repulsive yet fascinating subject. The fundamentals of the orderly microscopic world the bacteriologist imagined thus reappeared in the macroscopic world of Africa or in the jungles of Bengal, where nature had not yet been tamed by civilization and hygiene. That Koch formulated very similar statements about nature as the source of dirt and infection over a distance of several decades suggests that this was a point where science and ideology reinforced each other. Koch's concept of pathology was informed by the idea of lurking, minuscule, or invisible dangers coming from the outside. These were characteristics he ascribed to bacteria as well as to nature as a whole. To borrow the words of Ludwik Fleck, not the least important purpose of such concrete and trivial images is to stabilize scientific thinking within a firm context of life experience: "The concreteness of a body of knowledge has a special effect. First applied by the expert in order to make a thought understandable to others (or as a kind of mnemotechnical device), this concreteness, initially a means, assumes the importance of a body of knowledge acquired. The image wins out over the specific proofs and often returns to the expert in this new guise."[293]

Robert Koch's imaged concept of African nature, his ideas about the natural history of epidemics in Bengal and similar matters, brought together ideology and scientific practice in a quite creative, if not necessarily sophisticated manner. They expressed the fundamental assumptions about the relations between inside and outside, cleanliness and health, dirt and infection that were as important to the scientific concept of infectious disease as they were for the popular ideas about cleanliness, hygiene, and the boundaries of the body.

## V.4. Robert Koch's Travels

Comparing the two research expeditions, one is struck right away by the passion Koch brought to this kind of research. In both cases it is impressive to observe how he attempted to change or accommodate to the given circumstances, which quite often were not suitable for scientific work. Indeed, he was capable of tossing out earlier plans without much ado in order to make use of the given situation. But whereas in the first case these efforts led to success, they produced considerable lurching in the second case. Both in Amani and in Muanza, Koch broke off studies that had not yet yielded satisfactory results. Subsequently the conditions on Sese were such that further work on unsolved questions was out of the question. Koch, to be sure, repressed these problems, which in the last analysis had to do with the twofold purpose of his expedition. He believed in atoxyl therapy and saw the emerging parasitological questions as mere "theoretical problems." "Hence I believe that everything has been done to avert the looming threat of having the epidemic spread in our colony."[294]

Of course there are things about the two epidemics that cannot be compared. Thus cholera directly threatened Germany, and sleeping sickness affected only its colonies. Nonetheless a comparison allows us to take a closer look at the major changes that occurred in the conditions of scientific work between 1883 and 1906. The two expeditions were connected by way of Koch's biography, the history of German medical bacteriology, and the history of the German Reich. The two diseases were so important that the measures to combat them shaped the development of important areas in the German health system. While cholera showed the importance of medicine and hygiene in the early stages of the welfare state, sleeping sickness, and especially its control, was part of the development that made medicine into a technological and ideological resource of colonialism. Cholera embodied the threat of epidemics in the Europe of the industrial revolution; it was the Great Scourge that fostered the building of the classic contemporary arsenal for combating epidemics. Sleeping sickness, together with malaria, was the epidemic that endangered Europe's dominion over faraway lands, and combating it became an integral part of the ideology and practice of colonialism. In both cases one therefore finds an ideological dimension characteristic of bacteriological hygiene: Koch's understanding of sleeping sickness not only showed connections to the colonial human economy, the expedition itself made sure that the problems of colonialization were conceived as problems that could be solved by means of medical technology. In the same manner, back in 1883, the discovery of the pathogen for cholera was not simply the solution to a medical problem; it was touted as a victory over the enemies of the human race. Bacteriological hygiene seemed to be capable of contributing to the solution of problems that lay far beyond the health of individuals.

The special status of cholera certainly was a major factor in the success of Koch's expedition of 1883–84, to the extent that one of the causes of its great resonance was precisely the close connection between the scientific construction and the public presentation of its subject matter. The usefulness of the knowledge produced seemed obvious to contemporaries, and scientists also felt that once the general context had been elucidated, practical consequences were bound to follow. If this had not been the case, Koch and his findings could not have stood up to Pettenkofer in the imperial cholera commission. And finally, the prestige of the epidemic operated at the personal level, where it afforded Koch a new prominent status along with a cultural experience.

In the same manner the problems of the sleeping sickness expedition had not arisen simply because of bad luck. They resulted from an attempt to follow the strategy of the cholera expedition under different conditions and, in particular, to link the study of the disease to the control of the epidemic. This attempt was thwarted not only by the diversification of bacteriological hygiene but also by changes in the political environment. After the turn of the century, a separation had occurred between basic and applied research. Henceforth there were differences between biological and medical research, between studying trypanosoma and combating sleeping sickness. In one sense Koch was perfectly aware of this and had actually used such arguments in order to cut off the annoying "protozoa specialist" Schaudinn. But he did not draw the conclusions for his own work and instead tried—as he had been able to do in the case of cholera—to keep exclusive control of the subject of his research. In doing so he set up the oddly lurching course of the undertaking. Moreover, he had guaranteed the success of his measures before he ever left Germany. The result was that the experimental studies remained incomplete for a time, and Koch prematurely announced success in controlling the disease. When problems surfaced—for instance when the etiology of sleeping sickness, which had looked consistent in the Berlin laboratory, was put into question by the appearance of nonpathogenic trypanosoma—these insights were suppressed. The self-imposed pressure to succeed that Koch experienced was compounded by the fact that almost simultaneously experiments with atoxyl were taking place elsewhere too. As his associate Kleine recalled decades later, Koch was at that time someone who precisely because of earlier successes saw himself as condemned to succeed and lived with the feeling that time was running out on him.[295] In terms of his biography, he had fallen victim to his own renown on the sleeping sickness expedition, allowing himself to cling stubbornly to questionable achievements. This behavior is strongly reminiscent of his conduct in the tuberculin affair and his work on east coast fever in East Africa.

In a certain sense it was impossible anyway to repeat the success of 1884. It had been part of the rise of bacteriological hygiene to a dominant medical discipline of the time. Presented as a breakthrough in 1884, it had by 1907 become nothing more than the application of what its inventor considered a successful model to a new subject. In his speech to the Berlin physicians at the banquet in his honor in February 1908, Koch addressed this issue: "This evening recalls to mind a similar celebration with which the Berlin physicians honored the cholera expedition on its return from India some 20 years ago. Then too I felt that the event celebrated not so much one person as the joyful satisfaction that a new building block had been added to the edifice of our medical knowledge—and this building block was not of an ordinary kind . . . What was involved here was not an individual case but a general principle."[296] Such a principle, however, could be founded only once. Hence the sleeping sickness expedition was, as Koch expressed it "the purposeful application of existing experience to a specific case."[297]

The irony of the situation of 1907 is unmistakable: the problems of specialization and professional competition from which he had wanted to escape by turning to tropical medicine also characterized this specialty. In this respect Koch's travels to tropical lands remind one of the race between the hare and the hedgehog, for the very difficulties for which he had left Berlin caught up with him on Lake Victoria. The evolution from heroic discovery to specialized research was characteristic of the development of science in that era, as exemplified in the more highly specialized careers of Koch's students Paul Ehrlich and Emil von Behring. Koch's attempt to escape such specialization failed when it came up against the interlocking demands of research protocol and colonial policy. Due to a certain obstinacy and the tendency to defend acquired scientific possessions at almost any price, a combination of circumstances finally brought about the unfortunate course of the expedition described here. The fact that Koch nonetheless enjoyed his travels suggests that as an escape from the problems and the competitors lurking in Germany, travel fulfilled its purpose in his life story.

At a different level, the two expeditions had one interesting aspect in common. Both had been planned as relatively short undertakings yet unexpectedly took much more time. The cholera expedition expanded enormously. For the sleeping sickness expedition, the extension of the original plan was insignificant, but it took longer than Koch had assumed it would, even while abroad—in November 1906 he announced his impending return. Nor was this expedition the recreational outing it had been planned to be in part. In certain respects, such complications, and especially their unpredictability, are remarkable. They make it clear that the intent of bacteriological hygiene to break out of the confines of the laboratory was much more difficult to realize than it had appeared to be.

What worked in Berlin could elsewhere literally dissolve—as it did in the case of Koch's nutrient gelatin. The relations among humans, the tsetse fly, trypanosoma, and sleeping sickness, so clearly structured in the laboratory, became more and more complicated and finally uncontrollable on Lake Victoria. In terms of structure, the problem is one of replicating experimental systems in a changed environment.[298] In the first case, it proved difficult to transport the laboratory to the place where the comma bacillus was found; in the second case, at least as far as therapeutic research was concerned, the decisive point was the control of the patients. In the first case, it was possible to transfer the laboratory to Calcutta. As a result, the bacterium traveled back to Berlin with the expedition and became a regular tool of scientific endeavor there. Actually, it had been present in Germany even before, but it remained unrecognizable as long as it remained impossible to identify it unambiguously as a species and to bring it into a relation of mutual definition with the disease. In the case of sleeping sickness, replication did not succeed. Far from Europe, where hospitals, insurance companies, and other factors made for controllable relations between medicine and the patient, the camp on Sese would have required nothing less than barbed wire and armed guards. In this case the questionable results were largely the consequence of Koch's refusal to accept the inadequate conditions of his experimental work.

The claim that bacteriological hygiene would be able to reproduce the results elaborated in the laboratory beyond its confines was far more difficult to realize than the elegance of scientific theory would lead us to suppose. A laboratory's ability to produce knowledge is not so much a matter of being isolated as of controlling its relations with the outside world and the conditions for experimental work obtaining there. Studying nature in the laboratory makes it possible, on the one hand, to control it outside the laboratory, yet in a sense it already presupposes such control. Rarely can this be seen as clearly as in the case of the traveling bacteriologist, whose life and work have been the subject of this book.

# A Perspective

In the preceding pages I have described the history of medical bacteriology and related it to the biography of Robert Koch. This has yielded the picture of a rapidly developing but also heterogeneous discipline. Among the important characteristics of this historical phenomenon and its development were certain tension-filled relationships within the discipline and with other fields. These included the simultaneous existence of biological and medical explanatory approaches, the difficult relations with other medical disciplines such as internal medicine, and a tension between Koch's personal life and the development of his science, which caused Koch, the pioneer in this field, a great deal of difficulty.

In these concluding reflections, I would like to turn our attention in a different direction and look into the connection between the theme of this book, the early history of the medical-biological laboratory, and larger historical developments. The main question to be asked here is how the early history of the medical-bacteriological laboratory fits into the subsequent history of bacteriology and hygiene that shaped science and public perceptions in the twentieth century.

By this I mean not so much the subsequent history of the discipline of bacteriology[1] as the role bacteriology and hygiene assumed in the public space of the twentieth century. Starting with what might be considered the popular aspect of this theme, one is led to ask how bacteriology as a science was related to a matrix studied by the newest historiography, namely, the perceived threat of invisible and omnipresent bacteria and its use for political ends. That the origin of a political language of bacteriology with its specific metaphors lies in the nineteenth century and in the area of bacteriological hygiene is undisputed,[2] as is the fact that this threat of invisible and omnipresent enemies became one of the great narratives of the twentieth century. Its manifestations reached from the relatively concrete "killer microbes" and the popular comics characters Karius and Bactus that were used in my childhood to promote dental hygiene to the Nazis' "Jewish parasites" and the murderous implications of such language.[3] To this day the image of "the enemy" is suffused with the martial metaphors of bacteriology. This narrative was reactivated after September 11, 2001, in a discourse of the threat from biological arms that has decidedly phantasmagorical aspects.[4] In stark contrast to the historical record of bacteriology, popular ideas have remarkable staying power, by and large reproducing—notwithstanding all manner of scientific rectifications[5]—the image of an antagonism between humans and microbes, sometimes enlarged to include the image of the heroic struggle of an army of immune cells, which was not yet available to people of the nineteenth century.[6] Historical investigations of the popular image of bacteriology show that German medical bacteriology of the late nineteenth century in particular was a semantic reservoir of a kind of "us against them" thinking that spawned specific metaphors in the twentieth century.[7] And this thinking has remained impervious to all scientific critique until the present.

But upon closer inspection of the scientific work of the bacteriologists of that time, such as I have attempted here with the example of Robert Koch, this brief assessment raises more questions. It is true that Koch made ample use of the warlike rhetoric I have described above and that his conceptions of the relation between human and microbe were based on the concept of an absolute opposition. His understanding of infectious disease as an invasion of the body by bacteria must be seen in this context. Yet it does not seem plausible to me to see these conceptions as resulting from his work. The scientific systems of significance were and have remained too unstable for that. They did not always yield overwhelming proof for such conceptions, as we have seen especially in the case of Koch's work on tuberculosis. Indeed, they were more apt to confuse them. The simplicity and clarity that characterized the popular image of disease-causing bacteria since the nineteenth century was also present in scientific practice, namely to the extent that it informed an ideological component that placed multiple and sometimes contradictory pieces of scientific information into larger contexts and united them in one worldview. Popularized science entailed statements about the relation between humans and microbes that were labeled the fruit of scientific work but that, subjected to closer historical scrutiny, rather look like its underlying assumption.

Good examples of this were Koch's speculations about the natural history of the cholera bacterium.[8] Although there was nothing to prove that the bacterium originated in the swamps of Bengal, Koch's emphasis on the hostile environment in that area—think tigers, swamps, and humid tropical heat—created a convincing metaphorical background for the scientific construct of the *Vibrio cholerae* as a hostile bacterium. Another interesting indication to bolster this argument is the fact that the metaphors of threat, war, and destruction that Koch and his associates used in their texts were clearly not always in keeping with their own science. As an example, here is a short passage from a lecture that Koch delivered in 1888 about the war epidemics: "Even in peacetime they slink about and sap the strength of armies, but when the torch of war is lit, they creep out of their hiding places, rear their heads to tremendous heights and destroy everything in their path. Proud armies have often been decimated, even destroyed by epidemics; wars, and with them the fate of nations, have been decided by them."[9] If one did not know otherwise, one would place the author of this and other texts in the prebacteriological era. The actors here are the war epidemics, not their pathogens. Koch speaks about typhus, not about the recently identified bacterium that causes it. Even if in the texts of the bacteriologists, and particularly in popularizing accounts, bacteria appear as the reification of disease, it is striking to see that the bacteria do not dislodge the traditional image of epidemics, and that, particularly in public lectures, traditional and modern metaphors for epidemics appear side by side. The bacteriological imagery of "man against the microbe," then, could easily accommodate the traditional representations of epidemics. Considering everything that has been said so far, it is likely that this accommodation was crucial, not only for popular imagery but also for scientific work.

This takes us back to the fundamental question about continuity and ruptures in the history of medical bacteriology that was raised at the beginning of this book and brings it to bear on the mutual relations between popularized and scientific bacteriology. What constituted the special and novel quality of medical bacteriology was not only the break with older systems of knowledge that bacteriologists liked to emphasize, but particularly the ability to provide the modern scientific underpinnings that enhanced the traditional anxieties about the threat of epidemics. With respect to popularized bacteriology, Elias Canetti has illustrated this change with the example of the disappearance of the formerly ubiquitous little devils:

> The devils in their familiar shape are no longer to be found anywhere, their former large numbers not withstanding. But they have left traces . . . They have now given up all the traits that might be reminiscent of human beings and have become even smaller. Much changed, therefore, and in even greater numbers, they have resurfaced in the nineteenth century as *bacilli*. Instead of the soul, they now attack the human body. And they can be quite dangerous to it. Very few people have ever looked into a microscope and actually seen them there. But everyone who has heard of them is always aware of their presence and tries not to come in contact with them, which is a somewhat vague endeavor, given their invisibility, their dangerousness and the concentration of outrageously large numbers of them in a very small space they have no doubt taken over from the devils.[10]

Taking this cue, we can now formulate more specifically what the medical bacteriology of the late nineteenth century contributed to the set of political metaphors in this area. It was the modernization and the updating—a kind of "scientification"—of the traditional imagery of epidemics. At the same time this imagery also served as an ideological complement to the discipline itself. What followed was a process of reformulation and enlargement. Alongside the Horsemen of the Apocalypse, the bacteria now became the reification of disease and, equally typical for modern society, held out the promise of a technologically possible control or even elimination of such dangers.

# Acknowledgments

The German original of this book was based on a postdoctoral thesis (Habilitationsschrift), which was accepted by the Faculty of Medicine of Heidelberg University in the winter semester of 2001–2. It was extensively revised for its initial publication in 2005 and has now been translated into English.

My interest in the topic of Robert Koch's bacteriology originated in 1996, and as I traveled the road to this book, many persons and institutions have been helpful to its completion. Above all, I am grateful to Wolfgang U. Eckart. By discussing this topic with me and providing help and stimulation on countless occasions, he has furthered my work more than anyone else. I also owe important insights to many years' contact with Michael Hagner and Hans-Jörg Rheinberger. My conversations with them as well as the atmosphere at the Max Planck Institute for the History of Science in Berlin have strongly marked the manuscript, especially with respect to its methodology. I especially thank these two colleagues for valuable suggestions for revisions of the book. A very special thanks is due to Alexandre Métraux, who during the revision stage gave the half-finished text a critical reading and suggested ways to sharpen its language and its arguments. I also thank Axel Bauer, Andrea Engel, Volker Hess, and Thomas Schlich for their critical readings. I am grateful to Hans-Günther Sonntag for his willingness to share with me his knowledge of the research on animal experiments in the area of infectious disease. Over many years Gabriel Neumann provided indispensable help in tracking down literature and sources and in transcribing manuscript material.

The kind and knowledgeable assistance I encountered in a number of institutions contributed greatly to the completion of this work. Here I particularly wish to single out Ulrike Folkens who, in charge of the Robert Koch papers at the Robert Koch Institute in Berlin, was always a helpful and competent partner for conversation. I am also grateful to the Rockefeller Foundation, which in

2001 provided me with a grant that allowed me to work with the Paul Ehrlich papers kept at the Rockefeller Archive Center. In addition, I also appreciate the suggestions I received in conversations with my colleagues Anne I. Hardy, Sarah Jansen, Andrew Mendelsohn, Oliver Mengersen, Cay-Rüdiger Prüll, and Jutta Schickore. I also wish to express my gratitude for those who helped the most to get this translation going: to Elborg Forster and to Flurin Condrau.

My daughter, Friederike, told me that it's great to have a daddy who writes books. I dedicate this book to her.

# Notes

CHAPTER 1: Introduction

1. Wells, *"The Stolen Bacillus" and Other Incidents*. In the Wells story the bacteria under the microscope are in fact no longer alive, thanks to a bacteriologist and his tools: " 'Those have been stained and killed,' said the bacteriologist" (2).

2. A fundamental study for this is Cunningham, *Transforming Plague*. A contemporary parallel might be the fate of those researchers who doubted the existence and the etiological significance of HIV: Epstein, *Impure Science*.

3. This point is made by Elias Canetti (*Masse und Macht*, 52; English translation, *Crowds and Power*). For recent work on the history of bacteriologically inspired notions of cleanliness, Brecht and Nikolow, "Displaying the Invisible"; Tomes, *Gospel of Germs*.

4. For a survey of the period, see Bynum, *Science and the Practice of Medicine*; Coleman, *The Investigative Enterprise*; Cunningham and Williams, *Laboratory Revolution in Medicine*. The following books convey an understanding of the basics of bacteriology: Bulloch, *History of Bacteriology*; Foster, *History of Medical Bacteriology*; Loeffler, *Die Entwicklung der Lehre von den Bakterien*; Brock, *Milestones in Microbiology*; Lechevalier and Solotorovsky, *Three Centuries of Microbiology*.

5. Fleck, *Genesis and Development*.

6. Good descriptions of German practices are in the studies by Evans, *Death in Hamburg*; Labisch, *Homo Hygienicus*; Weindling, *Health, Race, and German Politics*, 1989.

7. For a list of the most important biographies using information available in the late 1980s, see Brock, *Robert Koch*. An important publication since then were the fragments of the second volume of Bruno Heymann's biography (*Robert Koch*, part 2). A survey of the literature up to 1981 is found in Holz, *Robert Koch*. The most recent literature is listed in Gradmann, "Robert Koch," and Vasold, *Robert Koch*.

8. An important guide to the use of laboratory notes as a source was Holmes, "Fine Structure." Research in the history of science that made use of such materials was largely influenced—though not necessarily in the case of Holmes—by sociology and ethnology. Fundamental here is Latour and Woolgar, *Laboratory Life*. An introduction to the present state of the discussion is furnished by Creager, *Life of a Virus*, and Rheinberger, *Experimentalsysteme und epistemische Dinge*.

9. Carter, "Koch's Postulates"; Mazumdar, *Species and Specificity*; Mendelsohn, *Cultures of Bacteriology*.

10. Some efforts in this directions are found in Schlich, "Repräsentationen von Krankheitserregern."

11. The Koch papers are archived in the Robert Koch Institute and in part at the Institute for Microbiology and Hygiene at Berlin University. An inventory of these materials was published recently (Münch, *Robert Koch und sein Nachlass*) but was not yet available during the writing of the present text. Hence it is based on my own research in the Koch papers and in other archives.

12. The bacteriological laboratory created a doctrine of disease "in whose name the systems bequeathed by the physicians of the eighteenth century and even the models invented by the physiologists of the late ninetenth century were banished to the realm of ideology." Canguilhem, "Beitrag der Bakteriologie," 127.

13. The extensive historiography on these postulates contrasts with their presumptive author's scanty statements on the subject. The designation "Koch's postulates" actually originated with Loeffler ("Zum 25jährigen Gedenktage"). For an introduction to this topic, Carter, "Causal Concepts of Disease."

14. Without pursuing epistemological ambitions of its own, my work follows Ludwik Fleck's model of a specific style of thought connecting a group of scientists in a "thought collective." Fleck, *Genesis and Development*.

15. For the issues in writing biographies in the history of science and medicine, see Shortland and Yeo, *Telling Lives in Science*, 1996. Useful surveys of the voluminous literature on biography in general can be found in the introductions to Szöllösi-Janze, *Fritz Haber*, and Goschler, *Virchow*. For the biography in the history of medicine, see Gradmann, "Helden in weissen Kitteln."

16. Geison, *Pasteur*, 7–16. Frederic Holmes used laboratory materials to work on such scientists as Lavoisier and Hans Krebs. An important methodological contribution is Holmes, "Scientific Writing and Scientific Discovery."

17. This was one of the things Max Perutz accused Geison of at the time. In this connection see also William Summer's reply to Perutz's attack (Summers, "Pasteur's 'Private Science.'")

18. These are in the Robert Koch Institute in Berlin. Other archives also contain some notes, but these rarely document the laboratory's experimental work.

19. Warner, "The History of Science."

20. Koch's letter of 22 Jan. 1904, probably to Carl Weigert (cf. Möllers, *Robert Koch*, 282.) SBPK Slg. Darmstadt, 3b 1882 (2) Koch, Robert, Koch an Ruge (?), Bulwago, 22 Jan. 1904.

21. For an overview see Pickstone, "Medical Innovations."

22. See, for instance, the race between Koch and Pasteur for the comma bacillus: Coleman, "Koch's Comma Bacillus." For Koch's identification of the tuberculosis germ as a public event, see Schlich, "Symbol medizinischer Fortschrittshoffnung." For the diffusion of bacteriology among the public: Brecht, "Bakterien in der Ausstellung"; Gradmann, "Bazillen, Krankheit und Krieg"; Otis, *Membranes*; and Tomes, *Gospel of Germs*, among other studies.

23. Cited from Pfohl's article "Und Naunyn hat's doch gesagt." Pfohl also lists the different variants of this quotation.

24. Gradmann, "Invisible Enemies."

25. Schlich, "Kontrolle notwendiger Krankheitsursachen."

26. Schlich, "Symbol medizinischer Fortschrittshoffnung." The public image of bac-

teriology is succinctly depicted in Gorsboth and Wagner, "Die Unmöglichkeit der Therapie." The triumphalist perception of science later found its way into the culture of commemoration and into historiography.

27. This is impressively documented in Worboys, *Spreading Germs*.
28. See, for instance, Bynum, *Science and the Practice of Medicine*, chapter 5. See also the studies by Condrau, *Lungenheilanstalt*, and Weindling, "Serum Therapy for Diphtheria."
29. Tomes and Warner, "Rethinking the Reception of the Germ Theory."
30. By way of introduction: Elkeles, *Das medizinische Menschenexperiment*, 1996; Maehle, "Assault and Battery"; Sauerteig, "Ethische Richtlinien, Patientenrechte."
31. For the time being Brock, *Robert Koch*, remains the best available biography. Anyone looking for a quick orientation in German might consult Elkeles, *Robert Koch*, or Vasold, *Robert Koch*. The best introduction to Koch's scientific work is found in Dolman, "Koch."
32. The best account of the early days of the Koch family and its most famous offspring is Heymann's *Robert Koch*. The present study is based on that book, unless otherwise noted, until 1882. For the years thereafter, the basic facts have been taken from Möllers, *Robert Koch*.
33. An interesting glimpse into the inner life of the fifteen-year-old Koch is provided by some letters he wrote to his aunt Marie Goedicke. They were edited by Kurt Kolle: *Koch, Briefe an Wilhelm Kolle*.
34. For the social history of the extended Koch family, Radday, *Auswanderung*.
35. Cf. Gradmann, "Helmholtz."
36. Koch, "Über das Entstehen der Bernsteinsäure."
37. Koch, Inaugural lecture at the Academy of Science (1909). See also chapter II.
38. Cf. Skrobacki, "Robert Koch in Wolsztyn"; Skrobacki, "Robert Koch und die Polen."
39. Cf. Bulloch, *History of Bacteriology*, 179–82. See chapter II.
40. Hoppe, *F. J. Cohn*.
41. Koch, "Die Ätiologie der Milzbrand-Krankheit."
42. Maulitz, "Rudolf Virchow, Julius Cohnheim."
43. Koch, "Verfahren zur Untersuchung"; for microscopy, and particularly for the collaboration with Abbe, cf. Hellmuth and Mühlfriedel, *Zeiss*; for microphotography in bacteriology, see Schlich, "Wichtiger als der Gegenstand selbst."
44. Koch, "Untersuchungen über die Ätiologie der Wundinfektionskrankheiten"; Koch, "Neue Untersuchungen über die Mikroorganismen." For an introduction to Koch's use of animal testing, Opitz, *Tierversuche und Versuchstiere*, 192–210; for Koch's postulates, particularly Carter, "Koch's Postulates."
45. For the criteria's classical description, see Koch, "Die Ätiologie der Tuberkulose" ([1884] 1912). See also chapter III.
46. For the history of that institution, see *Das Reichsgesundheitsamt*.
47. Koch, "Zur Untersuchung von pathogenen Mikroorganismen."
48. Brock, *Robert Koch*, chapter 16; Carter, "Koch-Pasteur Dispute"; Mendelsohn, *Cultures of Bacteriology*, chapters 2 and 3.
49. See chapter II.3.
50. Koch, "Die Ätiologie der Tuberkulose" ([1882] 1912); cf. Schlich, "Symbol medizinischer Fortschrittshoffnung."

51. For the influence of experimental pathology and bacteriology on the concept of infectious disease in the case of plague, see in particular Carter, "Transforming Plague."

52. Möllers, *Robert Koch*, 133

53. For Koch's cholera expedition, see Gradmann, "Die Entdeckung der Cholera in Indien"; and chapter V.2; for Koch's work in the context of the history of science, cf. Coleman, "Koch's Comma Bacillus"; Ogawa, "Uneasy Bedfellows."

54. An account of the contrast between Koch and Pettenkofer and their clash in the cholera commission is found in Evans, *Death in Hamburg*.

55. Opitz, "Robert Kochs Ansichten."

56. Koch's career was thus part of the celebrated Althoff System. See vom Brocke, *Wissenschaftsgeschichte und Wissenschaftspolitik* (of particular interest is Eckart's contribution concerning Althoff and medicine). For the history of the Institute for Hygiene, Eschenhagen, *Hygiene-Institut*; Heinicke and Heinicke, *Hygiene an der Universität zu Berlin*.

57. The tuberculin affair has been the subject of numerous investigations, which are listed in chapter III of the present study. For an introduction see Gradmann, "Anatomie eines Fehlschlags."

58. In a diary that has now disappeared, Hedwig Freiberg wrote about her acquaintance with Koch and about the time of the tuberculin affair. Parts of this diary are known through an article by Schadewald: "Entdeckung des Tuberkulins."

59. Evans, *Death in Hamburg*.

60. Mendelsohn, *Cultures of Bacteriology*, chapter 10.

61. For a survey of this topic, see the general history of German colonial medicine in Eckart, *Medizin und Kolonialimperialismus*.

62. Cranefield, *Science and Empire*; Dwork, "Koch and the Colonial Office."

63. Eckart, "Malaria and Colonialism." For criticism leveled against Koch by other experts in tropical medicine, cf., for instance, Bynum and Overy, *Beast in the Mosquito*.

64. For an introduction to Koch's sleeping-sickness expedition, Eckart, *Medizin und Kolonialimperialismus*, 340–49; see also chapter V.3.

65. Cf. Maulitz, "Robert Koch in the United States."

66. Heymann, "Robert Koch als Patient."

67. Geison, *Pasteur*, 260. Koch's former student and eventual rival Emil von Behring was also buried in a mausoleum on the grounds of his own institute (Zeiss and Bieling, *Behring*, 547).

68. Robert Koch's name was first added to that of the Institute for Infectious Diseases in 1912. In 1918 the institute was renamed the Robert Koch Institute.

69. For Koch's posthumous reputation, see the recent study "Culture of Commemoration," by George E. Haddad. For the National Socialist cult of Koch, see Reim, "Der Robert Koch—Film."

CHAPTER II: Lower Fungi and Diseases

1. Koch, "Antrittsrede in der Akademie der Wissenschaften am 1. Juli 1909," 3.
2. Koch, "Die Ätiologie der Milzbrand-Krankheit."
3. Koch, Inaugural Lecture, 3.
4. Ibid., 1.
5. Cited by Heymann, *Robert Koch*, 151.

6. For Cohn, Hoppe, "F. J. Cohn"; cf. Drews, "Ferdinand Cohn." For the connection between Cohn's microbiology and Koch's bacteriology, see Mazumdar, *Species and Specificity*.

7. Cohn, "Ein Brief über Koch."

8. Koch, Review of *Die niederen Pilze*.

9. For the history of bacteriology, the following titles provide a point of entry: Bulloch, *History of Bacteriology*; Loeffler, *Die Entwicklung der Lehre von den Bakterien*; Mazumdar, *Species and Specificity*; Foster, *History of Medical Bacteriology*. For the history of the concept of disease in that period: Diepgen, "Krankheitswesen und Krankheitsursache"; Faber, *Nosography;* Berghoff, *Krankheitsbegriff*.

10. It is confirmed, for instance, in the memoirs of one of Koch's teachers, Karl Ewald Hasse. Hasse freely admitted that "unfortunately we here in Göttingen cannot boast to have taught him [Koch] on the path he chose to take. In the early sixties I myself still had a rather skeptical attitude toward the thesis of the importance of microorganisms" (Quoted in Heymann, *Robert Koch*, 55). On the other hand, research on what was called vegetal parasites in humans at the time did go on in Göttingen (Heymann, *Robert Koch*, 55). And finally, there are statements by Koch in which he explicitly acknowledged his debt to the earlier work of others. Most important, he mentioned Davaine and Pollender (Koch, "Die Ätiologie der Milzbrand-Krankheit").

11. Cf. Pickstone, "Medical Innovations."

12. Wilson, *The Invisible World*. The early microscopes, such as Antoni van Leeuwenhoek's, had only one lens and therefore were magnifying glasses rather than microscopes.

13. Leven, *Geschichte der Infektionskrankheiten*, 36–38.

14. As an introduction see Conrad and Wujastyk, *Contagion*.

15. Heninger, "Leeuvenhoek."

16. Jahn, *Biologiegeschichte*, 352.

17. Ibid., 351–64.

18. For this and the following paragraphs, see also Bulloch, *History of Bacteriology*, chapter 7.

19. Farley, *Spontaneous Generation*. Gerald Geison has made an exemplary study of the problems with which Pasteur had to contend in the experiments he designed to refute the doctrine of spontaneous generation. (Geison, *Pasteur*, 110–42).

20. For a survey, see Bleker, *Naturhistorische Schule*.

21. Bynum, *Science and the Practice of Medicine*, chapter 2; Ackerknecht, *Paris Hospital*.

22. Hess, "Disease as Parasite"; Cf. Bleker, *Naturhistorische Schule*.

23. Faber, *Nosography*, 60. For an introduction to the history of the medical laboratory, Bynum, *Science and the Practice of Medicine*; Cunningham and Williams, *Laboratory Revolution in Medicine*; Lenoir, *Politik im Tempel der Wissenschaft*.

24. Hess, *Der wohltemperierte Mensch*; Cf. Hess, *Normierung der Gesundheit*.

25. Faber, *Nosography*, 59–94.

26. For this passage and the following, see Diepgen, "Krankheitswesen und Krankheitsursache," 107–8.

27. Henle, *Miasmen und Contagien*.

28. For instance, Foster, *History of Medical Bacteriology*, 7–8; Leven, *Geschichte der Infektionskrankheiten*, 171–72.

29. Most clearly in Evans, *Causation and Disease*. For a more differentiated discussion, Carter, "Koch's Postulates."
30. Henle, *Miasmen und Contagien*.
31. Heymann, *Robert Koch*. 33. Koch did take one course in medical botany, which was taught by Hofrat Griesebach (ibid., 49).
32. In 1865 he won a prize offered by the [medical] faculty for the best paper about whether the nerves of the uterus contain ganglion cells. Subsequently he wrote his second experimental physiological study about the origin of butric acid in the human organism. Both papers are reprinted in vol. 2.1 of Koch's complete works.
33. Koch to Biewend, 12 Dec. 1865. Cited in Heymann, *Robert Koch*, 46–49. Koch even declined to accept a travel supplement of 50 thaler.
34. For Birch-Hirschfeld, Bulloch, *History of Bacteriology*; Birch-Hirschfeld, *Lehrbuch der pathologischen Anatomie*.
35. Birch-Hirschfeld, "Vorkommen und Bedeutungen niederer Pilzformen (Bakterien)," 223.
36. Cf. Birch-Hirschfeld's reports on new research: "Krankmachende Schmarotzerpilze"; "Vorkommen und Bedeutungen niederer Pilzformen (Bakterien)."
37. Klebs, *Über die Umgestaltung der medicinischen Anschauungen*, 35.
38. For Henle, Hintzsche, "Henle"; Carter, "Koch's Postulates"; Carter, "Jacob Henle's Views."
39. Henle, *Miasmen und Contagien*, 17.
40. Ibid., 17, 24–26.
41. Jahn, *Biologiegeschichte*, 377.
42. Henle, *Miasmen und Contagien*, 46. See also Bulloch, *History of Bacteriology*, 159–61.
43. Henle, *Miasmen und Contagien*, 76. See also Bulloch, *History of Bacteriology*.
44. Henle, *Miasmen und Contagien*, 68.
45. Farley, *Spontaneous Generation*, chapters 4 and 5.
46. Henle, *Miasmen und Contagien*, 67.
47. In the extensive literature about Koch's postulates, this view was expressed most forcefully by Evans: *Causation and Disease*. For a critical assessment by Carter, see "Koch's Postulates," 363–64.
48. Henle, *Miasmen und Contagien*, 52.
49. Cf. Klebs, *Über die Umgestaltung der medicinischen Anschauungen*, 31–32. For all his respect for Henle, Klebs clearly outlined this incompatibility. For the bacteriological understanding of disease, see Cunningham, "Transforming Plague"; King, *Medical Thinking*, 60–64; Schlich, "Repräsentationen von Krankheitserregern."
50. Diepgen, "Krankheitswesen und Krankheitsursache," 309–12.
51. Overview in Bauer, *Krankheitslehre*, 95–115 (using the example of teachings about inflammation).
52. Gradmann, "Tuberculosis and Tuberculin," 4–5.
53. Richter, "Die neueren Kenntnisse," 1867, 1868, 1871, 1873, 1875.
54. Richter, "Die neueren Kenntnisse," 1867.
55. Ibid.
56. Ibid.
57. Bulloch, *History of Bacteriology*, 52–54. For Pasteur's microbiology, Geison, *Pasteur*.
58. Farley, *Spontaneous Generation*.

59. Mazumdar, *Species and Specificity*, part 1.
60. Richter, "Die neueren Kenntnisse," 90–91. Italics in original.
61. Mazumdar, *Species and Specificity*, 15–45.
62. Cohn, "Untersuchungen über Bacterien," 133.
63. Mazumdar, *Species and Specificity*, 46–67. For Cohn: Hoppe, "F. J. Cohn."
64. Bulloch, *History of Bacteriology*, 187; Mendelsohn, *Cultures of Bacteriology*, chapters 2 and 3.
65. Farley, *Spontaneous Generation*, 92–120.
66. Hallier no doubt obtained his position in Jena in part thanks to the advocacy of his uncle Mathias Schleiden, who also taught there. For Hallier, Jahn, "Hallier"; Théodoridès, "Hallier, Ernst Hans"; Loeffler, *Die Entwicklung der Lehre von den Bakterien*, 75–85.
67. Jahn, *Biologiegeschichte*, 1990, pp. 352–54; Loeffler, "Die Entwicklung der Lehre von den Bakterien," 76–77.
68. Richter, "Die neueren Kenntnisse."
69. Bulloch, *History of Bacteriology*, 177.
70. This is described most clearly in Hallier, *Gärungserscheinungen*. For Hallier's microbiology see also Bulloch, *History of Bacteriology*, 190–91; Loeffler, *Die Entwicklung der Lehre von den Bakterien*, 76–85.
71. For a summary of Hallier's theory, see Richter, "Die neueren Kenntnisse"; and Eidam, *Mycologie*, 165–71 (rather critical).
72. Richter, "Die neueren Kenntnisse," 103.
73. The individual morphs are listed in Richter, "Die neueren Kenntnisse," 103; cf. also the summary in Eidam, *Mycologie*, 165–71, followed by a detailed critique.
74. Hallier, *Gärungserscheinungen*, 15.
75. Description in Hallier, *Gärungserscheinungen*, 10–16; cf. Bulloch, *History of Bacteriology*, 190–91; Richter, "Die neueren Kenntnisse," 101–2.
76. Survey in Richter, "Die neueren Kenntnisse"; cf. Loeffler, *Die Entwicklung der Lehre von den Bakterien*, 82.
77. Hallier, *Gärungserscheinungen*, 76–91.
78. Evans, *Death in Hamburg*, 260–61.
79. Hallier, *Das Cholera-Contagium*.
80. Loeffler, *Die Entwicklung der Lehre von den Bakterien*, 82–85.
81. Both quotations in Bary, "Referat," 244.
82. The illustration is found in Bary, "Referat," 250.
83. Ibid., 249.
84. Ibid., 251.
85. Hallier, *Das Cholera-Contagium*, 37–38.
86. Loeffler, *Die Entwicklung der Lehre von den Bakterien*, 85.
87. Quoted in Bulloch, *History of Bacteriology*, 192. It was Brefeld who in 1868, (that is, immediately after Hallier's experiments), called for sterile culture media and formulated a procedure for producing pure cultures of bacteria.
88. Bulloch, *History of Bacteriology*, chapter 9, esp. 221–22, on Oscar Brefeld's fundamental work in this area.
89. Birch-Hirschfeld, "Krankmachende Schmarotzerpilze," 107.
90. Cf. Heymann, *Robert Koch*, 138. The protocols of Koch's anthrax experiments are reprinted in the appendix. The first entry dates from March 1873.

91. Heymann, *Robert Koch*, 58, 61–68
92. Evans, *Death in Hamburg*, 324.
93. Biewend, "Familienchronik von Robert Koch," 308.
94. Heymann, *Robert Koch*, 66–68.
95. Ibid., 346. The allusion is found in a passage about the sources of error in culturing experiments: Robert Koch, "Die Ätiologie der Milzbrand-Krankheit," 1912, 9.
96. Roudolf, "Die wissenschftliche Bibliothek Robert Kochs," 448. The works in question are *Die pflanzlichen Parasiten des menschlichen Körpers* (1866); *Gärungserscheinungen, Untersuchungen über Gärung, Fäulnis und Verwesung, mit Berücksichtigung der Miasmen und Contagien sowie der Desinfektion* (1867); and *Parasitologische Untersuchungen bezüglich auf die pflanzlichen Organismen bei Masern, Hungertyphus, Darmtyphus, Blattern, Kuhpocken, Schafpocken, Cholera nostras, etc.* (1868).
97. RKI/Koch papers, as/WI/015.2
98. Hallier never received a professorship. In 1870 he moved away from botany and instead worked for the propagation of a "scientific world view." Cf. Gradmann, "Geschichte als Naturwissenschaft."
99. Cohn, "Untersuchungen über Bacterien," 127.
100. Ibid.
101. Loeffler, *Die Entwicklung der Lehre von den Bakterien*, 86–89.
102. Koch, "Untersuchungen über die Ätiologie der Wundinfektionskrankheiten."
103. Billroth, *Coccobacteria septica*; cf. Loeffler, *Die Entwicklung der Lehre von den Bakterien*, 142–49.
104. For Cohn: Hoppe, "F. J. Cohn"; Mazumdar, *Species and Specificity*, 46–67.
105. Schroeter, "Über einige durch Bakterien gebildete Pigmente." Cf. Loeffler, *Die Entwicklung der Lehre von den Bakterien*, 111.
106. Loeffler, *Die Entwicklung der Lehre von den Bakterien*, 213–15. Cf. Schlich, "Wichtiger als der Gegenstand selbst"; Travis, "Science as Receptor of Technology." Loeffler, *Die Entwicklung der Lehre von den Bakterien*.
107. Mazumdar, *Species and Specificity*, 66–67.
108. Richter, "Die neueren Kenntnisse," 1867, 1868, 1871, 1873.
109. Birch-Hirschfeld, "Krankmachende Schmarotzerpilze"; Birch-Hirschfeld, "Vorkommen und Bedeutung niederer Pilzformen (Bakterien)."
110. Birch-Hirschfeld, "Krankmachende Schmarotzerpilze," 107.
111. In traumatic infections, the systemic form was referred to as septicemia, and the localized one as pyemia. Cf. Diepgen, "Die Lehre von der Entzündung," 82.
112. Loeffler, *Die Entwicklung der Lehre von den Bakterien*, 91–103.
113. Bynum, *Science and the Practice of Medicine*, 132–37.
114. Billroth, *Coccobacteria septica*; Billroth also described the phenomenon of transforming bacterial cultures by varying the culturing techniques. Cf. Loeffler, *Die Entwicklung der Lehre von den Bakterien*, 1887, p. 144.
115. Hueter, *Allgemeine Chirurgie*, 8–54. Cf. Loeffler, *Die Entwicklung der Lehre von den Bakterien*, 101–3
116. Klebs, *Schusswunden*. For Klebs, Bulloch, *History of Bacteriology*, 376; Köhler and Mochmann, "Edwin Klebs"; Stürzbecher, "Klebs, Edwin."
117. Klebs, "Cellularpathologie und Infectionskrankheiten." Cf. Ackerknecht, *Virchow*, 95–96. Bauer, *Krankheitslehre*, 114–15.
118. Carter, "Koch's Postulates," 1985; Carter, "Klebs' Criteria." Koch explicitly

named Klebs's work as a model for his own studies: Koch, "Über die Ätiologie der Tuberkulose," 446.

119. Klebs, *Schusswunden*, 106. The only distinction Klebs still made here was between the systemic phenomenon of septicemia and the localized one of pyemia. Cf. Diepgen, "Die Lehre von der Entzündung," 82.

120. Klebs, *Schusswunden*, 106.

121. Loeffler, *Die Entwicklung der Lehre von den Bakterien*, 94.

122. Klebs, *Schusswunden*, 105.

123. Ibid., 119.

124. In control experiments with secretions containing fungi, the animals had sickened and then died. Klebs, *Schusswunden*, 119. Cf. Loeffler, *Die Entwicklung der Lehre von den Bakterien*, 94, which also has bibliographical data about the work of Klebs's students Zahn and Tiegel. For Klebs's student Friedrich Wilhelm Zahn, Benaroyo, "Friedrich Wilhelm Zahn"; cf. Carter, "Edwin Klebs's *Grundversuche*."

125. Klebs, *Schusswunden*, 118.

126. Loeffler, *Die Entwicklung der Lehre von den Bakterien*, 97–98; cf. Köhler and Mochmann, "Edwin Klebs," 1039.

127. Klebs, *Schusswunden*, 120–22.

128. Cf. Rather, "Virchow und die Entzündungsfrage."

129. Bauer, *Krankheitslehre*, 107–12. For Cohnheim's concept of inflammation, Diepgen, "Die Lehre von der Entzündung," 73–79; Doerr, "Cohnheims Entzündungslehre."

130. For the contemporary theories of inflammation: Bauer, *Krankheitslehre*, 95–115; Diepgen, "Die Lehre von der Entzündung"; Rather, "Virchow und die Entzündungsfrage."

131. Birch-Hirschfeld, "Krankmachende Schmarotzerpilze," 108. Klebs's descriptions of pathological anatomy did indeed open new paths; see Bulloch, *History of Bacteriology*, 146–47; Diepgen, "Die Lehre von der Entzündung," 82.

132. Köhler and Mochmann, "Edwin Klebs."

133. Klebs, *Über die Umgestaltung der medicinischen Anschauungen*, 46–47.

134. Ibid., 45.

135. Carter, "Klebs' Criteria," 84–86.

136. Birch-Hirschfeld, "Vorkommen und Bedeutung niederer Pilzformen (Bakterien)," 169.

137. Ibid., 198.

138. Virchow, *Fortschritte der Kriegsheilkunde*, 28.

139. Ibid., 33–34. Virchow also emphasized the need to go beyond morphology in defining a species.

140. Ackerknecht, *Virchow*, 95–100. Virchow's criticism of bacteriology became increasingly sharper as the medical public heaped uncritical praise on medical bacteriology. Cf. Virchow, "Der Kampf der Zellen und Bakterien."

141. Klebs, *Über die Umgestaltung der medicinischen Anschauungen*. For the stages of the debate, Bauer, *Krankheitslehre*, 144–45.

142. Klebs, *Über die Umgestaltung der medicinischen Anschauungen*. The text reproduces a lecture given in September 1877.

143. Loeffler, "Die Entwicklung der Lehre von den Bakterien," 146–48.

144. Cohn, "Untersuchungen über Bacterien II."

145. Sander, "Die Bakterienfrage zu London und Berlin," 9–10.

146. Billroth, *Coccobacteria septica*. Billroth's objections, to be sure, applied to the medical field, whereas in the area of biology, Naegeli was no doubt the more important opponent, as Mazumdar has shown in *Species and Specificity*, 98–103. While Koch sharply attacked Naegeli (Review of *Die niederen Pilze*), he expressly acknowledged the importance of Billroth. Decades later, he confessed to Billroth that "when I began my investigations . . . I was profoundly marked by your coccobacteria" (Koch to Billroth, 24 April 1890, in Möllers, *Robert Koch*, 109).

147. Billroth, cited in Loeffler, *Die Entwicklung der Lehre von den Bakterien*, 148

148. Loeffler, *Die Entwicklung der Lehre von den Bakterien*, 181–82.

149. Ibid., 154. Birch-Hirschfeld, "Vorkommen und Bedeutung niederer Pilzformen (Bakterien)," 210–11.

150. We now know that there are several forms of relapsing fever, which are transmitted by different borellia. Since this is a vector-dependent disease, the failure of Obermeier's animal trials is easy to understand today. Hardy, "Relapsing Fever."

151. Cohn, "Untersuchungen über Bacterien II," 196–99. Loeffler, *Die Entwicklung der Lehre von den Bakterien*, 154–55.

152. For the history of animal experimentation, see Holmes, "Martyr of Science"; Logan, "Before There Were Standards"; Maehle, *Kritik und Verteidigung des Tierversuches*; Opitz, *Tierversuche und Versuchstiere*; Rupke, *Vivisection*; for the pathological animal experiments of the early nineteenth century, see Bynum, "Animal Models."

153. Henle, *Miasmen und Contagien*, 52.

154. Rather, "Virchow und die Entzündungsfrage."

155. Predöhl, *Geschichte der Tuberkulose*, 169–74. Cf. Opitz, *Tierversuche und Versuchstiere*, 197.

156. For Cohnheim: Maulitz, "Rudolf Virchow, Julius Cohnheim," 1773–74. Cf. Diepgen, "Die Lehre von der Entzündung."

157. Opitz, *Tierversuche und Versuchstiere*, 200; cf. Predöhl, *Geschichte der Tuberkulose*, 223–25.

158. Bulloch, *History of Bacteriology*, 179–81. The structures had already been shown to exist by Brauell, Pollender, and Rayer.

159. For this and the following paragraphs, see Birch-Hirschfeld, "Vorkommen und Bedeutung niederer Pilzformen (Bakterien)," 205–7.

160. Ibid., 205.

161. Cohn, "Untersuchungen über Bacterien II"; cf. Hoppe, "F. J. Cohn," 168.

162. Cohn, "Untersuchungen über Bacterien II," 188–90.

163. Ibid., 177.

164. Loeffler, *Die Entwicklung der Lehre von den Bakterien*, 167.

165. Bollinger, "Infectionen durch thierische Gifte," 459–68. Cohn criticized this view, which questioned the existence of the genus bacillus. In this connection he gave his own opinion about the still hypothetical spores in 1875: "[I] have to adhere to my earlier concept of the anthrax bacteria as a species of *bacillus*, and will not pronounce myself on their connection with the spherical bacteria (micrococcus) for the time being. Since the *bacilli*, as we mentioned above, usually propagate through spherical durable spores, perhaps we can expect such spores for the rods of anthrax as well, and suppose that they are the germs of infection in the seemingly rod-free blood and in dried-up contagia." Cohn, "Untersuchungen über Bacterien II," 200.

166. Koch, "Die Ätiologie der Milzbrand-Krankheit." Heymann has described

Koch's procedure in detail and reprinted the protocols of his experiments. Their originals are kept in the RKI. A short descripion is found in Loeffler, *Die Entwicklung der Lehre von den Bakterien,"* 169–73.

167. Koch, "Die Ätiologie der Milzbrand-Krankheit," 7.
168. Ibid., 6.
169. Koch's notes cited in Heymann, *Robert Koch*, 345.
170. Koch expressly emphasized this. Koch, "Die Ätiologie der Milzbrand-Krankheit," 8.
171. Koch, "Die Ätiologie der Milzbrand-Krankheit," 21. Here Koch referred to the fact that his insights had been gained in animal experiments with rodents.
172. Ibid., 21–22.
173. Ibid., 24.
174. Heymann, *Robert Koch*, 147–54.
175. Koch's diary of the trip to Munich, RKI/Koch papers, as/L3/001 (cited in full in Heymann, *Robert Koch*, 131–34). The visit to Pettenkofer's laboratory is noted under 19 May, and it does not appear that Koch, who kept a careful record of the people he met, made the acquaintance of the hygienist.
176. Heymann, *Robert Koch*, 153–54, about Koch's first visit to Breslau, or ibid., 210–14, about his visit in October 1977.
177. General hygiene, let alone bacteriological hygiene, was not an academic field at the time. Eulner, *Spezialfächer*, 130–58.
178. Heymann, *Robert Koch*, 155.
179. Loeffler, *Die Entwicklung der Lehre von den Bakterien*, 213–14. This visit is also documented in a diary entry (RKI/Koch papers, as/L3/001, cited in Heymann, *Robert Koch*, 153–54). For Weigert: Wohlrab and Henoch, "Carl Weigert." Cohnheim and Weigert also acquainted Koch with the special technique for pathological animal experimentation they had jointly developed.
180. In this connection, see Schlich, "Wichtiger als der Gegenstand selbst," 1995.
181. Heymann, *Robert Koch*, 200–1; Loeffler, *Die Entwicklung der Lehre von den Bakterien*, 173–74.
182. Bollinger, "Thierkrankheiten." The *Deutsche Medizinische Wochenschrift* 3 (1877) 24 carried a brief review of this work.
183. Billroth to Cohen, 10 Nov. 1876. Cited in Heymann, *Robert Koch*, 1932, 172.
184. Koch, "Verfahren zur Untersuchung."
185. Heymann, *Robert Koch*, 179–82. For Frisch's microphotography see Hagner, "Mikro-Anthropologie und Fotografie."
186. Schlich, "Wichtiger als der Gegenstand selbst."
187. Koch to Cohn, 15 Nov. 1876, cited in Heymann, *Robert Koch*, 174.
188. Ibid.
189. Ibid.
190. They were published in Cohn's *Beiträgen*: Koch "Verfahren zur Untersuchung." Cf. Heymann, *Robert Koch*, 184–85.
191. Koch to Cohn, cited in Heymann, *Robert Koch*, 185.
192. For Cohnheim's experimental pathology, Maulitz, "Rudolf Virchow, Julius Cohnheim."
193. Loeffler, "Die Entwicklung der Lehre von den Bakterien," 213–15; cf. Bulloch, *History of Bacteriology*, 214–17.

248   Notes to Pages 54–58

194. Koch's diary entries for October 1877 (RKI/Koch papers, as/L3/001). Cited in Heymann, *Robert Koch*, 212–13.

195. Koch to Cohn, 24 Nov. 1877, cited in Heymann, *Robert Koch*, 222.

196. In his survey of the relevant literature for the years 1872 to 1875, Felix Victor Birch-Hirschfeld listed forty-six studies: Birch-Hirschfeld, "Vorkommen und Bedeutung niederer Pilzformen (Bakterien)," 171–99.

197. Panum, "Das putride Gift." Cf. Loeffler, "Die Entwicklung der Lehre von den Bakterien," 200.

198. Birch-Hirschfeld, *Lehrbuch der pathologischen Anatomie*, 1225.

199. Hiller, *Lehre von der Fäulnis*; cf. Loeffler, "Die Entwicklung der Lehre von den Bakterien," 190–98.

200. Davaine, "Recherches." Cf. Bulloch, *History of Bacteriology*, 179–82; Opitz, "Tierversuche und Versuchstiere," 196–97.

201. Bulloch, *History of Bacteriology*, 141–44; Loeffler, "Die Entwicklung der Lehre von den Bakterien," 183.

202. Birch-Hirschfeld, "Vorkommen und Bedeutung niederer Pilzformen (Bakterien)," 198. Cf. Birch-Hirschfeld, *Lehrbuch der pathologischen Anatomie*, 1225, where the author laments that the variation in methods and experiments was largely responsible for the contradictory findings of the many investigators.

203. Some of Koch's notes have survived (RKI/Koch papers, as w1/010, 011, 012). The earliest experiment in this area is dated 24 October.

204. Koch, "Neue Untersuchungen über die Mikroorganismen." It seems to have been at Cohnheim's suggestion that Koch attended the Natural History Convention, where he could personally follow the debate between Klebs and Virchow. Heymann, *Robert Koch*, 262–63; cf. Bauer, *Krankheitslehre*, 114–15. Cohnheim, whom Koch visited that autumn in Leipzig, may also have encouraged the Leipzig publisher F. C. W. Vogel to publish Koch's book.

205. Koch, *Untersuchungen über die Ätiologie der Wundinfektionskrankheiten*.

206. Koch, "Die Ätiologie der Milzbrand-Krankheit," 13.

207. RKI/Koch papers, as/w1/15.1.

208. Koch, *Untersuchungen über die Ätiologie der Wundinfektionskrankheiten*, 72. Koch was fully aware of the controversial nature of this conviction and presented the extant literature on the subject to his readers.

209. Ibid., 72.

210. Ibid., 74.

211. Ibid., 75.

212. In doing so, Koch clearly indicates his opposition to Billroth's view and explicitly refers to Virchow's conclusion of 1874: "If the same forms bring about completely different effects, they must be internally different" (Virchow, *Fortschritte der Kriegsheilkunde*, 33).

213. Koch, *Untersuchungen über die Ätiologie der Wundinfektionskrankheiten*, 75.

214. Ibid., 62–63; Birch-Hirschfeld, "Vorkommen und Bedeutung niederer Pilzformen (Bakterien)," 170–72.

215. Cf. Mazumdar, *Species and Specificity*, 66–67.

216. In experimental pathology Koch thus assigned to the animal experiment the central role that Georges Canguilhem has described in general as the essential hallmark

of the development of medicine and biology into laboratory sciences in the nineteenth century (Canguilhem, *Experimentieren in der Tierbiologie*).

217. Koch, *Untersuchungen über die Ätiologie der Wundinfektionskrankheiten*, 75.
218. Ibid., 61; Bulloch, *History of Bacteriology*, 179–82.
219. RKI/Koch papers, as/w1/012. Unfortunately only the descriptions of a series of experiments have been preserved; bacteriological notes—for instance on the identification of microorganisms—are lacking.
220. Koch, *Untersuchungen über die Ätiologie der Wundinfektionskrankheiten*, 95–96.
221. Ibid., 89.
222. Here Koch chose to compare the pathological process with the growth of peat mosses. Koch, *Untersuchungen über die Ätiologie der Wundinfektionskrankheiten*, 90.
223. Koch, *Untersuchungen über die Ätiologie der Wundinfektionskrankheiten*, 100. For the further elaboration of Koch's experiments, for instance by the use of human material, see Bulloch, *History of Bacteriology*, 149–51.
224. Koch, *Untersuchungen über die Ätiologie der Wundinfektionskrankheiten*, 101. Translation from Thomas Brock, ed. *Milestones in Microbiology* Englewood Cliffs, N.J.: Prentice Hall, 1961.
225. Ibid.
226. Using the case of plague, Andrew Cunningham has provided an exemplary analysis of this argument and the resulting conflicts between pathological anatomy and medical bacteriology: Cunningham, *Transforming Plague*.
227. The reviewer in Virchow and Hirsch's annual survey about studies on "vegetal and animal parasites" opened his report with Koch's study, which he characterized as "of outstanding importance for the further development of our understanding of the pathogenic character of vegetal organisms" (Ponfick, "Pflanzliche und thierische Parasiten," 288). Cf. Heymann, *Robert Koch*, 269. It should be noted that Ponfick also taught at Breslau and had taken over Cohnheim's professorship that very year. (Fischer, *Lexikon der hervorragenden Ärzte*, 1234); Klebs, "Sepsis," esp. 277.
228. Koch to Cohn 22 Dec. 1877, cited in Heymann, *Robert Koch*, 222.
229. Koch, Review of *Die niederen Pilze*.
230. Koch, *Untersuchungen über die Ätiologie der Wundinfektionskrankheiten*, 105.
231. Ibid.
232. Koch to Cohn, 24 Nov. 1877, cited in Heymann, *Robert Koch*, 219.
233. Mendelsohn, *Cultures of Bacteriology*; cf. Carter, "Koch-Pasteur Dispute."
234. Dolman, "Brefeld," 221–23. Note that Brefeld and others understood "pure culture" to mean the complete life cycle of a single microorganism, whereas Koch, in the procedure he eventually developed, attached particular importance to cultivating only organisms of one species together (cf. Schlich, "Repräsentationen von Krankheitserregern").
235. Koch, *Untersuchungen über die Ätiologie der Wundinfektionskrankheiten*, 102.
236. Ibid.
237. Ibid., 103.
238. None of these are mentioned in the published study, *Untersuchungen über die Ätiologie der Wundinfektionskrankheiten*, or in the extant laboratory notes.
239. Koch to Flügge, 10 July 1879, Robert Koch correspondence, Henry Barton Jacobs Collection.
240. RKI/Koch papers, as/w1/011.

241. In 1876, for instance, he remarked, "For the most part I use mice, which are the most convenient and easy to handle subjects to inoculate." Koch, *Untersuchungen über die Ätiologie der Wundinfektionskrankheiten*, 6.

242. Koch, *Untersuchungen über die Ätiologie der Wundinfektionskrankheiten*, 539 and 550.

243. Koch, *Zur Untersuchung von pathogenen Mikroorganismen*, 127–28.

244. Koch, *Untersuchungen über die Ätiologie der Wundinfektionskrankheiten*, 103.

245. Ibid.

246. Ibid., 104.

247. Koch, *Zur Untersuchung von pathogenen Mikroorganismen*, 127–28; cf. Schlich, "Repräsentationen von Krankheitserregern."

248. Koch, "Die Ätiologie der Tuberkulose" ([1884] 1912), 521.

249. In the case of tuberculosis, Koch, "Die Ätiologie der Tuberkulose" ([1884] 1912), 520–21.

250. Logan, "Before There Were Standards."

251. Some remarks to this effect are found in the study on traumatic infections (Koch, *Untersuchungen über die Ätiologie der Wundinfektionskrankheiten*) and in the polemic against Pasteur (Koch, *Experimentelle Studien*). A systematic discussion of the problem is found only in 1881 (*Zur Untersuchung von pathogenen Mikroorganismen*, 127–31). A special piece on the methodology of the animal experiment promised on this occasion was never published.

252. Koch, *Zur Untersuchung von pathogenen Mikroorganismen*, 127.

253. A similar argument was advanced by the author of the first textbook of medical bacteriology, Ferdinand Hueppe (*Methoden der Bakterien-Forschung*). For the concept of the model, see Churchill, "Life Before Model Systems," 266–67. Churchill proposes that in the description of a historical context, the concept of the model organism should be brought in only if such an organism has become the focus of the work of an entire institute; if it has been standardized by means of domestication, selection, and interbreeding; and finally, if researchers who use it make explicit statements about its universality. Churchill's criteria were developed for biological research, and the first criterion makes little sense for research on infection. Still, Koch's experimental animals did not meet the second and third criteria.

254. During Koch's work on anthrax at the IHO,, financial considerations therefore ruled out the use of bovine cattle (Opitz, *Tierversuche und Versuchstiere*, 198). For the work with sheep, which were apparently affordable, "suitable localities were not available at the agency, so that a stable had to be built for the purpose on the grounds of the Central Police Station" (Koch, *Experimentelle Studien*, 233).

255. Tröhler and Maehle, "Antivivisection." The authors find a weakening of the antivivisectionist movement in the 1880s and describe this movement's fixation on the physiological animal experiment of the preceding years.

256. Koch, "Die Ätiologie der Tuberkulose" ([1884] 1912), 531.

257. Heymann, *Robert Koch*, 155–56.

258. Koch, "Die Ätiologie der Tuberkulose" ([1884] 1912), 512.

259. Koch, *Untersuchungen über die Ätiologie der Wundinfektionskrankheiten*, 88.

260. Ibid.

261. One would really like to know the subject of the "rather heated debate with Klebs" that Koch notes in his diary for 22 July 1878 (Heymann, *Robert Koch*, part 2,

258–59.) The two men met in Leipzig, after Koch had discussed his work on traumatic infections in Breslau. Conversations with Klebs are mentioned for the following days as well. A letter to Cohn of 27 June indicates that Koch had gone to Leipzig in part to present his findings to Klebs (Heymann, *Robert Koch*, part 2, 257–58).

262. Diepgen, "Krankheitswesen und Krankheitsursache," 312–15; Faber, *Nosography*, 96–97.

263. In this sense, according to Georges Canguilhem, disease is not a gradually measurable deviation from the norms of health but rather a "functional innovation" facing the body (Canguilhem, *Das Normale und das Pathologische*, 96; Canguilhem, "Beitrag der Bakteriologie," 121; Canguilhem, *Experimentieren in der Tierbiologie*, 15).

264. This goal was implied most clearly in Koch's work on tuberculosis. Here the bacteriological diagnosis led to a new definition of the disease, which covered a whole gamut of affections, such as phthisis and lupus.

265. Virchow, "Der Kampf der Zellen und Bakterien."

266. For bacteriology and immunology, the idea of nontheoretical empirical work that carries embedded theories is found in Canguilhem, "Beitrag der Bakteriologie."

CHAPTER III: Tuberculosis and Tuberculin

1. Loeffler, "Zum 25jährigen Gedenktage," 449.
2. Möllers, *Robert Koch*, 113.
3. Ehrlich, in *Frankfurter Zeitung*, 2 June 1910, cited in Möllers, *Robert Koch*, 133.
4. Johne, *Geschichte der Tuberkulose*, 4. In the book's preface we read that the author had participated in a contest offering a prize for the best history of tuberculosis!
5. Brock, *Robert Koch*, chapters 14 and 15.
6. For a first orientation, see Rosenkrantz, "Bovine Tuberculosis," 1985.
7. Brock, *Robert Koch*, chapter 18.
8. Möllers, *Robert Koch*, 593; Brock, *Robert Koch*, 198; Foster, *History of Medical Bacteriology*, 62. See also the recent study by Kaufmann, "Höhen und Tiefen," 14.
9. Fleck, *Genesis and Development*, 27–38.
10. Silverstein, *A History of Immunology*, chapter 9.
11. Foster, *History of Medical Bacteriology*, 62.
12. Gerald Geison, for example, has analyzed Pasteur's work on anthrax and rabies in these terms (Geison, *Pasteur*, 145–256).
13. As late as 1909, Koch wrote about extensive experimental work in a letter to Wilhelm Kolle: Koch to Kolle, Berlin, 26 Nov. 1909, SBPK. Eli Metchnikoff reported on a visit to Koch's laboratory in 1909, saying that he had found Koch working on improvements to tuberculin (Brock, *Robert Koch*, 302).
14. Schlich, "Symbol medizinischer Fortschrittshoffnung." Today the building is used by the Institute of Microbiology and Hygiene of the Humboldt University and also shelters a Robert Koch Museum.
15. Koch, "Die Ätiologie der Tuberkulose" ([1882] 1912).; Koch, "Über die Ätiologie der Tuberkulose"; Koch "Kritische Besprechung."
16. Koch, "Die Ätiologie der Milzbrand-Krankheit."
17. Koch, *Untersuchungen über die Ätiologie der Wundinfektionskrankheiten*; Koch, *Neue Untersuchungen über die Mikroorganismen*.
18. For an introduction to the extensive literature on the history of tuberculosis,

Rosenkrantz, *From Consumption to Tuberculosis*. Two important comprehensive studies are Dubos et al., *The White Plague*; Smith, *The Retreat of Tuberculosis*. For an introduction to the present state of the research, see Barnes, *The Making of a Social Disease*; Condrau, *Lungenheilanstalt*. Tuberculosis was also the disease that was most intensely reflected in the art and literature of the era: Engelhardt, *Medizin in der Literatur*, 73–89.

19. Bochalli, *Tuberkuloseforschung*; King, *Medical Thinking*; Myers, "Development of Knowledge of Unity of Tuberculosis"; Seiffert, "Die Tuberkulose als übertragbare Krankheit." An older and most useful account is Predöhl, *Geschichte der Tuberkulose*.

20. King, *Medical Thinking*, 34–35; Duffin, *Laënnec*, chapters 8 and 9.

21. Faber, *Nosography*, 76–78.

22. For the history of contemporary etiological thinking, see Carter, "Koch's Postulates"; Diepgen, "Krankheitswesen und Krankheitsursache"; Schlich, "Kontrolle notwendiger Krankheitsursachen."

23. Niemeyer, *Lehrbuch der speziellen Pathologie und Therapie*, 17.

24. Predöhl, *Geschichte der Tuberkulose*.

25. Ibid.

26. Weigert, "Zur Lehre von der Tuberculose."

27. Cohnheim, *Die Tuberkulose*. Cf. Doerr, "Cohnheims Entzündungslehre"; Maulitz, "Rudolf Virchow, Julius Cohnheim."

28. Cf., for the case of plague, Cunningham, "Transforming Plague."

29. Koch, "Die Ätiologie der Tuberkulose" ([1884] 1912), 428.

30. Ibid., 469.

31. Ibid., 470–71.

32. Möllers, *Robert Koch*, 535.

33. For the history of the Imperial Health Office, *Das Reichsgesundheitsamt*; Huentelmann, *Das Kaiserliche Gesundheitsamt*, 2003; Ritter, *Grossforschung und Staat*, 19–20.

34. Koch, "Die Ätiologie der Tuberkulose" ([1882] 1912), 429.

35. Koch, "Verfahren zur Untersuchung," 27–50. Cf. Bulloch, *History of Bacteriology*, 213–17; Heymann, *Robert Koch*, 168. For the history of bacteria staining, see Clark and Kasten, *History of Staining*, 91–101; Bracegirdle, *History of Microtechnique*, 70–74; for Ehrlich, Travis, "Science as Receptor of Technology." Cf. Schlich, "Repräsentationen von Krankheitserregern."

36. According to Koch's own words, "all efforts to find bacteria or other microorganisms in these preparations [initially] . . . remained unsuccessful." Koch, "Die Ätiologie der Tuberkulose" ([1884] 1912), 472.

37. Ibid. In a lecture of summer 1882 Koch espoused a variant of this thesis, stating that one could also see the bacteria without staining, provided one had first used staining to establish their presence. Koch, "Die Ätiologie der Tuberkulose" ([1882] 1912), 448. In March 1882 he did not broach the subject.

38. Koch, "Die Ätiologie der Tuberkulose" ([1882] 1912), 429.

39. Ibid., 472 ff. Cf. Möllers, *Robert Koch*, 552–53.

40. Koch, "Die Ätiologie der Tuberkulose" ([1882] 1912), 430.

41. Brock, *Robert Koch*, 119.

42. "Given the regular ocurrence of tubercle bacilli, it must be considered strange that so far no one has seen them. But this is explained by the fact that the bacilli are extremely small structures and usually so few in number . . . that without a very special

staining reaction they are bound to escape the notice of the most attentive observer" (Koch, "Die Ätiologie der Tuberkulose" ([1882] 1912), 432–33).

43. Koch, "Die Ätiologie der Tuberkulose" ([1884] 1912), 479.

44. Baumgarten, "Tuberkelbakterien." Baumgarten had, however, brightened his preparations with sodium and potassium solutions and not undertaken cultivation and inoculation experiments. His lecture took place earlier than Koch's but was published later. Predöhl, *Geschichte der Tuberkulose*, 347, emphasizes the simultaneous discovery and provides a detailed description of Baumgarten's work. Cf. Brock, *Robert Koch*, 133.

45. RKI/Koch papers, folder "versch. Tuberkulose—S. 1881/82," which contains a list of the animal experiments conducted. These continued until October 1882, and it appears that until early March very few infection experiments with cultures were successfully completed. The dates of the listed experiments and the fact that many of them can be identified in Koch's article of 1884 suggest that the list was drawn up in the context of Koch's work on the major article "Die Ätiologie der Tuberkulose" ([1884] 1912).

46. For the first time, on June 12, 1881, five guinea pigs and two rabbits were inoculated with cultures. Four of the guinea pigs and one rabbit died between December 17 and February 1, and the other two were killed on February 4 and January 3. RKI/Koch papers, folder "versch. Tuberkulose—S. 1881/82," as/w2/001. By the time of his lecture, Koch had already carried out six series of experiments involving the testing of innumerable animals of various species (ibid.).

47. Möllers, *Robert Koch*, 550.

48. Koch, "Zur Untersuchung von pathogenen Mikroorganismen," 122. Cf. Schlich, "Wichtiger als der Gegenstand selbst"; Schlich, "Repräsentationen von Krankheitserregern."

49. In "Die Ätiologie der Tuberkulose" ([1884] 1912), 484, for instance, Koch provides precise measurements and compares the bacterium with others that he had been able to photograph.

50. Loeffler, "Zum 25jährigen Gedenktage," 451.

51. This information is largely based on Johnston, "Tuberculosis." For Carl Fränkel in 1890 the existence of spores was an unsolved question (Fränkel, *Bakterienkunde*, 308). In Kolle and Wassermann's *Handbuch der pathogenen Mikroorganismen* of 1903 as well, the spores are discussed with a great deal of reserve: Cornet and Meyer, "Tuberkulose," 78–177.

52. See chapter II.3.1.

53. "Die Ätiologie der Tuberkulose" ([1884] 1912), 551.

54. Ibid., 554.

55. On p. 509 of his article, Koch discusses the case of infectious content in tubercles in which bacteria cannot be documented; on p. 502 he describes spores in organic tissue; and on p. 526 spores in pure cultures.

56. Koch, "Die Ätiologie der Tuberkulose" ([1884] 1912), 485.

57. Descriptions to be found in Koch, "Die Ätiologie der Tuberkulose" ([1882] 1912), 431; Koch, "Die Ätiologie der Tuberkulose" ([1882] 1912), 491.

58. Ibid., 485.

59. Ibid., 537. The laboratory notes (RKI/Koch papers "Versch. Tuberkulose—S. 1881/2" as/w2/001) show Koch's modus operandi: he began in October by transmitting animal forms of tuberculosis, such as bovine pulmonary tuberculosis known as Perlsucht, to other animals, for the most part guinea pigs. Beginning in late November, human

forms were transmitted to animals. Retransmission to humans could not of course be performed as a demonstration. Instead, Koch conducted twenty-eight experiments in which he inoculated various forms of human tuberculosis into a total of 179 guinea pigs, 35 rabbits, and 4 cats. Koch, "Die Ätiologie der Tuberkulose" ([1884] 1912), 537. In the laboratory notes, page 7 mentions monkeys instead of cats.

60. Koch, "Die Ätiologie der Tuberkulose" ([1884] 1912), 512. cf. Opitz, *Tierversuche und Versuchstiere*, 200–4.

61. RKI/Koch papers, "Versch. Tuberkulose—S. 1881/2" as/w2/001. The subsequent documents indicate that the underlined numbers refer to animals that have been killed.

62. "The surest way to produce a pure culture is to seed [the medium] with a bacilli-filled tubercle or the same substance from inside the not yet caseated lymph glands of a killed tubercular guinea pig." Koch, "Die Ätiologie der Tuberkulose" ([1884] 1912), 520.

63. Ibid., 522.

64. Ibid., 435.

65. Ibid., 525.

66. Ibid., 524. Koch refers (p. 525) to experiments conducted by Klebs, who, probably because of a different composition of the culture medium, observed small beads at the edges of the cultures. He considered these to be micrococci, while Koch thought they were elements of the nutrient substance.

67. Koch used thirteen different species, among them such varied ones as sparrows, gold fish, and turtles (ibid., 539).

68. Koch also conducted control experiments, both with noninfectious materials and with nonsusceptible animals.

69. Koch, "Die Ätiologie der Tuberkulose" ([1884] 1912), 538.

70. Koch, "Die Ätiologie der Tuberkulose" ([1882] 1912), 438.

71. Ibid., 441.

72. Ibid., 442.

73. Even in 1884, when he described the tubercular tissue from which he had grown his pure culture, he added standard information indicating possible hereditary susceptibility of the deceased from whom the tissue had been taken (Koch, "Die Ätiologie der Tuberkulose" ([1884] 1912), 504–5).

74. Virchow, "Der Kampf der Zellen und Bakterien," 1885. However, Virchow also criticized an over-reliance on the knowledge of pathogens, which dominated "not only the thinking but also the dreams of many older and almost all young physicians" (8).

75. Koch, "Kritische Besprechung," 457.

76. Koch to Justi, 27 Sept. 1882, SBPK Slg. Darmst., 3b 1882 (2) Koch, Robert. Cf. Koch, "Die Ätiologie der Tuberkulose" ([1882] 1912), 442.

77. Koch, "Die Ätiologie der Tuberkulose" ([1884] 1912), 467.

78. Koch, "Die Ätiologie der Tuberkulose" ([1882] 1912), 442.

79. King, *Medical Thinking*, 59. For a survey of the experimental studies stimulated by Villemin, see Predöhl, *Geschichte der Tuberkulose*, 163–349.

80. Koch, "Die Ätiologie der Tuberkulose" ([1884] 1912), 442.

81. Virchow, for instance, considered the caseous mass inside the tubercles—which Koch, following Villemin, identified as a product of tuberculosis—to be a "caseous pneumonia" not specific to tuberculosis. Cf. Faber, *Nosography*, 99–101. Faber emphasizes the

connection of Koch's concept of tuberculosis to that of Laënnec and Villemin. Cf. also Orth, *Lungenschwindsucht*, 4: "Koch's discovery was, as it were, but the capstone of the etiological edifice, which in the main was already completed."

82. Koch, "Die Ätiologie der Tuberkulose" ([1884] 1912), 531.
83. Ibid., 469–70.
84. This information is based on Loeffler, "Zum 25jährigen Gedenktage."
85. Koch, "Die Ätiologie der Tuberkulose" ([1884] 1912), 550.
86. Tomes and Warner, "Rethinking the Reception of the Germ Theory."
87. Koch, "Die Ätiologie der Tuberkulose" ([1882] 1912), 444.
88. Cf., for instance, Frevert, *Krankheit als politisches Problem*; Göckenjahn, *Kurieren und Staat machen*; Huerkamp, *Der Aufstieg der Ärzte*; Labisch, *Homo Hygienicus*. Also Evans, *Death in Hamburg*, 330–63; Weindling, *Health, Race, and German Politics*; Weindling, *Epidemics and Genocide*; the health education of the time is treated from a similar perspective in Tomes, "Gospel of Germs."
89. Gorsboth and Wagner, "Die Unmöglichkeit der Therapie," 142.
90. Bruno Latour has presented the most convincing account of this connection. Concerning the influence of Pasteur's microbiology on French society, he writes, "When Pasteur and the hygienists introduced the notion of a microbe as the essential cause of infectious disease, they did not take the society to be made up of rich and poor, but rather of a different list of groups: sick contagious people, healthy but dangerous carriers of microbes, immunised people, vaccinated people, and so on" (Latour, *Science in Action*, 115–16; cf. Latour, *The Pasteurisation of France)*.
91. Koch, "Die Ätiologie der Tuberkulose" ([1882] 1912), 444–45.
92. The authoritative study is Carter, "Koch's Postulates." See also Evans, *Causation and Disease*; Grafe, "Kochsche Postulate"; Harden, "Koch's Postulates"; Lennox, "Those Deceptively Simple Postulates"; Maiwald, "Charakterisierung 'neuer' Infektionserreger"; Schlich, "Kontrolle notwendiger Krankheitsursachen"; Thagard, *How Scientists Explain Disease*.
93. Carter, "Koch's Postulates."
94. Mendelsohn also claims that in the studies of the 1870s and early 1880s, Koch developed an etiological model. At that time he made some fundamental decisions, for instance, not to regard virulence as an important phenomenon. Mendelsohn, *Cultures of Bacteriology*, esp. chapter 3.
95. Carter, "Koch's Postulates."
96. Koch, "Die Ätiologie der Tuberkulose" ([1884] 1912), 469. Cf. Carter, "Koch's Postulates," 361.
97. Mazumdar, *Species and Specificity*, 66–67.
98. Koch, "Die Ätiologie der Tuberkulose" ([1884] 1912), 516.
99. Ibid., 558.
100. Ibid., 498.
101. For Koch's concept of disease as bacterial invasion, Mendelsohn, *Cultures of Bacteriology*, 255–63. For the popular understanding, Gradmann, *Bazillen, Krankheit und Krieg*; Gradmann, "Invisible Enemies."
102. Fleck, *Genesis and Development*, 43. Cf. Carter, "Klebs' Criteria," 88. Koch himself began to correct this view after 1892–93: Mendelsohn, *Cultures of Bacteriology*, chapter 7.
103. Koch, "Die Ätiologie der Tuberkulose" ([1884] 1912), 551 ff.

104. As an introduction, see Myers, "Development of Knowledge of Unity of Tuberculosis." Cornet argued that tuberculosis spread through dust; Flügge incriminated sputum; and Behring thought it was foodstuffs (milk). Koch, "Die Ätiologie der Tuberkulose" ([1884] 1912), 554, wrote, " A phthisic [is] very apt to provide ample amounts of infectious material for his closest companions, and in a form that will most easily cause an infection." One of the consequences of this was that it now seemed necessary to treat people suffering from pulmonary tuberculosis in special institutions rather than at home (Wilson, "Decline of Tuberculosis," 381).

105. Koch, "Die Ätiologie der Tuberkulose" ([1884] 1912) 491.

106. Ibid., 485.

107. Ibid., 487.

108. Ibid., 490.

109. Ibid., 546.

110. The notes summarizing these experiments clearly show this modus operandi: RKI/Koch papers, folder "versch. Tuberkulose-S. 1881/82."

111. Koch, "Die Ätiologie der Tuberkulose" ([1884] 1912), 560.

112. David, *Rudolf Virchow*, chapter 3.

113. Diepgen, "Krankheitswesen und Krankheitsursache," 315.

114. Koch, "Die Ätiologie der Tuberkulose" ([1884] 1912), 483. Koch had obtained these deceased patients from two Berlin hospitals, Friedrichshain and Moabit.

115. RKI/Koch papers, folder "versch. Tuberkulose—S. 1881/82" as/w2/001. At this point, "Dr. L." no doubt refers to Friedrich Loeffler.

116. Diepgen, "Krankheitswesen und Krankheitsursache." Engelhardt, "Kausalität und Konditionalität." Georges Canguilhem considers rationalistic simplification and the refusal to engage in systematic discussions of pathology a hallmark of bacteriological thinking about disease. Canguilhem, "Beitrag der Bakteriologie," 123.

117. Koch, "Kritische Besprechung," 455.

118. Carter, "Klebs' Criteria"; Klebs, "Cellularpathologie und Infectionskrankheiten." Virchow, "Krankheitsursachen."

119. Koch, "Über bakteriologische Forschung." For a more detailed discussion of the tuberculin question, see Elkeles, "Tuberkulinrausch"; Gradmann, *Money, Microbes and More*; Opitz and Horn, "Die Tuberkulinaffäre."

120. Coleman, "Koch's Comma Bacillus"; Gradmann, "Die Entdeckung der Cholera in Indien." See also chapter V.2.

121. Opitz, "Koch's Ansichten."

122. Eschenhagen, *Hygiene-Institut*.

123. For the significance of the bacteriology courses, see Eschenhagen, *Hygiene-Institut*; Mendelsohn, *Cultures of Bacteriology*, 285–99; cf. D'Orazio, "Scienza Tedesca"; Gossel, "Standard Methods."

124. Möllers, *Robert Koch*, 192, 224.

125. Koch to Flügge, 26 Dec. 1888, Koch correspondence, Jacobs Collection.

126. Pasteur's rabies vaccine of 1885 was generally seen as a sensational achievement and led to the founding of the Institut Pasteur: Löwy, "Networks and New Disciplines," 1994; Weindling, "Scientific Elites and Laboratory Organisation."

127. This point is most clearly made by Foster, *History of Medical Bacteriology*. Foster sees Koch's later work as a tragic attempt to repeat the successes of his early years. The newest biography by Brock takes a similar position.

128. Mendelsohn, *Cultures of Bacteriology*. This view is shared by Briese, *Angst in den Zeiten der Cholera*, 1:358–83.
129. Koch, "Die Ätiologie der Tuberkulose" ([1882] 1912), 444. Cf. Faber, *Nosography*, 108–11.
130. Koch, "Die Ätiologie der Tuberkulose" ([1882] 1912), 445.
131. Mendelsohn, *Cultures of Bacteriology*, chapter 3; Carter, "Koch-Pasteur Dispute."
132. Koch, "Die Ätiologie der Tuberkulose" ([1884] 1912), 546.
133. Ibid., 552.
134. Schill and Fischer, "Über die Desinfektion des Auswurfes."
135. Brock, *Robert Koch*, 105–13.
136. Schill and Fischer, "Über die Desinfektion des Auswurfes."
137. Koch, "Die Ätiologie der Tuberkulose" ([1884] 1912), 540.
138. Ibid., 543.
139. Fräntzel's report at the meeting of the Society for Internal Medicine: Leyden and Pfeiffer, *Verhandlungen des Congresses für innere Medicin*, 46–49.
140. Möllers, *Robert Koch*, 556.
141. This view is shared by Eschenhagen, *Hygiene-Institut*, 113. Eschenhagen speaks of a "focused search over several years."
142. Cornet, "Tuberkelbacillen im thierischen Organismus," 1889. For the contemporary drug-based treatment of tuberculosis, see Bochalli, *Tuberkuloseforschung*, 1958, pp. 79–83; Smith, *The Retreat of Tuberculosis*, 1987, chapter 5.
143. Cornet, "Tuberkelbacillen im thierischen Organismus," 100–101.
144. In this case, three more infected animals were kept in Berlin as controls.
145. Koch to Farbwerke Höchst, 23 May 1888, SBPK, Slg. Darms., 3 b 1882 (2) Koch, Robert. Cf. Koch, "Über bakteriologische Forschung," 659. The recipient of the letter, Arnold Libbertz, later became an important associate in the development of tuberculin.
146. In RKI/Koch papers, as/w2/003.
147. Note of June 1888, RKI-Archiv, as/w2/003. Auramin is the chlorhydrate of Amidotetrametyldiamidophenylmethan (*Merck's Index*, 35).
148. RKI/Koch papers, as/w2/008.
149. Koch, "Über bakteriologische Forschung." For the sumptuous setting of the lecture, see Gorsboth and Wagner, "Die Unmöglichkeit der Therapie"; Winau, "Serumtherapie."
150. Koch, "Über bakteriologische Forschung," 659.
151. Ibid.
152. It would seem that Ferdinand Hueppe was the only contemporary who immediately noticed this change in research strategy. While Koch, he wrote, had initially looked for ways to combat the bacteria, he was now working on the tissues (Hueppe, "Die Heilung der Tuberculose").
153. Koch, "Weitere Mitteilungen über ein Heilmittel." Koch claimed that this failure had to do with difficulties in the production of the medication. However, secrecy also protected his financial goals. On November 24 the Prussian minister of culture Gossler assured the Prussian parliament that he assumed responsibility for this secrecy (Stenographic transcripts, 1890, session of 19 Nov. 1890). Cf. Möllers, *Robert Koch*, 197.
154. Koch, "Fortsetzung der Mitteilungen über ein Heilmittel."
155. Elkeles, "Tuberkulinrausch," 1731.

156. In any case, Koch disclosed his information to the scientific public only piecemeal, publishing yet another article about tuberculin in the summer of 1891: Koch, "Weitere Mitteilung über das Tuberkulin." For the production and the ingredients of tuberculin, cf. Möllers, *Robert Koch*, 556-93.

157. Hueppe and Scholl, "Die Natur der Koch'schen Lymphe." Hueppe eventually wrote a critique of Koch's publications on this subject: Hueppe, "Koch's Mitteilungen über Tuberkulin." Following Hueppe and Klebs, Martin Hahn worked on the subject as well but was also unable to go beyond the recognition that he was probably dealing with a protein that could not be isolated. Hahn, "Über die chemische Natur."

158. Klebs, "Die Zusammensetzung des Tuberkulin."

159. In the summer of 1891, Koch acknowledged that all his attempts to produce a pure form had been unsuccessful. Koch, "Weitere Mitteilungen über das Tuberkulin," 678.

160. Klebs, "Über die Wirkung des Koch'schen Mittels."

161. Baumgarten, "Tuberculinwirkung"; Grawitz, "Versuche mit dem Koch'schen Mittel."

162. Pfuhl, "Behandlung tuberculöser Meerschweinchen."

163. The documents are to be found in the RK/Koch papers, as/w2/003. Pfuhl, "Behandlung tuberculöser Meerschweinchen."

164. Ibid., 242-44.

165. Compare this with Koch's account: Koch "Fortsetzung der Mitteilungen über ein Heilmittel," 670.

166. Koch, "Weitere Mitteilungen über das Tuberkulin," 664.

167. Ibid., 672.

168. Ibid., 670.

169. RK/Koch papers, as/w2/003.

170. Koch, "Fortsetzung der Mitteilungen über ein Heilmittel," 670.

171. Koch's original extraction method used alcohol and glycerin (Koch, "Fortsetzung der Mitteilungen über ein Heilmittel," 671). Extracts made with alcohol, however, rapidly lost their effectiveness. In the course of the year 1891, Koch's associates at the Institute for Infectious Diseases tested a wide variety of other extraction methods (Koch, "Weitere Mitteilungen über das Tuberkulin," 675).

172. Koch, "Weitere Mitteilungen über das Tuberkulin," 664.

173. Ibid., 662.

174. Ibid., 663.

175. Koch's description of the reaction, ibid. The overwhelmingly positive evaluation of tuberculin as a diagnostic tool is visible in the official reports: "Die Wirksamkeit."

176. It is no coincidence that the euphoria over tuberculin is related to the spread of bacillus metaphors in the public at large: Gradmann, "Popularisierte Bakteriologie," 45-50.

177. For Pasteur's vaccines exemplified by the case of rabies, see Geison, *Pasteur*; For Koch's motivation, cf. Foster, *History of Bacteriology*, 59.

178. For additional details, see Gradmann, "Ein Fehlschlag und seine Folgen."

179. Koch to Althoff, 31 Oct. 1890, GStAPK, I HA, Rep. 76 VIII B, 2892.

180. Eschenhagen, *Hygiene-Institut*, 131.

181. GStAPK, I HA, Rep. 92 Althoff, A I, Nr. 256. For Althoff, see vom Brocke,

*Wissenschaftsgeschichte und Wissenschaftspolitik*, especially Eckard's article on Althoff and medicine.

182. Cf. The report on the session of the (Prussian) *Landtag* on 29 Nov. 1890 in the *Vossische Zeitung*: "Die Interpellation über das Koch'sche Heilverfahren."

183. See the very good account of these negotiations by Opitz and Horn, "Die Tuberkulinaffaire," 733. Cf. GStAPK, I HA, Rep. 76 VIII B 2937. Although Koch declared his willingness to accept an endowment, he did want the contract to stipulate a share of the profits for himself.

184. Cf. GStAPK I., Rep. 76 VIII B, Nr. 2937. Cf. Ibid., Rep. 92 Althoff, B 95, vol. 1: This is the letter of December 9 to Althoff, to which Koch appended the contract signed by him, Libbertz, and Pfuhl, characterizing it as "a difficult birth." It calls for a payment of 1 million marks for Koch and 250,000 each for Libbertz and Pfuhl. Cf. Eschenhagen, *Hygiene-Institut*, 132 ff.

185. Gossler had made it clear to Chancellor Caprivi back in November that he too was opposed to an endowment, which was a way of asking for the chancellor's veto. See Gossler's note to Caprivi before the presentation to the Prussian King on November 9 "I hope that His Majesty will be convinced by the last report that it is not yet time to decide the matter of a reward or distinction for Geheimrat Koch. In my opinion a premature compensation would be unfortunate from a personal and an objective point of view." (Cf. GStAPK, I HA, Rep. 92 Althoff, 233a). However one judges Koch's financial intentions in the autumn of 1890, he was correct when he felt put off, indeed tricked by the bureaucrats of the Ministry of Culture. As he put it to Althoff, he thought they wanted to "gyp" him (GStAPK I., Rep. 76 VIII B, Nr. 2937, cited by Eschenhagen, *Hygiene-Institut*).

186. Caprivi's *Votum* of 25 Dec. 1890: GStAPK I., Rep. 76 VIII B, Nr. 2937.

187. Koch, "Über bakteriologische Forschung."

188. Virchow, "Ueber die Wirkung des Koch'schen Mittels." Within the medical community, Virchow's article represented a kind of turning point. See also the editorial statement of the *Berliner klinische Wochenschrift*, Jan. 1891, no. 4, p. 96.

189. Koch, "Fortsetzung der Mitteilungen über ein Heilmittel."

190. One detail in Koch's application for a leave of absence clearly indicates his weariness with his work situation: "I am taking the liberty of informing you that I wish to use this leave for recreational travel." Having immediately obtained the leave—granted by the Minister of Culture in person!—Koch left town.

191. See chapter IV.

192. Schadewaldt, "Entdeckung des Tuberkulins"; Virchow, "Ueber die Wirkung des Koch'schen Mittels." For contemporaries, Virchow's article was a turning point in the discussion: see the editorial statement of the *Berliner klinische Wochenschrift*, Jan. 1891, no. 4, p. 96; Elkeles, "Tuberkulinrausch." For the international discussion: Chauvet, "Cenetaire"; Hansen, "Images of a New Medicine"; Leibowitz, "Scientific Failure"; Smith, *The Retreat of Tuberculosis*, 17–62.

193. See chapter IV.2.

194. Schadewaldt, "Entdeckung des Tuberkulins." Subsequently some of the assistants, such as August von Wassermann and Shibasaburo Kitasato, were also allowed to act as experimental subjects. Koch, "Weitere Mitteilungen über ein Heilmittel," 679.

195. Hansemann, "Erfahrungen über die Koch'sche Injectionsmethode."

196. Opitz and Horn, "Die Tuberkulinaffaire," 731 ff.

197. As early as November 30, 1890, the *Neue Preussische Zeitung* reported on the

priority claims for a tuberculosis serum registered by Héricourt and Richet in the *Revue Rose*. See Bulloch, *History of Bacteriology*, 261.

198. Mendelsohn, *Cultures of Bacteriology*, 310–16.

199. Althoff's memorandum of 21 March 1981. GStAPK, 1 HA, Rep. 92 Althoff, A1, Nr. 233a. Cf. Eschenhagen, *Hygiene-Institut*, 1983, p. 136.

200. GStAPK, 1 HA, Rep. 92 NI Althoff, Nr. 256, of 20 April 1891.

201. The expertise that Gossler submitted to the governing board at the ministry was produced by Professor Köhler. Gossler's request and a copy of the expertise are in BA, R 09.01 AA, Nr. 21037, of 5 Dec. 1891.

202. GStAPK, 1 HA, Rep. 76 VIII B, Nr. 2937. Koch's statement is dated June 29, 1891, but contains the remark "as repeatedly stated earlier." This business arrangement had been the subject of long drawn-out disagreements.

203. GStAPK, 1 HA, Rep. 76 VIII B, 9 June 1891.

204. GStAPK, 1 HA, Rep. 76 VIII B, Nr. 2937.

205. For Koch's statement on this point of 26 June 1891, see GStAPK, 1 HA, Rep. 76 VI-IIB, Nr. 2904.

206. Excerpt of the minutes of the meeting of the Ministry of Culture, 15 May 1891.

207. Külz was one of the pioneers of physiological chemistry in Germany (Walter, "Külz"). For the "Althoff system" and Külz's role as advisor to Althoff, see vom Brocke, *Wissenschaftsgeschichte und Wissenschaftspolitik*.

208. Külz to Althoff, 21 April 1891, GStAPK I, HA, Rep, 76 VIII A, Nr. 2955.

209. Koch, "Weitere Mitteilungen über ein Heilmittel"; Koch "Weitere Mitteilung über das Tuberkulin," 678.

210. Koch to Althoff, 39 Oct. 1891, GStAPK I, HA, Rep, 76 VIII A, Nr. 2956.

211. Letter to Althoff, of 25 Oct. 1891. The name of the signer is illegible, but there is no doubt that the document is a report of the Marburg working group. GStAPK I, HA, Rep, 76 VIII A, Nr. 2955.

212. Accordingly, Koch was requested to provide, free of charge, considerable quantities of tuberculin.

213. Külz to Ministry of Culture, 9 May 1892, GStAPK I, HA, Rep, 76 VIII A, Nr. 2955.

214. Koch to Ministry of Culture, 9 May 1892, GStAPK I, HA, Rep, 76 VIII A, Nr.2955.

215. Ibid.; cf. for Proskauer, Möllers *Robert Koch*, 1950, 396. The expression "Tuberculin swindle" was apparently coined by Virchow's former student Johannes Orth.

216. Schadewaldt, "Entdeckung des Tuberkulins." For the public's view of tuberculin, cf. Gorsboth and Wagner, "Die Unmöglichkeit der Therapie."

217. The demand that Koch publish his animal experiments was voiced at the time by such scientists as Czerny: Czerny, "Erster Bericht," 68. Koch, "Über neue Tuberkulinpräparate"; Koch, "Über die Behandlung der Lungentuberkulose"; Brock, *Robert Koch*, 302. For the history of tuberculin therapy: Schmidt, "Geschichte der Tuberkulin-Therapie."

218. Johnston, *Tuberculosis*.

219. Nolte, "Autovakzine."

220. In the relevant article in the *Handbuch der pathogenen Mikroorganismen*, which was the standard work in bacteriology around the turn of the nineteenth century, Koch's

spores have become "vacuoles." The authors, bacteriologists of the Koch school all, were very doubtful about the existence of spores. Cornet and Meyer, "Tuberkulose," 81.

221. Fleck, *Genesis and Development*, 29–30.

222. Ibid., 40.

223. Koch, "Weitere Mitteilungen über ein Heilmittel," 662. Koch injected healthy guinea pigs with a full 2cc. of tuberculin without triggering a reaction.

224. Silverstein, *A History of Immunology*, 229.

225. Koch, "Weitere Mitteilungen über ein Heilmittel," 663.

226. Schreiber, "Uber das Koch'sche Heilverfahren," 18. Cf. Schreiber's contribution in the official reports: "Die Wirksamkeit," 650–59.

227. Peiper, "Über die Wirkung des Koch'schen Mittels."

228. Cf. Hueppe, "Uber Erforschung der Krankheitsursachen."

229. See the description of Koch's work on traumatic infections in chapter II.

230. Cf. Canguilhem, "Beitrag der Bakteriologie."

231. Buchner, "Tuberculin Reaction." Cf. Bulloch, *History of Bacteriology*, 258. For Buchner's studies: Mazumdar, *Species and Specificity*, 78–85.

232. Lahmann, *Koch und die Kochianer*.

233. Hueppe, "Ueber Erforschung der Krankheitsursachen"; Martius, "Krankheitsursachen und Krankheitsanlage." Cf. Diepgen, "Krankheitswesen und Krankheitsursache," 316–27.

234. Rosenbach, *Aufgaben und Grenzen der Therapie* (reprinted in Rosenbach, *Arzt contra Bakteriologe*). For Rosenbach, see especially Maulitz, "Physician versus Bacteriologist."

235. For orientation, see Engelhardt, "Kausalität und Konditionalität."

236. For orientation, see Mendelsohn, "Von der 'Ausrottung' zum Gleichgewicht."

237. Mendelsohn, *Cultures of Bacteriology*, part 2.

238. Rosenkrantz, "Bovine Tuberculosis."

239. Hetsch, "Tuberkulose," 443–46.

240. GStAPK I, HA, Rep, 76 VIII A, Nr. 2937.

241. The first medication to be protected by a patent was antipyrin in 1890. Although this was not the purpose of the law, patent protection gradually began to take the place of secrecy—which was used by Koch—as a means of securing property rights (Wimmer, "*Wir haben fast immer was Neues*," 85–106).

242. GStAPK I, HA, Rep, 76 VIII A, Nr. 4163, accompanying letter from Paul Ehrlich to the Prussian Ministry of Culture, concerning the monetary aspect of the expertise.

243. Koch, "Über neue Tuberkulinpräparate."

244. Koch to Ministry of Culture, 12 Dec. 1898 (GStAPK I, HA, Rep, 76 VIII B, Nr.4163). Cf. Möllers, *Robert Koch*, 375–79; Zeiss and Bieling, *Behring*, 238–50.

245. Koch to Ministry of Culture, 12 Dec. 1898 (GStAPK I, HA, Rep, 76 VIII B, Nr.4163). Cf. Koch's letter to Libbertz of 22 Jan. 1899, in which he relates his objection. (Möllers, *Robert Koch*, 379.)

246. Behring to Ministry of Culture, 30 Dec. 1898; Koch to Ministry of Culture, 12 Dec. 1898 (GStAPK I, HA, Rep, 76 VIII B, Nr.4163).

247. GStAPK I, HA, Rep, 76 VIII B, Nr.4163. Interestingly, the first item in the file is Koch's declaration going back to June 1891. Cf. Zeiss and Bieling, *Behring*, 249–51.

248. Zeiss and Bieling, *Behring*, 1940, 234–36. Modern authors, such as Thomas Brock, have often followed Behring's interpretation: Brock, *Robert Koch*.

249. SBPK Slg. Darmst., 3 b 1882 (2) Koch, Robert. Koch's letter from Bulawayo, 22 Jan. 1904. According to the catalog of the Prussian State Library, the letter was sent to Ruge. Möllers, (*Robert Koch*, 282) who cites a slightly shorter printed version of the letter, assumes it was addressed to Weigert, which seems plausible.

250. On this point I agree with a few authors who also describe the limiting programs of medical-biological research: Amsterdamska, "Medical and Biological Constraints"; Helvoort, "Bacteriological and Physiological Research"; Helvoort, "Bacteriological Paradigm." Both authors are concerned with the history of early virology. Kohler, "Bacterial Physiology," has provided an entry into the history of bacterial physiology.

251. Cf. Hueppe's reflections on the relationship between pathology and bacterial proteins: Hueppe, "Ueber Erforschung der Krankheitsursachen." For two theories on the effect of tuberculin, see Hertwig, *Grundlage der Tuberkulinwirkung*; Köhler and Westphal, "Eine neue Theorie."

CHAPTER IV: Of Men and Mice

1. Koch, "Die Ätiologie der Tuberkulose" ([1884] 1912), 467. For the relation between clinical medicine and bacteriology, a good introduction is Bynum, *Science and the Practice of Medicine*, 118–41; also, Faber, *Nosography*, 95–111.
2. "Über bakteriologische Forschung," 659.
3. Gradmann, "Invisible Enemies."
4. Schlich, "Kontrolle notwendiger Krankheitsursachen."
5. Koch, "Die Ätiologie der Tuberkulose" ([1882] 1912), 444.
6. Virchow, "Der Kampf der Zellen und Bakterien," 8.
7. Korb-Döbeln, *Liederbuch*, 8.
8. See chapter V.2.4. for the public reaction to Koch's cholera expedition. Cf. Hansen, "Images of a New Medicine."
9. L. "Der letzte Arzt."
10. Best known is Rosenbach, *Arzt contra Bakteriologe*. See also Maulitz, "Physician versus Bacteriologist." Cf. Dinges, *Medizinkritische Bewegungen*.
11. Brock, *Robert Koch*, chapter 12.
12. "In the future it will not be difficult to decide what is tubercular and what is not tubercular," Koch wrote in 1882. As a diagnostic marker he advocated "the demonstration of tubercle bacilli, either in the tissue by means of a reaction to dye [Farbenreaktion] or by culturing on congealed blood serum." Koch, "Die Ätiologie der Tuberkulose" ([1882] 1912), 442. Cf. Bochalli, *Tuberkuloseforschung*, 53–65. At the same time, it is unclear to what extent this diagnosis actually took hold. We do know, however, that in lung sanatoria the bacteriological examination of the sputum was standard procedure by the turn of the century (Condrau, *Lungenheilanstalt*, 221; Martin, "Bedeutung und Funktion").
13. Geison, *Pasteur*, 206–32.
14. Weindling, "Serum Therapy for Diptheria."
15. A recent study of the topic is Sauerteig, "Ethische Richtlinien, Patientenrechte."
16. Lenoir, *Politik im Tempel der Wissenschaft*, 107–45. For the history of antibiotics, Wainwright, *Miracle Cure*.

17. By way of introduction: Elkeles, *Das medizinische Menschenexperiment*, 104–13; Sauerteig, "Ethische Richtlinien, Patientenrechte."
18. Elkeles, *Das medizinische Menschenexperiment*, 104–13.
19. Ibid., 105–7.
20. Lederer, *Subjected to Science*, 11–12.
21. Elkeles, *Das medizinische Menschenexperiment*, 112; cf. Lederer, *Subjected to Science*, 3, which also notes such an expansion.
22. Tröhler and Maehle, "Antivivisection."
23. Geison, *Pasteur*, 206–33.
24. Elkeles, *Das medizinische Menschenexperiment*, 180–217.
25. For a view of the contemporary discussion see Tröhler and Maehle, "Antivivisection"; Maehle, "Assault and Battery"; Winau, "Medizin und Menschenversuch."
26. Koch, "Über bakteriologische Forschung," 658.
27. For an introduction to the history of the chemical-pharmaceutical industry and its products, see Bäumler, *Farben, Formeln, Forscher*; Müller-Jahnke and Friedrich, *Geschichte der Arzneimitteltherapie*, 1996; Wimmer, "*Wir haben fast immer was Neues*."
28. Lutz Sauerteig has convincingly done this by using the example of salvarsan therapy: Sauerteig, "Ethische Richtlinien, Patientenrechte."
29. Koch, "Über bakteriologische Forschung," 659.
30. Overviews in Bochalli, *Tuberkuloseforschung*; Smith, *The Retreat of Tuberculosis*; Klebs, *Tuberkulose*.
31. Gaffky and Koch, together with Oscar Fräntzel of the Berlin Charité, had subjected humans to experimental treatments, which have been described above. In their experiment, a total of twenty-seven patients inhaled substances that had proved effective in vitro. Such substances as camphor, creosote, and carbolic acid had been known for some time and proved to be completely ineffective. Cf. Fränzel's report in Leyden and Pfeiffer, *Verhandlungen des Congresses für innere Medicin*, 46–49.
32. RKI/Koch papers, as/w2/003. See chapter III.4.
33. Koch, "Über bakteriologische Forschung," 659.
34. This according to Schadewaldt, "Entdeckung des Tuberkulins," who cites passages from Hedwig Koch's (née Freiberg) diary. Unfortunately, this diary has since been lost. In his first publication about tuberculin, Koch refers to the experiments on himself in this sense. Koch, "Weitere Mitteilungen über ein Heilmittel," 483. Cf. Altman, *Who Goes First?*
35. Koch, "Weitere Mitteilungen über ein Heilmittel," 679.
36. Guttman had assigned some tuberculosis cases to Koch for study purposes. Koch, *Die Ätiologie der Tuberkulose* 1912 [1891] p. 483.
37. Koch, "Weitere Mitteilungen über ein Heilmittel," 662
38. Koch, "Weitere Mitteilung über das Tuberkulin," 679.
39. He started on September 11. "Die Wirksamkeit," 78.
40. Starting on September 11 or 12. Ibid., 78, 91. In October, Koch ("Weitere Mitteilungen über ein Heilmittel," 661) cited other physicians and clinics that had tested the medication, among them the private clinic of Dr. Levy and Bardeleben's department at the Charité. As far as one can establish today, however, treatment in these places was almost always initiated later. In the end these physicians, who had started trying out Koch's medication on October 11 (in the case of Westphal and Köhler) or September 22

(in the case of Levy), became loyal followers of Koch. Köhler and Westphal, "Ueber die Versuche"; Levy, "Methode des Herrn Geheimrath Dr. Koch," 30.

41. "Die Wirksamkeit," 89. The figure of thirteen is taken from Fräntzel's report, 90–101

42. Fräntzel and Runkwitz, "Anwendung des Koch'schen Specificums."

43. Ibid., 19. The government report (Amtliche Berichte) states the patient's full name and also the fact that his treatment was begun on September 13. "Die Wirksamkeit," 93.

44. Fräntzel and Runkwitz, "Anwendung des Koch'schen Specificums," 18. A comparison with Fräntzel's account in the government report shows that at the time of the first publication, all the patients were already deceased ("Die Wirksamkeit," 90–93).

45. Fräntzel and Runkwitz, "Anwendung des Koch'schen Specificums," 26.

46. Levy, "Methode des Herrn Geheimrath Dr. Koch," 33. According to Levy, Koch himself determined the timing and the dosage of the injections.

47. Ibid., 30.

48. Ibid., 32–33.

49. Ibid., 35.

50. Ibid., 36.

51. Cornet, like Levy, was later sharply attacked for his interest in financial gain. A paper he wrote in his defense in 1891 provides an interesting glimpse into the tuberculin euphoria: Cornet, *Rückblick*. For Cornet's career, see Möllers, *Robert Koch*, 384–85.

52. Köhler and Westphal, "Über die Versuche." It is striking that the observations gathered at the university clinic, where tests had also begun very early, were not included in the special issue of the *Deutsche Medizinische Wochenschrift (DMW)*. The director of that clinic, Senator, eventually became one of the first critics of tuberculin. He noted as early as December 1890 that the "intensity of the reaction, as well as its duration and the rapidity of its onset, is not related to the intensity of the spreading of the tubercular process." Senator, "Mittheilungen über das Koch'sche Heilverfahren."

53. For Ehrlich's illness and this traveling, see Dolman, "Ehrlich." Koch's papers contain, close to the protocols of the self-experiments of other associates and colleagues, a note on the treatment of "Ehrl.," which under the circumstances no doubt refers to Ehrlich (RKI/Koch papers, as/w2/009).

54. Koch, "Weitere Mitteilungen über ein Heilmittel." These contributions appeared in issue 47 of November 20. In a letter to Virchow of November 7, 1890, Koch wrote that "for the time being" he intended to provide "only an abbreviated report." (Koch to Virchow, 7 Nov. 1890, Akademiearchiv, NL Virchow). Apparently Virchow had invited Koch to give a lecture, but Koch declined.

55. This was Senator, "Mittheilungen über das Koch'sche Heilverfahren." Senator had tested the remedy since September 11. "Die Wirksamkeit."

56. Koch, "Weitere Mitteilungen über ein Heilmittel," 661.

57. Ibid., 664, 666.

58. It should be noted that tuberculin was usually given diluted 1:10, so that 0.1 cc corresponds to approximately 1 mg.

59. For more details about Koch's theory of the effect of tuberculin, see chapter III.4.

60. Koch, "Weitere Mitteilungen über ein Heilmittel." According to Koch, the lower limit of effectiveness in healthy subjects was 0.01 cc (p. 662). This was also the initial dose

for the treatment of lupus (p. 665). For phthisics, who reacted much more strongly, a 10 times lower initial dose (0.001cc.) was indicated, but it could be rapidly increased.

61. Koch, "Weitere Mitteilungen über ein Heilmittel," 665.

62. Ibid., 668.

63. Ibid., 661. The Prussian Minister of Culture, Gossler, speaking to the Prussian Diet, employed the same reasoning when he assumed responsibility for disclosing the nature of the medication (*Stenographische Berichte*).

64. "Die Wirksamkeit," 78. Cf. also the rather critical statement issued as early as December 1890 by the director of this department: Senator, "Mittheilungen über das Koch'sche Heilverfahren." Observers might also have noted and discussed the conspicuous fact that of the four very ill patients whom Fräntzel injected, three died within a few days ("Die Wirksamkeit," 91–93).

65. Bergmann, "Die mit dem Koch'schen Heilverfahren gewonnenen Ergebnisse," 63.

66. For Bergmann: Genschorek, *Wegbereiter*.

67. Bergmann, "Die mit dem Koch'schen Heilverfahren gewonnenen Ergebnisse," 64.

68. Ibid., 66. Bergmann chose an initial dose of 1 cc. in a solution of 1:100.

69. Ibid., 66–67.

70. Ibid., 81–82.

71. For this topic, Elkeles, "Tuberkulinrausch"; Gorsboth and Wagner, "Die Unmöglichkeit der Therapie"; Gradmann, "Ein Fehlschlag und seine Folgen." Further literature in Gradmann, "Fehlschlag."

72. Geison, *Pasteur*, 145–76; cf. Cooter and Pumfrey, "Separate Spheres and Public Places."

73. See also, for instance, the report on the special session of the Society of Charité physicians in the *DMW* of 20 November (*DMW*, "Robert Koch's Heilmittel," 1890/91, fascicle 1, 94–98), or the case, reported from Heidelberg, of a tuberculosis of the larynx cured within two days (ibid. fascicle 2, 37–38).

74. *Kölnische Zeitung* and *Neue Preussische (Kreuz-) Zeitung*.

75. Cf. Gorsboth and Wagner, "Die Unmöglichkeit der Therapie."

76. In his memoirs Theodor Brugsch recalls the redesignation of a coffee house into a tuberculin clinic. The end of this clinic was equally sudden. Brugsch, *Arzt seit fünf Jahrzehnten*, 47–48.

77. "Der neue Ritter St. Georg."

78. "Robert Koch," *Vossische Zeitung*, Nov. 16 1890.

79. Koch had agreed to this condition: "Die Wirksamkeit," preface.

80. By December the Brehmer Institute, where Georg Cornet worked, already seems to have been using the remedy for three weeks. ("Robert Koch's Heilmittel," *DMW*, 1890/91, fascicle 2, 53). In his memoir on the tuberculin euphoria, Cornet talks about its use in private clinics such as that of William Levy, and about the criticism that was soon leveled at these institutions (Cornet, *Rückblick*). In other countries the euphoria usually took a few more days to set in. For the international reaction to tuberculin, see Brock, *Robert Koch*; Burke, "Postulates and Pecadilloes"; Chauvet, "Cenetaire"; Hansen, "Images of a New Medicine"; Leibowitz, "Scientific Failure."

81. The figure sixty-nine is taken from the bibliography in *Schmidtsche Jahrbücher der gesammten in- und ausländischen Medicin*, volumes 230 and 231.

82. *DMW* 1890–91, fasc. 2, 73–74, "Robert Koch's Heilmittel." (Cited in a report on the meeting of the Hamburg medical society of 2 Dec. 1890.) The "barks" in the text refer to the scabs mentioned by Koch.

83. Grünwald, "Bericht." For other dramatic cures see *DMW*, 1890–91, fasc. 1, 97; fasc. 2, 38.

84. Cornet, *Rückblick*, 3–4.

85. Dehmel, "Ein Dankopfer." This poem gives voice to a mother suffering from tuberculosis who may not kiss her child, fearing that she might infect it.

86. Möllers, *Robert Koch*, 193–200.

87. Alfred Grotjahn, *Erlebtes und Erstrebtes. Erinnerungen eines sozialistischen Arztes*, Berlin, 1938. Cited in Gorsboth and Wagner, "Die Unmöglichkeit der Therapie," 131. According to the official report of the director of the Greifswald medical clinic, Mosler ("Die Wirksamkeit," 502), tuberculin was brought to Greifswald on November 20 through the good offices of Loeffler. It was then tested on fifteen patients in a special session of the local medical society meeting in the auditorium of the medical clinic.

88. Oppenheimer, "Aus dem St. Josephshaus in Heidelberg."

89. Helfrich, "Erfolge," 51.

90. *DMW* 1890–91, fasc. 1, 98, "Robert Koch's Heilmittel."

91. Israel, "Anatomische Befunde."

92. Senator, "Mittheilungen über das Koch'sche Heilverfahren."

93. Köhler, "Mittheilungen über das Koch'sche Heilverfahren," 1127.

94. Hertel, "Mittheilungen über die Einwirkungen des Kochschen Mittels." The author describes a number of cases that initially showed no evidence of tuberculosis where the diagnostic injections led to a finding of tuberculosis of the larynx.

95. Fräntzel, "Bemerkungen."

96. Jürgens et al., "Mittheilung über das Koch'sche Heilverfahren." Cf. in "Die Wirksamkeit," the report of Leyden's clinic, where on pages 15 and 53 the patient's full names and histories can be found.

97. "Die Wirksamkeit," 53.

98. Ibid., 15.

99. Jürgens et al., "Mittheilungen über das Koch'sche Heilverfahren," 128.

100. Henoch, "Mittheilungen über das Koch'sche Heilverfahren," 1169.

101. Ibid.

102. Ibid., 1171.

103. "Die Wirksamkeit," 201.

104. Henoch, "Mittheilungen über das Koch'sche Heilverfahren," 1171.

105. Rosenbach, "Reactionserscheinungen," 16.

106. Hueppe, "Heilung der Tuberculose."

107. Leyden, "Bericht," 1145.

108. Leichtenstern, "Mittheilungen über das Koch'sche Heilverfahren," 41.

109. *DMW* 1890–91, fasc. 4, 35, "Robert Koch's Heilmittel."

110. GStAPK, 1 HA, Rep. 76 VIII A, Nr. 2955.

111. Evidently Libbertz—along with many of his contemporaries—had worked with a 1:10 diluted solution. If he had not done this, 2 milligrams would have been twice the initial dose. This contradicts Libbertz's claim that 2½ to 10 times as much was usually given for diagnostic purposes.

112. *DMW* 1890–91, fasc., 73, "Robert Koch's Heilmittel."

113. Ibid., 60 (Brehmer'sche Anstalten); ibid., 16 ff. (Rosenlaub).

114. Rosenbach, "Reactionserscheinungen," 16.

115. "Die Wirksamkeit," 251. By spring 1891 Bergmann joined the camp of the outspoken critics: Genschorek, *Wegbereiter*, 175. Cf. Elkeles, *Das medizinische Menschenexperiment*, 145.

116. Cf., for instance, the discussion of the Berlin Clinical Society of 21 Jan. 1891, *DMW* 1890–91, fasc. 6, 101, "Robert Koch's Heilmittel."

117. Ibid, fasc. 5, 92, or Gerhard in "Die Wirksamkeit," 68. Several of the accounts mention the discontinuation of the treatment at the patient's request.

118. The publication of the collective reports of the Prussian university clinics, including a statistical analysis, did not appear until early 1891: "Die Wirksamkeit."

119. Czerny, "Erster Bericht über die Koch'schen Impfungen," 97.

120. Ibid., 68. Even though Czerny considered it unlikely that tuberculin would cure pulmonary tuberculosis, he was convinced that it would be effective against other forms of the disease and felt that this remedy represented a major step forward.

121. Fürbringer, "Vierwöchige Koch'sche Behandlung," 97.

122. Deviating reactions are described by such authors as Hahn, *DMW* 1890–91, fasc. 4, 352, "Robert Koch's Heilmittel," and Leichtenstern (Ibid., 39ff). The observation that healthy subjects can also react is found in Bäumler (ibid., 92). Erich Peiper later devised a series of experiments in this area: Peiper, "Über die Wirkung des Koch'schen Mittels."

123. Maulitz, "Physician versus Bacteriologist."

124. Rosenbach, "Körpertemperatur."

125. Ibid., 30.

126. Schwimmer, "Die Behandlung mit Koch'scher Lymphe," 65.

127. Ibid., 66.

128. Virchow, "Über die Wirkung des Koch'schen Mittels."

129. Hansemann, "Erfahrungen über die Koch'sche Injectionsmethode," 80.

130. Lahmann, *Koch und die Kochianer*.

131. Overview in Baumgarten, "Tuberculinwirkung." Koch's son-in-law Pfuhl was able to replicate Koch's experiments to the extent that guinea pigs treated with tuberculin clearly survived longer than untreated ones (Pfuhl, "Behandlung tuberculöser").

132. Schadewaldt, "Entdeckung des Tuberkulins."

133. Koch, "Fortsetzung der Mitteilungen über ein Heilmittel."

134. Hueppe and Scholl, "Ueber die Natur der Koch'schen Lymphe."

135. In autumn 1891 Koch finally issued a more detailed description: Koch, "Weitere Mitteilungen über das Tuberkulin."

136. RKI/Koch papers, as/w2/009. This bundle consists of ninety-seven sheets and obviously refers to cases reported in the published literature.

137. See chapter III.5.

138. GStAPK, I. HA, Rep. 76 VIII B, Nr. 4162.

139. Ibid., evaluation of 30 Dec. 1890. For the drug legislation of the time, Wimmer, "Wir haben fast immer was Neues," chapter 1; cf. Fleischer, *Patentgesetzgebung*.

140. GStAPK, I. HA, Rep. 76 VIII B, Nr. 4162. Shortly thereafter, using the remedy on prison inmates against their will was outlawed: Winau, "Medizin und Menschenversuch," 20–21.

141. "Die Wirksamkeit." The decree is mentioned in the preface.

142. There are corrections that point to February 1891. "Die Wirksamkeit," 651.

143. The number of persons treated in the Department of Internal Medicine alone was 196 ("Die Wirksamkeit," 844). Added to these were 61 in the Department of Surgery. For Koch's participation, see p. 794. Second to the Moabit Hospital were the departments of the Berlin University clinic headed by von Leyden (131) and Bergmann (111).

144. "Die Wirksamkeit," 795.

145. Ibid., 5

146. Ibid., 798

147. Together with the Department of Surgery, which treated 61 patients and performed about 1,000 injections, the Moabit Hospital used "about 175 gr." tuberculin. ("Die Wirksamkeit"). Some of the other physicians whom Koch respected also stood out for their excessive consumption of tuberculin: Fräntzel, with his 37 cases, used at least 20 grams!

148. "Die Wirksamkeit," 808.

149. Sonnenburg, "Aus der chirurgischen Abteilung."

150. Ibid., 109.

151. "Die Wirksamkeit," 904–5.

152. Of these, 50 percent (884) showed no improvement, and the rest were improved (431) or greatly improved (319). "Die Wirksamkeit," 904–5.

153. Leichtenstern, "Mittheilungen über das Koch'sche Heilverfahren," 40.

154. Ibid., 40.

155. "Die Wirksamkeit," 33, reports on the case of a maidservant who as early as November 29, 1890, refused further treatment because of severe pain.

156. This was at Medical Clinic II of the Charité. Its director, Gerhard, reported that he had personally obtained the consent of prospective patients. If the patients later withdrew this consent, the treatment was "immediately discontinued" ("Die Wirksamkeit," 62).

157. In his experiments with nontubercular subjects, Erich Peiper contrasted twenty-one tested subjects whom he identified by name with a certain "R. , cand. Med., 22 years old, healthy." Peiper, "Über die Wirkung des Koch'schen Mittels," 18.

158. Schreiber, "Ueber das Koch'sche Heilverfahren." This was a lecture given on January 19. The continuation of his investigation can be found in the official reports, in which Schreiber added corrections as late as February ("Die Wirksamkeit," 657).

159. Schreiber, "Ueber das Koch'sche Heilverfahren," 18.

160. "Die Wirksamkeit," 657–58.

161. Peiper, "Über die Wirkung des Koch'schen Mittels." The stark contradiction to Schreiber's findings was lost on their contemporaries.

162. "Die Wirksamkeit," 515.

163. Peiper, "Über die Wirkung des Koch'schen Mittels," 30.

164. Paul Fürbringer (Fürbringer, "Vierwöchige Koch'sche Behandlung") established a whole catalog of divergent fever reactions, but this did not bring him to a fundamental critique.

165. Czerny, "Erster Bericht über die Koch'schen Impfungen," 68.

166. Korach, "Ueber die mit dem Koch'schen Heilmittel . . . erzielten Resultate," 95.

167. "Die Wirksamkeit," 6–7. For the history of internal medicine as a discipline, see Eulner, *Spezialfächer*, 180–201.

168. "Die Wirksamkeit," 9.

169. Koch, "Weitere Mitteilungen über ein Heilmittel," 661 (footnotes).
170. "Die Wirksamkeit," 7.
171. This premise is shared by, among others, Elkeles, *Das medizinische Menschenexperiment*; Lederer, *Subjected to Science*; Sauerteig, "Ethische Richtlinien, Patientenrechte."
172. For drug-based therapies, cf. Bochalli, *Tuberkuloseforschung*, 79–80. For an overview of the period after 1900, see Cornet, "Tuberkulose," 21002–52.
173. Thus it was still recommended in a respected handbook as late as 1919. The author pointed out that the remedy had recovered from its discreditation in the early 1890s (Hetsch, "Tuberkulose," 843–46). Cf. Schmidt, "Geschichte der Tuberkulin-Therapie"; Worboys, "Sanatory Treatment for Consumption."
174. Wolff, "Über die Anwendung des Tuberkulins," 156. Cf. Wolff, "Die Reaktion der Lungenkranken."
175. Rosenbach, *Aufgaben und Grenzen der Therapie*, 156.
176. Ibid., 156.
177. Ibid., 157. Italics in original.
178. Cf., for instance, the report of the director of the Medical Clinic II, Gerhard, in the government report ("Die Wirksamkeit," 61–77). Paul Fürbringer and his assistants had carried out a series of examinations at the municipal hospital in Friedrichshain, which he directed. These are described in "Koch's Heilmittel," *DMW* 1890–91.
179. Guttmann and Ehrlich, "Anfangs-Behandlung," 5.
180. Wolff, "Ueber die Anwendung des Tuberkulins," 97. Here the author refers the reader to the work of Biedert, Fürbringer, Ewald, and Bäumler.
181. Lewin, "Die Nebenwirkungen der Arzneimittel," 191.
182. A minor controversy between Virchow and Neisser concerned this very point: Neisser had criticized Virchow for obtaining his samples, which militated against tuberculin therapy, from patients who, because of their poor condition "had not received it for therapeutic reasons at all" ("Robert Koch's Heilmittel," *DMW* 1890–91, fasc. 6, 87). The samples therefore gave no indication about their effect in less severe cases. In his rebuttal Virchow toyed with the inevitable suspicion that these patients had been used for human experiments.: "On the other hand, Herr Virchow feels compelled to state that in every individual case good intentions were no doubt present and rejects the idea that the treating physicians gave injections *experimenti causa* or for extraneous reasons." Ibid., fasc. 7, p. 95.
183. Lichtheim, "Das Koch'sche Heilverfahren."
184. "Robert Koch's Heilmittel," *DMW* 1890–91, fasc. 7, p. 98. Minutes of the meeting of the Berlin Medical Society, 28 Jan. and 4 Feb. 1891.
185. Schultze, "Weitere Mittheilungen." According to the text, this was the publication of a lecture given on February 28. It was printed in the *DMW* of March 23, 1891.
186. Ibid., 16–17.
187. Ibid., 18.
188. Ibid., 20
189. That winter Schultze initially stated that he would treat with tuberculin only on request (Schultze, "Weitere Mittheilungen," 21). Then, at the Congress for Internal Medicine in April, he declared that "since February he had not dared" treat anyone with this medication ("Robert Koch's Heilmittel," *DMW* 1890–91, fasc. 11, p. 51).
190. Baumgarten, "Impftuberculose," 465. Since Baumgarten also succeeded in pro-

voking a general tuberculin reaction in his experimental animals, he had refuted Koch on this point as well. But then he explained the contradiction between his and Koch's findings by the different virulence of the cultures used. Cf. Grawitz, "Versuche mit dem Koch'schen Mittel," 1891.

191. The responsible government expert, Skrezka, proposed on January 27, 1891, to have tuberculin sold in pharmacies only, a rule that was adopted in Prussia effective March 1 (GStAPK, I, HA, Rep. 76 VIII B, Nr. 4162).

192. GStAPK, I, HA, Rep. 92 NL Althoff, A1, Nr. 233a.

193. "Robert Koch's Heilmittel," *DMW* 1890–91, fasc. 11, 22–58.

194. From such practitioners as Ziemssen (pp. 45–46), Sonnenburg (pp. 43–44), Cornet (p. 51)

195. "Robert Koch's Heilmittel," *DMW* 1890–91, fasc. 11, 23.

196. Ibid., 50.

197. "Robert Koch's Heilmittel," *DMW* 1890–91, fasc. 11, 51–52

198. "Robert Koch's Heilmittel," *DMW* 1890–91, vol. 5, fasc. 11, 57–58.

199. Stenographic minutes, 1891, vol. 5, p. 2261 (Delegate Broemel).

200. Ibid., 2258.

201. Ibid.

202. Ibid., 2260.

203. Ibid., 2263.

204. Ibid.

205. Cf. Sauerteig, "Ethische Richtlinien, Patientenrechte." The author finds an increasing consciousness of the problem of human experiments on hospital patients in the 1890s.

206. Siegmund, "Die Stellung des Arztes zur Tuberculinbehandlung," 415.

207. RKI/Koch papers, as/ws/009, italics in original. Patient Nr. 74. Admitted 6 October 1890, deceased 7 May 1891. Cf. The report on Gustav Nitschke in "Die Wirksamkeit," 98–99.

208. Siegmund, "Die Stellung des Arztes zur Tuberculinbehandlung," 420.

209. Ibid.

210. Wimmer, *"Wir haben fast immer was Neues,"* 87. This was done by circumventing a legal prohibition that expressly forbade the patenting of medical drugs and their manufacturing.

211. In this connection, see the remembrances of August Laubenheimer, a chemist at Hoechst, who provided the following interesting reason for Hoechst's becoming involved in the production of tuberculin: "At the time the enthusiasm with which tuberculin was greeted at first had pretty much died down, its sales had fallen to a low level, and so the gentlemen who were in charge of production were amenable to the idea of moving the production to Hoechst... Even though at first tuberculin had little financial success, it did serve to show that the course of an infectious disease can be influenced by specific bact. products." Hoechst-Archiv, Hoechst C/1/2/c, ms. "Zur Geschichte der Serumherstellung in den Farbwerken," signed "Laubenheimer, 1904."

212. Koch and his associates Pfuhl and Libbertz sold the rights to the remedy to the firm for a period of twenty years. The fact that in the contract the company also agreed to take over any remaining quantities is a hint that sales had already declined sharply (Wimmer, *"Wir haben fast immer was Neues,"* 157–58; Gradmann, "Ein Fehlschlag und seine Folgen," 51).

213. Throm, *Diphterieserum*; Weindling, "Serum Therapy for Diphteria"; Winau, "Serumtherapie." Similar consequences are also known for Great Britain: Smith, *The Retreat of Tuberculosis*, 62.

214. I have in my possession a number of instruction sheets for its correct therapeutic use from before World War I.

215. Whitrow, "Wagner-Jauregg"; Whitrow, *Julius Wagner-Jauregg*, 151–59.

216. Smith, *The Retreat of Tuberculosis*, 59–62; Burke, "Postulates and Peccadilloes"; Chauvet, "Cenetaire"; Hansen, "Images of a New Medicine"; Leibowitz, "Scientific Failure."

217. Koch worked on publications to this effect into the twentieth century.

218. Koch's report on tuberculocidin to Minister of Culture Bosse of 27 Nov. 1892. GStaPK, I, HA, Rep. 76 VIII A. Nr. 2956.

219. Klebs, "Die Zusammensetzung des Tuberkulin."

220. Ibid.

221. Ibid.

222. The ditty ran: "May you give us soon to help us / A remedy against the fraud bacillus. Happy New Year 1891."

223. Lewin, "Die Nebenwirkungen der Arzneimittel," 390.

224. Ibid., 391.

225. For the change in the concept of fever in the course of the nineteenth century, see Hess, *Der wohltemperierte Mensch*, 61–67.

226. Moll, *Ärztliche Ethik*, 1. Cf. ibid., 305–6, 366.

227. Ibid. Cf. Elkeles, *Das medizinische Menschenexperiment*, 180–217.

228. Wimmer, *"Wir haben fast immer was Neues,"* 87.

229. Ibid., 74–85.

230. Liebenau, "Paul Ehrlich"; Weindling, "Serum Therapy for Diptheria."

231. Cf. Sauerteig, "Ethische Richtlinien, Patientenrechte," 309–10. The author points out that a certain caution shown by Ehrlich, Behring, and others contrasted sharply with the testing of tuberculin.

232. Ibid., 310–20.

233. Elkeles, *Das medizinische Menschenexperiment*, 180–217.

234. Cited verbatim in Elkeles, *Das medizinische Menschenexperiment*, 209.

235. Ibid., 2010–17; Sauerteig, "Ethische Richtlinien, Patientenrechte," 307–12.

236. Moll, *Ärztliche Ethik*, 504; Cf. Maehle, "Paternalismus und Patientenautonomie."

237. Kolle, "Volksvertretung und Medicin."

238. Sauerteig, "Ethische Richtlinien, Patientenrechte," 317–18.

239. Ibid., 319.

240. See chapter V.4.1.

241. Bäumler, *Paul Ehrlich*, 174–206.

242. Lyons, *The Colonial Disease*.

243. See chapter V.3.

244. For the contemporary state of knowledge, see Scheube, *Die Krankheiten der warmen Länder*. Today untreated sleeping sickness is considered to be 100 percent lethal (Lyons, "African Trypanosomiasis").

245. GStAPK 1, HA, Rep. 76 VIII B, Nr. 4118. The institute's director was trying, apparently in agreement with the ministerial officials, to keep the matter secret. Cf. Gaffky's report of 24 Aug. 1906.

246. Möllers, *Robert Koch*, 317. Gaffky's first report of August 4, 1906, however, mentions mice.

247. Möllers cites, for instance, a letter from Koch to Gaffky (20 Aug. 1906), in which Koch makes detailed suggestions for diagnosis and treatment (Möllers, *Robert Koch*, 314–16).

248. Ibid. Koch accounted for the atoxyl resistance that Schmidt had developed by the fact that the dosage given in Berlin was different from that given in Africa.

249. Möllers, *Robert Koch*, 317, 333 (Koch to Dönitz, 20 Aug. 1906)

250. Ibid., 317. Cf. also Schmidt's personnel file (RKI/Koch papers, PA Schmidt), which indicates that Schmidt retired on March 31, 1949.

251. GStAPK I, HA, Rep. 76 VIII B, Nr. 4118 and 4119. Gaffky's first report is dated 4 Aug. 1906, the last one I was able to consult is of 9 Sept. 1908. The series of reports that Gaffky had started in August 1908 had to be deposited at the ministry under a decree of 2 Jan. 1907. The personnel file is found in RKI/Koch papers, PA Schmidt. It indicates that Schmidt was born in 1874 and had served in the dragoons before coming to the institute. The file also contains several of Gaffky's handwritten reports. In the RKI/Koch papers are the patient's charts, where fever curves, pulse rates, indications, and similar data were recorded.

252. Letter from Gaffky to "Herr Geheimrat," 9 August 1906 (GStaPK I, HA, Rep. 76 VIII B, Nr. 4118) reporting Schmidt's transfer.

253. Handwritten addition to the report of 4 Aug. Signed "Abel, 14.8.1906."

254. Gaffky to Ehrlich, 11 Jan. 1907, Rac, RU 650.3 Eh89 Martha Marquand Collection, folder 52. Cf. Gaffky's letter to the minister of education of 24 Aug. 1906, in which he reports on the deception. "As for the patient himself, we let him continue to believe that he too, like the prosector Bobbermin, is suffering from a recurrent fever infection."

255. Gaffky to the cashier of the Institute for Infectious Diseases, 4 August 1906. RKI/Koch papers, PA Schmidt.

256. "Nor was Privy Councelor Prof. Dr. Ehrlich, whom we consulted, able to propose a more promising treatment. In case Privy Councelor Prof. Dr. Koch should by now have garnered new experiences with trypanosoma in humans in East Africa, I have written to him asking for such information" (Gaffky's report of 4 August ). For the history of atoxyl, see Riethmiller, "Ehrlich, Bertheim, and Atoxyl"; in the therapy of syphilitic skin diseases, this arsenic compound with the nontoxic name was already widely used.

257. "Report of Privy Councilor Prof. Dr. Dönitz, Department Head in the Institute for Infectious Diseases, on the further course of the illness of the laboratory attendant B. Schmidt," undated, submitted to Gaffky, and appended to Gaffky's report of 14 August.

258. Koch to Gaffky, 15 September 1906, Möllers, *Robert Koch*, 315. Gaffky had parts of this letter copied and marked them "confidential" (RKI/Koch papers, as/b2/095).

259. For instance, Dönitz mentions in his report of December 13, by way of a reason for changing the treatment protocol, that "the patient had become rather sensitive to the frequent and painful injections."

260. Koch to Gaffky, 15 September 1906, Möllers, *Robert Koch*, 315. Koch recommended an initial dose of 0.2 grams, to be increased to 0.4 grams (in two doses of 0.2).

261. Wilhelm Dönitz to Paul Ehrlich, 3 March 1907, StA Frankfurt, NI Paul Ehrlich. Cf. Koch to Gaffky, 15 September 1906, Möllers, *Robert Koch*, 314.

262. Dönitz's report of 13 Dec.
263. Wassermann's report of 6 Oct. 1906.
264. Dönitz's report of 13 Dec. Another report of 2 November cannot be found.
265. See chapter V.4.3.
266. Gaffky to Ehrlich, 11 Jan. 1907, Rac, RU 650.3 Eh89, Martha Marquard Collection, Rockefeller Archive Center. Dönitz's report of December 13 speaks of improvement, not of a cure.
267. Koch to Gaffky 15 September 1906, Möllers, *Robert Koch*, 316. The main reason Koch gave was concern for Möllers' health.
268. Koch to Möllers 20 Dec. 1906, Möllers, *Robert Koch*, 323. Cf. Koch to Gaffky, ibid., 322, where Koch informs Gaffky of his consent and mentions that he had already written to Möllers about this matter.
269. Gaffky's report of 9 March 1907. Cf. Ehrlich, "Trypanosomenstudien."
270. Bornemann, "Erblindung nach Atoxylinjektionen."
271. See Ehrlich's lecture delivered on February 13, 1907, in which he reported on his research based on animal experiments: Ehrlich, "Trypanosomenstudien."
272. Dönitz to Ehrlich, 3 March 1907, StA Frankfurt, NI Ehrlich.
273. One gathers from Dönitz's letter that Schilling had sidestepped Gaffky, the Institute's director, and tried to obtain dyes from Ehrlich.
274. Ehrlich to Dönitz, 12 Aug. 1906, BA, R 86, Nr. 2631. In the course of this letter, Ehrlich also talks about the problems of dye therapy with trypan red, such as its unclear dosology.
275. Ehrlich to Dönitz, 23 Febr. 1907, RAC RU650, EH89 Paul Ehrlich copy books, box 24, CB XXI.
276. It appears that for several days previously, the treating physicians had run out of atoxyl and were therefore reduced to using rosaniline alone (Gaffky's report of 9 March 1907).
277. Gaffky's report of 23 April 1907.
278. This was proved by injecting mice with Schmidt's blood (report of 23 April 1907).
279. Gaffky's report of 10 July 1907.
280. Gaffky's report of 16 July 1907.
281. Schmidt had tolerated rosaniline in an olein base very badly, and Ehrlich commented to Dönitz as follows. "What you tell me about the appearance of abdominal pain is interesting; I myself, as I said in my lecture, have not yet had the opportunity to try out Fuchsin in humans." Ehrlich to Dönitz, 25 March, RAC RU650, EH89 Paul Ehrlich Copy books, box 24, CB XXI.
282. The spike of fever occurred on September 8 (Gaffky report of 16 Sept. 1907). The reappearance of the parasites was observed in October (Gaffky's report of 17 Dec. 1907).
283. Koch to Gaffky 20 April 1907, Möllers, *Robert Koch*, 333.
284. Ehrlich, "Trypanosomenstudien." Cf. Koch to Gaffky, 19 May 1907, Möllers, *Robert Koch*, 335.
285. "After his return, His Excellency Koch, at my request, took charge of the treatment of Schmidt" (Gaffky's report of 17 Dec. 1907).
286. Gaffky's report of 20 Febr. 1908.
287. Möllers, *Robert Koch*, 345.

288. Schulz, "Quecksilber," 132.
289. Gaffky's report of 20 Feb. 1908.
290. Scheube, *Die Krankheiten der warmen Länder*. In his presentation of atoxyl treatment, the author expressly mentions the futility of combining it with the use of mercury.
291. At this time Schmidt was again feverish, but according to Gaffky this was related to a bronchial affection. Since, moreover, trypanosoma could no longer be seen, interpreting this occurrence was difficult (Gaffky's report of 4 June 1908).
292. Koch to Möllers 22 May 1908 and 11 July 1908, Möllers, *Robert Koch*, 1950, pp. 147–48.
293. Gaffky's report of 9 Sept. 1908.
294. Lyons, "African Trypanosomiasis," 7.
295. RKI/Koch papers, PA Schmidt. In this case, as in all the following ones, the Institute bore the cost of the stay at the sanatorium.
296. A bibliographical search yielded no results.
297. Möllers, *Robert Koch*, 390. Cf. Kolle, *Robert Koch*, 1959.
298. Cf. Eckart and Cordes, "People Too Wild."

CHAPTER V: Traveling

1. For scientific biography and history of science, cf. Shortland and Yeo, *Telling Lives in Science*; Gradmann, "Helden in weissen Kitteln."
2. SBPG, Slg Darmstadt, 3b 1882 (2) Koch, Robert; Koch to Ruge (?) Bulawayo 22 Jan. 1904. Reprinted in Möllers, *Robert Koch*, 282–83. According to the catalog of the Prussian National Library, the letter was sent to Ruge. According to Möllers, the letter was written to Carl Weigert, which is plausible, given its tone and its content.
3. Koch to Gaffky, 10 Oct. 1903, cited in Möllers, *Robert Koch* 272.
4. For an overview, see Brock, *Robert Koch*, chapter 20. Cf. Jusatz, "Robert Kochs Bedeutung."
5. Heymann, *Robert Koch*, 331–33.
6. This quote comes from Koch's mother (Heymann, *Robert Koch*, 60). Short biographies of the siblings can be found in the appendix to Heymann, ibid., 331–33. Cf. also p. 60, where Heymann reports on Emmy Koch's negative attitude toward renewed emigration plans in 1866.
7. Traveling before that time—for instance the trip from Wollstein to Breslau (see above)—was more educational in nature. Koch's activities as a physician in the war of 1870 should probably be seen in this context as well.
8. The deliberations of the government's cholera commission in 1884 led to an open confrontation between Koch and Pettenkofer: Möllers, *Robert Koch*, 160.
9. Opitz, "Robert Kochs Ansichten." To some extent the call to a chair in pathological anatomy at Leipzig University was also involved here, although Koch declined the appointment. Möllers, *Robert Koch*, 163. For academic chairs in hygiene, see Eulner, *Medizinischen Spezialfächer*, 154–55.
10. For the cholera courses, see Möllers, *Robert Koch*, 161. For the Hygiene Institute, see Eschenhagen, *Hygiene-Institut*. For the courses at the Institute for Infectious Diseases, see D'Orazio, "Scienza Tedesca."
11. Hueppe, *Methoden der Bakterien-Forschung*.

12. One of them was Alexandre Yersin, the future discoverer of the plague bacillus, *Yersinia pestis*. Mendelsohn, *Cultures of Bacteriology*, 280–85; Métraux, "Reaching the Invisible."

13. For instance Flügge, *Grundriss der Hygiene*; Hueppe, *Methoden der Bakterien-Forschung*.

14. An exemplary understanding of the process of dissemination is shown in D'Orazio, "Scienza Tedesca."

15. An overview of the development of experimental hygiene is found in Weindling, *Epidemics and Genocide*, 3–72.

16. Engelhardt, "Kausalität und Konditionalität."

17. SBPK Slg. Darmstadt, 3 b 1882, Koch Robert. Letter by Koch from Bulawayo, 22 Jan. 1904. Möllers, *Robert Koch*, 161.

18. This remark seems to have been made at the (belated) celebration of Koch's sixtieth birthday on July 27, 1904. *Ehrung Robert Kochs, 1904*, 1247–8, reprinted in Möllers, *Robert Koch*, 289.

19. See, for instance, the mordant commentaries about Koch and his studies in tropical medicine in the correspondence between Patrick Manson and Ronald Ross in Bynum and Overy, *Beast in the Mosquito*.

20. Dinges, *Medizinkritische Bewegungen*.

21. Hubenstorf, "Die Genese der Sozialen Medizin."

22. Zeiss and Bieling, *Behring*.

23. Eckart, "Von der Idee eines 'Reichsinstituts.'"

24. It would not be difficult to make an impressive list of former associates of Koch's who were not involved in any conflict with him. First and foremost, this was the case for Georg Gaffky. See Wolf, *Georg Gaffky*.

25. Koch to Pfeiffer, 23 Aug. 1899, cited in Möllers, *Robert Koch*, 242.

26. Koch to Dönitz 15 Jan. 1899, RKI/Koch papers as/b2/141. Reprinted in Möllers, *Robert Koch*, 280–81.

27. Möllers, *Robert Koch*, 280–91.

28. Mendelsohn, *Cultures of Bacteriology*, chapter 7.

29. Cranefield, *Science and Empire*.

30. For Koch's malaria studies, see Eckart, "Malaria and Colonialism."

31. Koch to Gaffky, 10 Oct. 1903, cited in Möllers, *Robert Koch*, 272. The remark referred to Koch's work on east coast fever.

32. Introductory reading on tropical hygiene: Arnold, *Warm Climates*.

33. Koch in a letter to Wilhelm Kolle of 9 July. 1898 (Kolle, *Robert Koch, Briefe an Wilhelm Kolle*, 41).

34. Koch to Flügge, 1 Dec. 1896, Koch -letters, Baltimore. Reprinted in Möllers, *Robert Koch*, 228.

35. Koch to Libbertz, 20 June 1903, cited in Möllers, *Robert Koch*, 253. Koch probably suffered from malaria himself.

36. Koch to Gaffky, 10 Oct. 1903, cited in Möllers, *Robert Koch*, 272. Koch was in Bulawayo in Rhodesia at the time.

37. Koch to Libbertz, 20 June 1900, cited in Heymann, *Robert Koch*, part 2, 100.

38. Heymann, *Robert Koch*, part 2, 100.

39. Koch to Wilhelm Kolle, 3 Aug. 1907, SPBK slg. Darmstadt, 3 b 1882, Koch, Robert.

40. Koch to Ehrlich, 24 Nov. 1907, Stadtarchiv Frankfurt am Main, NL Paul Ehrlich.

41. Koch to Flügge, 12 Dec. 1907, Koch correspondence. Reprinted in Möllers, *Robert Koch*, 344.

42. A sweeping assessment is not possible in any case. Koch's techniques for controlling typhus and the epidemiological research connected with it were highly successful in the eyes of the contemporaries (see Mendelsohn, *Cultures of Bacteriology*, chapter 9). Yet quite a few of them considered Koch's malaria studies insignificant, as was shown in the critiques of some British experts in tropical hygiene. For Patrick Manson's criticism, Bynum and Overy, *Beast in the Mosquito*.

43. By way of introduction, see Cunningham and Williams, *Laboratory Revolution in Medicine*; Rheinberger and Hagner, *Die Experimentalisierung des Lebens*.

44. Using the example of Louis Pasteur's microbiology, Bruno Latour has studied the special dynamic between the laboratory and the world at large: Latour, "Give Me a Laboratory." This differentiation is also valid if one keeps in mind that recent historiography uses such concepts as the experimental system to define the laboratory as an entity beyond its simple spatial dimensions. Rheinberger and Hagner, *Die Experimentalisierung des Lebens*; Rheinberger, Hagner, and Wahrig-Schmidt, *Räume des Wissens*; for the theoretical background: Rheinberger, *Experimentalsysteme und epistemische Dinge*, 18–34. It should also be mentioned that the relationship between experimental medicine and urban culture has at last aroused some interest: Dierig, "Feinere Messungen"; Sarasin and Tanner, eds. *Physiologie und industrielle Gesellschaft*. A fundamental study is Warner, *The History of Science*; a recent case study is Hess, *Der wohltemperierte Mensch*.

45. The example of control measures against sleeping sickness in Togo is examined by Eckart and Cordes, "People Too Wild."

46. More on this in Rupke, *Medical Geography*.

47. Large numbers of such studies can be found in Koch, *Gesammelte Werke*, vol. 2.1.

48. Helen Power has documented this in her history of the Liverpool School of Tropical Medicine: Power, *Tropical Medicine*; cf. Eckart, "Malaria and Colonialism."

49. The efforts to combat sleeping sickness in the Congo can serve as an example here: Lyons, *The Colonial Disease*. Cf. Eckart and Cordes, "People Too Wild."

50. As an introduction, see Fisch, "Forschungsreisen." For an introduction to the genre of travel literature, see Brenner, "Der Reisebericht." The role of travel in the history of medicine and science is touched on in Bourguet, Licoppe, and Sibum, *Instruments, Travel, and Science*; and Revill, *Pathologies of Travel*.

51. For the public resonance of Humboldt as a traveler, Daum, *Wissenschaftspopularisierung*, 269–79. Cf. Brenner, "Humboldts Reiserwerk."

52. For an introduction to the history of the scientific travel report, see Fisch, "Forschungsreisen." Cf. Daum, *Wissenschftspopularisierung*, 329–31.

53. Cf. the impressive study *Tropenfieber*, by Johannes Fabian.

54. For scientific work while traveling, see Cittadino, *Nature as Laboratory*; Fisch, "Forschungsreisen"; Kirchberger, "Deutsche Naturwissenschaftler"; Mackenzie, *Imperialism and the Natural World*; Miller and Reill, *Visions of Empire*; Osterhammel, "Forschungsreise und Kolonialprogramm." Cf. also the 1998 meeting of the History of Science Society dedicated to the theme "Science and Travel." (Eckart, "Wissenschaft und Reisen.")

55. Fisch, "Forschungsreisen," 396.

56. For an overview of science and colonialisms, see Arnold, *Warm Climates*. Cf. also the history of the Liverpool School of Tropical Hygiene, which features a great deal of travel: Power, *Tropical Medicine*.

57. Cf. Evans, *Death in Hamburg*. Calling cholera "the Great Scourge" is the invention of Thomas Nipperday.

58. An overview: Worboys, "Comparative History of Sleeping Sickness."

59. Schlich, "Symbol medizinischer Fortschrittshoffnung."

60. For the history of the IHO, see *Das Reichsgesundheitsamt*; cf. Lundgreen et al., *Staatliche Forschung*; Ritter, *Grossforschung und Staat*.

61. Cited in Heymann, *Robert Koch*, part 2, 52.

62. Pfuhl, "Privatbriefe von Robert Koch" (cited in Heymann, *Robert Koch*, part 2, 53).

63. Heymann, *Robert Koch*, part 2, 40.

64. Brock, *Robert Koch*, chapter 16; Carter, "Koch-Pasteur Dispute." An interesting—although partisan—documentation of the controversy is found in BA, R 15.01 RMDI, No. 11130.

65. For an introduction to the history of cholera, see Evans, *Death in Hamburg*. For a cultural history of cholera in Germany, see Briese, *Angst in den Zeiten der Cholera*; Cf. the work of Bourdelais and Dodin, *Visages du Cholera*; also Briese, "Defensive, Offensive, Strassenkampf"; Goltz, "Eine fatale Geschichte."

66. According to a letter of August 4, 1883 by the director of the IHO, Struck, the cholera outbreak offered a welcome opportunity to bring the work of the IHO to the public's attention. "After the experience with other infectious diseases gained over the last decade, there is every reason to believe that the causes and the manner of spreading of cholera will be sufficiently known, thanks to the new and extremely reliable methods and improved tools now available, to implement more effective preventive measures than were hitherto possible." Struck to RMDI, 4 August 1883, BA, R 15.01.Nr. 11334.

67. For more on the history of cholera, see Watts, *Epidemics and History*, 167–212. For Germany: Evans, *Death in Hamburg*. Cf. Bourdelais and Dodin, *Visages du cholera*; also Briese, "Defensive, Offensive, Strassenkampf"; Goltz, "Eine fatale Geschichte."

68. Jahn, *Cholera*.

69. Heymann, *Robert Koch*, 58. See chapter I.2.2.

70. Koch, "Erste Konferenz zur Erörterung der Cholerafrage," 23. To be sure, Koch's remark must be seen in its context. Made at the cholera meeting of 1884, it also served to contest the findings of older research.

71. See chapter II.2.2.

72. Heymann, *Robert Koch*, 68. The Italian physician Pacini had probably identified the pathogen for *Cholera asiatica* as early as 1854 (Franceschini, "Pacini," 1974). The pathologist Carl Thiersch had conducted successful feeding experiments with cholera materials. But then he did not interpret the success of his feeding experiments as the demonstration of an etiology but rather as the effect of cholera toxin. Bulloch, *History of Bacteriology*, 131.

73. Hallier, *Das Cholera-Contagium*. See chapter II.2.2.

74. Koch, *Gesammelte Werke*, vol. 2.2, 849–50.

75. The members of the four-man expedition were the pathologists Isidore Strauss and Edmond Nocari and two of Pasteur's associates, Emile Roux and Louis Thuillier. (Coleman, "Koch's Comma Bacillus," 118.)

76. The plan for the expedition (dated 4 Aug. 1883): BA, R 15.01, No. 11334. A few days later, an official request was submitted by Crown Prince Friedrich Wilhelm. Cf. Heymann, *Robert Koch*, part 2, 55.
77. Gaffky, *Bericht über die Thätigkeit*, 2.
78. A complete list of the equipment is found in Gaffky, *Bericht über die Thätigkeit*. As Heymann noted (*Robert Koch*, part 2, 56), the scientists took along all manner of tools, but not a single scientific book!
79. Heymann, *Robert Koch*, part 2, 59.
80. The German scientists received help for this problem from the city's Greek hospital and two physicians, Kartulis and Zankeröl, who worked there. They not only provided them with materials for study but also allowed them to use their premises (Koch's first report from Alexandria, *Gesammelte Werke*, vol. 2.2, 851).
81. Koch, "Erforschung der Cholera," 2.
82. The team of the French expedition was looking for the bacillus in the blood and was unable to obtain unambiguous results (Coleman, "Koch's Comma Bacillus," 323).
83. Koch, "Erforschung der Cholera," 3.
84. Ibid.
85. Gaffky, *Bericht über die Thätigkeit*, 6.
86. Koch, "Erforschung der Cholera," 5.
87. Ibid.
88. Koch to Hugo, 11 Nov. 1883, cited in Heymann, *Robert Koch*, part 2, 72.
89. Biewend, "Familienchronik von Robert Koch," 1891–92. Cited here from Heymann, *Robert Koch*, part 2, 70.
90. Koch to his daughter, Gertrud, 10 Nov. 1883, cited here from Heymann, *Robert Koch*, part 2, 70.
91. Koch to Hugo, 11 Nov. 1883, cited in Heymann, *Robert Koch*, part 2, 73.
92. Coleman, "Koch's Comma Bacillus," 323.
93. Ibid., 318. A British team was also on its way.
94. "Politische Tagesübersicht. Die deutsche Cholera-Kommission."
95. The reports are stored in BA, R 15.05 RMDI, Mo. 11334/5.
96. For the genre of the scientific travel report in the nineteenth century, see Brenner, "Reisebericht," 443–90.
97. Koch, "Erforschung der Cholera," 5.
98. Koch, *Gesammelte Werke*, vol. 2.2., 850–51. The original can be found in BA, R, 15.01 RMDI, No. 11334/5.
99. Heymann, *Robert Koch*, part 2, 74–75. The reasons for the long voyage were several interruptions in Ceylon and Madras, as well as bad weather during the crossing from Ceylon to India.
100. For the endemic cholera around Calcutta, see Gaffky, *Bericht über die Thätigkeit*, 183–220. For Cunningham, see Coleman, "Koch's Comma Bacillus," 326–27. For cholera in India, see Harrison, "Cholera in British India," 1996.
101. In a letter to his wife on December 24 Koch reported that at noon it was "relatively warm," and that in the evening temperatures were between 14 and 16 degrees C (Heymann, *Robert Koch*, part 2, 77).
102. Koch, "Erforschung der Cholera," 10–11.
103. Ibid., 12. Cf. Heymann, *Robert Koch*, part 2, 76.
104. Koch, "Erforschung der Cholera," 12.

105. Ibid., 13.
106. Ibid.
107. Ibid.
108. Coleman, "Koch's Comma Bacillus," 337; cf. Ogawa, "Uneasy Bedfellows."
109. Coleman, "Koch's Comma Bacillus," 334.
110. Koch, "Erforschung der Cholera," 15. This statement is also interesting because the problem of the asymptomatic carriers of infectious disease, which Koch was to encounter in Hamburg in 1892–93, is not broached at all. In the early 1880s Koch adhered to a simple model of bacterial invasions, in which there is indeed no room for this problem (Mendelsohn, *Cultures of Bacteriology*, chapter 7).
111. See chapter III.3.
112. Koch, "Erforschung der Cholera," 16. Koch was subsequently able to correct this: in the context of the Hamburg cholera epidemic he observed for the first time that the presence of pathogenic germs in the human organism does not necessarily have to lead to an infection (Mendelsohn, *Cultures of Bacteriology*, chapter 7).
113. Koch, "Erforschung der Cholera," 16. This argument is not conclusive, even by Koch's own standards. After all, in the case of tuberculosis, he had succeeded in infecting guinea pigs with the disease, even though he stated that these animals never suffered from tuberculosis outside the laboratory. In that context he realized that only animal experiments with pure cultures could truly prove a causal connection between the bacterium and the disease.
114. Koch, "Die Ätiologie der Tuberkulose" ([1884] 1912), 469–70. Cf. Carter, "Koch's Postulates," 360–62.
115. Koch, "Erforschung der Cholera," 16.
116. Ibid., 19.
117. Koch, "Erforschung der Cholera," 18–19.
118. Koch, "Erforschung der Cholera," 19.
119. Heymann, *Robert Koch*, part 2, 82.
120. If they had wished, they could have stayed longer. The Interior Ministry had expressly left this decision up to Koch. But in a telegram received in Berlin on March 11, 1884, Koch asked for permission to return (BA, R 15.01, RMDI No. 11335).
121. Heymann, *Robert Koch*, part 2, 77.
122. Ibid., 77, 78.
123. Ibid., 74.
124. Ibid., 82.
125. Ibid., 78.
126. Ibid.
127. Ibid., 82–83.
128. Ibid., 77. Koch's remark refers to the servants at the hotel where he lived.
129. Ibid., 78.
130. Gaffky, *Bericht über die Thätigkeit*, 265–71.
131. "Willkommen, Ihr Sieger!"
132. Both quotes from "Willkommen, Ihr Sieger!"
133. For the topos of German science, Kollenbrock-Netz, "Wissenschaft als nationaler Mythos."
134. "Willkommen, Ihr Sieger!"
135. L. G. "Das Aerzte-Bankett."

136. BA, R 1501 RMDI No. 11335. Draft memorandum Dr. Köhler. As indicated by the different initials on the document, it was written no later than April 27, 1884.
137. Möllers, *Robert Koch*, 152–56.
138. Möllers, *Robert Koch*, 152–54, 57.
139. Quoted from Möllers, *Robert Koch*, 149.
140. Coleman, "Koch's Comma Bacillus," 667. Cf. Ogawa, "Uneasy Bedfellows."
141. A summary of Pettenkofer's cholera theory is presented in Jahn, *Cholera*, 37–47.
142. If one believes K.B. Lehmann's "Frohe Lebensarbeit"—as Heymann does—Koch even told Pettenkofer that he had conducted successful animal experiments, even though this was not the case until summer 1884. None of the reports from the expedition mention successful experiments, and Koch publicly reported on such work only at the cholera conference in July 1884. On this occasion he referred to the studies of Nicati and Riehl, which he had learned of only in the summer of 1884 in Toulon (Koch, "Erste Konferenz zur Erörterung der Cholerafrage," 36–37).
143. Pettenkofer, "Die Entdeckung des Cholerapilzes," *Münchener Neueste Nachrichten*, March 1884 (BA, R 15 01 RMDI no. 11335).
144. The inquiry of the editors of the *Reichsanzeiger* reached the Ministry of the Interior on March 22, 1884. After consulting the Imperial Health Office, the ministry advised the editors not to publish the piece (BA, R 15 01 RMDI no. 11335).
145. F. "Der Entdecker des Cholerapilzes."
146. Valerius, "Kommabacillus." Cf. Schlich, "Repräsentationen von Krankheitserregern."
147. Valerius, "Kommabacillus." Cf. Koch, "Erste Konferenz zur Erörterung der Cholerafrage," 36–37, from which the citation is taken. The only difference is that Koch's text speaks not of "malignant" (bösartigen) but of "pernicious" (perniziösen) fevers.
148. Fleck, *Genesis and Development*, 148–50.
149. Schlich, "Repräsentationen von Krankheitserregern."
150. Valerius, "Kommabacillus."
151. Valerius, "Cholera-Gefahr."
152. F. "Der Entdecker des Cholerapilzes."
153. Referring the reader to the parallel cases of anthrax and malaria, the article pointed out that such an etiology had already been demonstrated.
154. For an overview, see Ritter, *Grossforschung und Staat*.
155. See the list of items furnished to the expedition in Gaffky, "Bericht über die Thätigkeit," appendix 1, and also the cost of the expedition in appendix 8.
156. Decades later, in East Africa, Koch also had to make do without autopsies, since these would have been seen as cannibalism by the local population (Koch, "Erforschung der Cholera," 18–19).
157. Subsequent travels, like the trip to Toulon in the summer of 1884 or the one to Hamburg in 1892, were essentially connected with epidemiological research and the control of epidemics. For microbiological work with the cholera pathogen, it was no longer necessary to leave Berlin.
158. Koch, "Über die Cholerabakterien."
159. Descriptions of these experiments to be found in Koch, "Über die Cholerabakterien," 64, as well as in Koch, "Erforschung der Cholera," 76–81. Cf. Colemann, "Koch's Comma Bacillus," 336. The problem was the need to perform a multiplicity of manipulations on the experimental animals. This involved, among other things, simultaneously

injecting a caustic soda solution into the stomach and partially paralyzing the intestinal muscles by means of a tincture of opium. Other manipulations might consist of devising an artificial intestinal opening, lowering the acidity level in the stomach, and so forth. The result was the proliferation of cholera bacilli in the intestine, not necessarily additional symptoms of cholera.

160. Möllers, *Robert Koch*, 152–54, 157.

161. Virchow, "Der Kampf der Zellen und Bakterien," 8.

162. Koch to Flügge, 18 Dec. 1884, Koch letters/Baltimore. Koch had to serve as acting director of the IHO in the interim between the two directors Struck and Köhler.

163. Brock, *Robert Koch*, chapters 14 and 15.

164. Introductory reading for the expedition is Beck, "Medicine in German East Africa"; Eckart, *Medizin und Kolonialimperialismus*, 340–49; Heymann, *Robert Koch*, part 2, 103–81; Möllers, *Robert Koch*, 304–6; Münch and Biel, "Expedition, Experiment und Expertise." Cf. also, for Koch's protozoa research, Kühn, *Robert Kochs Bedeutung*.

165. Documents from the IHO indicate that in early 1903, both the Berlin hygienists and the colonial administrations in Africa approached the Foreign Office with proposals in this direction. BA, R 86 KGA, No 2622.

166. Worboys, "Comparative History of Sleeping Sickness"; Lyons, *The Colonial Disease*, 64–75. Subsequently, sleeping sickness not only became the object of large-scale campaigns to control it but also gave rise to international conferences on ways to combat it. Unlike malaria, this disease affected only Africa. Its kinship to the South American Chagas' disease was not yet known at the time. See Perleth, "The Discovery of Chagas' Disease."

167. Lyons, *The Colonial Disease*, 76–101.

168. BA, R 86 KGA, No 2613: Minutes of the meeting of 1 April 1905. Added to this were several reports from physicians who, like Koch, were in East Africa for other reasons. BA, R 86 KGA, No 2622: copy of a letter to the Foreign Office from the government office for East Africa, which refers to consultations with Koch at Dar-es-Salaam on the topic of sleeping sickness. (Cf. Koch's report of 30 July 1904, GStAPK, I. HA. Rep. 76 VIII B, No. 4117).

169. As an overview, see Gründer, *Geschichte der deutschen Kolonien*.

170. See the letter of June 17, 1903, from the Foreign Office to the Interior Ministry, expressing the negative stance of the Foreign Office ( BA, R 86 KGA, No 2622) and the appended letter of the Foreign Office to the Health Office, which indicates that the Prussian Ministry of Culture had supported the project.

171. See the draft letter of June 16, 1903, from the director of the IHO, Köhler, to the Interior Ministry (BA, R 86 KGA, No 2613), which expresses this attitude. See also the marginal notes dated August 31and September 7, 1908, made in preparation for formulating a new opinion, as requested by the Interior Ministry on August 30, 1904. Both documents explicitly refer to the hesitant attitude of the previous year.

172. Lyons, *The Colonial Disease*, 73–74.

173. Copy of letter Foreign Office to Interior Ministry, 17 June 1903 in BA, R 86 KGA, No. 2622.

174. Response of the IHO (signed "Paul" ) to a situation report of Koch's concerning sleeping sickness, dated 30 July 1904 . BA, R 86 KGA, No. 2613.

175. Cf. Report of the imperial administration for East Africa to the Foreign Office

of 16 Febr. 1903 in BA, R 86 KGA, No. 2622. Koch had spent time in East Africa in 1903.

176. Lyons, *The Colonial Disease*, 73–74. For the priority contest between Bruce and Castellani, see Boyd, "Sleeping Sickness."

177. Copy of Foreign Office to Prussian Ministry of Culture, 30 Nov. 1904 ( BA, R 86 KGA, No 2613). The rejection was supported by a letter from the Foreign Office to KUMI (the Ministry of Culture) included in the letter to the Ministry of Culture. (GStAPK, I, HA, Rep. 76 VIII B, No. 4117).

178. GStAPK, I, HA, Rep. 76 VIII B, No. 4117. However, the Foreign Office did hold out the prospect of a financial subsidy and in that eventuality recommended Friedrich Fülleborn as the participating physician.

179. Koch to Prussian Ministry of Culture, undated. The letter is a reply to an inquiry of 5 April 1902 and is stamped "received, 15 April." (GStAPK, I, HA, Rep. 76 VIII B, No. 4117). Cf. Koch, *Gesammelte Werke*, vol. 2.2, for an almost complete copy of the letter, which is dated 12 April 1904.

180. Dönitz to the Prussian Ministry of Culture, 29 May 1903 (GStAPK, I, HA, Rep. 76 VIII B, No. 4117). This letter was Dönitz's reply to an inquiry of the Ministry of 9 May of that year.

181. To be found in GStAPK, I, HA, Rep. 76 VIII B, No. 4117, these reports were also published in the *Deutsche Kolonialzeitung*.

182. Koch, "Über die Trypanosomenkrankheiten."

183. Koch's letter to the Ministry of Culture, "Re: Sleeping sickness," of 30 July 1904 (GStAPK, I, HA, Rep. 76 VIII B, No. 4117). This report indicates that the necessary papers had been sent to Koch at Dar-es-Salaam. He submitted his report after his return to Germany in the summer of 1904. Cf. a similar report of the same date in Koch, *Gesammelte Werke*, vol. 2.2, 926–28.

184. Koch's letter to the Ministry of Culture, "Re: Sleeping sickness," of 30 July 1904 (GStAPK, I, HA, Rep. 76 VIII B, No. 4117).

185. "There can be no doubt that the fly was brought in by boat from the north shore near Ikuru Island. In view of the facts reported here I have come to the conviction that the epidemic is spreading." Koch's letter to the Ministry of Culture, "Re: Sleeping sickness," of 30 July 1904 (GStAPK, I, HA, Rep. 76 VIII B, No. 4117).

186. Koch's letter to the Ministry of Culture, "Re: Sleeping sickness," of 30 July 1904 (GStAPK, I, HA, Rep. 76 VIII B, No. 4117): "I would therefore once again urgently recommend sending a German commission as soon as possible." On September 23, 1904, Koch repeated this recommendation in a letter to the Prussian Ministry of Culture (Ibid.).

187. BA, R 86 KGA, No. 2613, inquiries from Ministry of Culture and Imperial Health Office. The questions make it clear that no one was thinking of Koch as the leader of such a commission. However, he had agreed to support the project.

188. "Notes on the Meeting held on 7 April at the Ministry of the Interior, Concerning the Sending of a Scientific Expedition for the Study of Sleeping Sickness" (BA, R 86 KGA, No. 2613) . Present were representatives of the Interior Ministry, the Foreign Office, the Colonial Department, the Imperial Navy, the Treasury Department, and the Imperial Health Office. According to the minutes, the representative of the Ministry of Culture was unable to attend.

189. The report mentioned a small focus of contagion for Togo, a few cases for Cameroon, and steady though not yet extensive spreading for Lake Victoria.

190. The only protozoologist present was Schaudinn. The plan also called for temporary stays in British Uganda, since it was expected that study materials would still be scarce in the German territory.

191. The planning process can be reconstructed, for instance, from a draft letter of the Health Office to the Interior Ministry of January 15, 1906 ("Re: Directorship of the prospective expedition for the study of sleeping sickness," BA, R 86 KGA, No.2613). The letter is not signed, but internal evidence indicates that this was a draft proposed to the director of the Health Office, Bumm, and corrected by him. Kleine mentions it in his memoirs (Kleine, *Ein deutscher Tropenarzt*, 394).

192. Neufeld had accompanied Koch on the second voyage to South Africa, when he was working on the east coast fever (Cranefield, *Science and Empire*; Möllers, *Robert Koch*, 394).

193. With Ehrlich's *Trypanrot* and Wendelstadt's *Malachitgrün*. Atoxyl, which he used later, was not yet mentioned here.

194. "However, a successful and comprehensive study of sleeping sickness appears to be possible only if other diseases caused by similar pathogens and transmitted in similar ways are included." BA, R 86 KGA, No. 2616, notes for the memorandum on sleeping sickness, 21 June 1905, typescript, 8 pp.

195. "The study of pathogenic protozoa has found . . . increased interest over the last few years . . . It has been found to be highly likely that a certain kind of spirochete is not, as was assumed until now, a bacterium but rather a protozoon and that it represents a stage in the development of other forms known as halteridia or trypanosoma" (BA, R 86 KGA, No. 2613, notes for the memorandum on sleeping sickness of 21 June 1905, typescript, 8 pp.).

196. Both Köhler (to Gaffky, 12 Sept. 1905, copy in BA, R 15 RMDI No. 9307) and Bumm ("Re: Directorship of the prospective expedition for the study of sleeping sickness," BA, R 86 KGA, No.2613, 15 Jan. 1906) emphasized that Koch changed his mind on this matter toward the end of the year.

197. Koch, "Über die Unterscheidung von Trypanosomenarten."

198. For Koch's stay in Amani, see Möllers, *Robert Koch*, 295–305. For Amani, Zepernik, "Zwischen Wirtschaft und Wissenschaft," 211–13; Bald and Bald, *Amani*.

199. Cf. Koch, "Ergebnisse einer Forschungsexpedition nach Ostafrika." The director of the research station, Stuhlmann, was also working on trypanosoma: Stuhlmann, "Beiträge zur Kenntnis der Tsetse-Fliege." Koch had encouraged him to publish these studies.

200. Koch to Gaffky, 25 June 1905, cited in Möllers, *Robert Koch*, 299.

201. Koch to Kolle, 1 Sept. 1905, cited in Möllers, *Robert Koch*, 303. In the same letter he criticized Schaudinn's thesis of the connection between spirochetes and trypanosoma.

202. Koch to Gaffky, 31 Aug. 1905, in Möllers, *Robert Koch*, 303

203. Koch to Kolle, 1 Sept. 1905, in Möllers, *Robert Koch*, 303: "I consider Schaudinn's claim that spirochetes are a stage in the development of trypanosoma erroneous, at least in the general formulation in which he presents his theory."

204. At the founding of the Berlin institute, Koch had tried in vain to obtain a more important role for himself. See Eckart, "Von der Idee eines 'Reichsinstituts.'"

205. Möllers, in *Robert Koch*, cites a letter from Metchnikoff indicating that his opinion in this matter had been solicited. According to the Archives of the Nobel Foundation in Stockholm, Schaudinn was never proposed for the Nobel Prize in Physiology and Medicine of 1905. Blanchard and Calmette only proposed him in 1906. Koch by contrast was nominated by colleagues again and again from the time the prize was established.

206. "Re: Directorship of the prospective expedition for the study of sleeping sickness," BA, R 86 KGA, No.2613, 15 Jan. 1906. The draft is marked "Urgent" and "Secret." Cf. Koch's letter to Libbertz of 22 Dec. 1905, in which he reports a conversation with Bumm.

207. Re: Expedition for the study of sleeping Sickness . . . Berlin, 20 Febr. 1906. Typescript, 5 pp. (BA, R86 KGA, No. 2613).

208. Appended to the report on the expedition (Koch, "Erforschung der Schlafkrankheit," 640–45), is the revised version.

209. Koch later (after the completion of the expedition) also distanced himself from this version of the goals, in which research and disease control had equal weight. In the official report he declared that disease control had been the sole goal and objective of the expedition: Koch, "Erforschung der Schlafkrankheit," 594. Ehrlich had supplied him with the necessary chemotherapeutica: "Notes on the meeting . . . of 14 February 1906 . . . ," GStAPK, I, HA, Rep. 76 VIII B, No. 4118. In the same vein Koch declared before the Imperial Medical Board (Reichsgesundheitsrat) that he had "focused his attention on finding effective ways and means of combating the epidemic." Minutes of the session of the Imperial Board of Health on 18 Nov. 1907, GStAPK, I, HA, Rep. 76 VIII B, No. 4118.

210. The institute in the form it was to keep had been founded against Koch's will (Eckart, "Von der Idee eines 'Reichsinstituts'").

211. Copy marked "Secret" of a letter from Köhler to Gaffky, 12 Oct. 1905, BA, R 1501 RMDI, No. 9703. Köhler suggested equipping another expedition for Schaudinn, an idea that was resuscitated by his successor Bumm in the memorandum cited above.

212. With good reason, considering that Beck later reported privately to the director of the IHO, Bumm (Münch and Biel, "Expedition, Experiment und Expertise," 20. Cf. also Beck's report of 6 May 1906 in BA, R 86, No. 2613). The colonial administration added two members to the expedition: the military physician Panse and the medical sergeant Sacher.

213. Heymann, *Robert Koch*, part 2, 110. Cf. Koch, "Über den bisherigen Verlauf der deutschen Expedition," 510.

214. Koch to Libbertz, 22 Dec. 1905, cited in Möllers, *Robert Koch*, 307.

215. Koch to Gaffky, 22 May 1906, cited in Möllers, *Robert Koch*, 311.

216. Heymann, *Robert Koch*, part 2, 110.

217. Koch to Bumm, 7 May 1906, BA, R 86, No. 2613.

218. Koch to Gaffky, 22 May 1906, cited in Möllers, *Robert Koch*, 311. At this point Libbertz left the expedition and returned to Germany.

219. Koch to Gaffky, 30 July 1906, cited in Möllers, *Robert Koch*, 312.

220. Koch, "Über den bisherigen Verlauf der deutschen Expedition," 512–13. The patient died and was autopsied.

221. Koch's notebook for Sese, ms. 11 pp. Undated, but no doubt written shortly before his departure from Sese, i.e., before August 6,1906. BA, R 86, KGA, No. 3782 . Cf. the objectively identical formulation in the report on the expedition: Koch, "Über

den bisherigen Verlauf der deutschen Expedition," 514. Feldmann, confronted with these facts, no longer defended his views. The handwritten originals of Koch's reports are found in BA, R 15.01, RMDI No. 9307.

222. View expressed later by Kleine: BA, R 86, KGA, No. 3515, letter of 23 Oct. 1909 to K [presumably Koch]. Feldmann had, for instance, mistaken leucocytes for trypanosoma.

223. "Report Dr. Feldmann from Bukoba," dated 22 Aug. 1905., GStAPK, I, HA, Pep. 76 VIII 86, No. 4118.

224. Koch, "Über den bisherigen Verlauf der deutschen Expedition," 514–15.

225. Worboys, "Comparative History of Sleeping Sickness," 90.

226. Koch's notebook for Sese. BA, R 86, KGA, No. 3782. Cf. identical formulations: Koch, "Über den bisherigen Verlauf der deutschen Expedition," 517; "Erforschung der Schlafkrankheit," 601.

227. Koch to Gertrud Pfuhl, 3 Aug. 1906, cited in Möllers, *Robert Koch*, 313–14.

228. Koch, "Erforschung der Schlafkrankheit," 589. Cf. Heymann, *Robert Koch*, part 2, 144–45.

229. BA, R 86 KGA, No. 3782.

230. As was mentioned above, forms pathogenic to nonhumans were rare in areas with endemic sleeping sickness. In his report on the expedition, Koch made a point of stressing the meager theoretical progress made on Sese: Koch, "Zweiter Bericht über die Tätigkeit der deutschen Expedition," 530. In a letter to Gaffky on Oct. 22, 1906 (RKI/Koch papers, as/b2), he reported: "They [the trypanosoma] are a big headache for me, for among them I have already found four different types, or rather species; added to these is one trypanosome found in crocodile blood."

231. Koch, "Über den bisherigen Verlauf der deutschen Expedition," 521.

232. Ibid.

233. Ibid., 520. To ascribe this decision to Koch's special humaneness (Münch and Biel, "Expedition, Experiment und Expertise") is speculative and unwarranted, given the fact that Koch subsequently recommended the sometimes brutal measures to combat sleeping sickness in the German colonial territory. Cf. the exemplary study by Eckart and Cordes "People Too Wild."

234. Koch, "Über den bisherigen Verlauf der deutschen Expedition," 524 (of 5 Nov. 1906).

235. Koch, "Erforschung der Schlafkrankheit," 591.

236. For atoxyl, see Lyons, *The Colonial Disease*, 108–9; Müller-Jahnke and Friedrich, *Geschichte der Arzneimitteltherapie*, 223–26; Riethmiller, "Ehrlich, Bertheim, and Atoxyl."

237. Koch, "Über den bisherigen Verlauf der deutschen Expedition," 521. Koch later justified the risk to the patient of increasing the dosage of a medication whose effect on humans was practically unknown by pointing out that these patients would have been irremediably lost if no remedy were found (Koch, "Erforschung der Schlafkrankheit," 590). He established the minimum dosis for effectiveness at 0.3–0.4 mg (Kleine, "Ein Tagebuch von Robert Koch," 216 (diary entry of 14 Sept.).

238. Koch, "Zweiter Bericht über die Tätigkeit der deutschen Expedition," 527.

239. Ibid., 528.

240. According to Koch's diary the first injections were administered on August 27, 1906. (Kleine, "Ein Tagebuch von Robert Koch," 185.)

241. Koch, "Über den bisherigen Verlauf der deutschen Expedition," 524 (of 5 Nov. 1906).
242. Ibid., 529.
243. Koch to Libbertz, 27 Nov. , cited in Möllers, *Robert Koch*, 321.
244. Koch to Bumm, 5 Nov. 1906, BA, R 86, No. 2616.
245. Koch to Libbertz, 27 Nov. 1906, cited in Möllers, *Robert Koch*, 320.
246. "The number of sleeping sickness patients seeking help is increasing all the time," Koch noted in his diary on August 29. (Kleine, "Ein Tagebuch von Robert Koch," 185.) On August 30, one hundred sick persons were in the camp; on September 3 there were already two hundred. In the letter of November 5, cited above, the figure given is "around 900."
247. Koch, "Über den bisherigen Verlauf der deutschen Expedition," 524. For Koch this was clearly a welcome opportunity to get rid of the military physician Beck, with whom he had no personal ties, by sending him off to the second camp at Bumangi. "Fortunately all members of the expedition now enjoy good health, and this is desirable, since there is much work for everyone. Regierungsrath Beck too, for whom I originally did not have much use, is doing good work now. He is in charge of a ward at Bumangi, where a large part of our severe cases is being observed and treated" (Koch to Bumm, 5 Nov. 1906, BA, R 86, No. 2616). According to Koch's diary, Beck left Bugalla on September 17 for Bumangi, where he established his own camp.
248. Möllers, *Robert Koch*, 321. Cf. Kleine, "Ein Tagebuch von Robert Koch," 216: diary entry of 13 Sept., when Koch "telegraphed and wrote Berlin for another shipment of atoxyl." Cf. BA, R 86 KGA, Np. 2616: Koch's letter to KGA of 27 Sept. ordering 1,550 marks' worth of atoxyl.
249. Koch to Gaffky, 7 Dec. 1906 (Möllers, *Robert Koch*, 321). "I feel that with this supply I will definitely be able to face all the demands I might have to face in the future. I shall even be able to give some of it to the two German stations on Lake Victoria and to the efforts to combat sleeping sickness on Lake Tanganyika."
250. Koch to Bumm, 6 Nov. 1906, BA, R 86, No. 2616. The visits of English doctors are noted in the diary on 18 Sept. (Kleine, "Ein Tagebuch von Robert Koch," 217).
251. *Neue Preussische Kreuzzeitung*, 18 Dec. 1906.
252. Möllers, *Robert Koch*, 340–41.
253. Koch, "Schlussbericht über die Tätigkeit der deutschen Expedition," 535.
254. Ibid.
255. Koch to Bumm, 18 Feb. 1907, BA, R 86, No. 3515.
256. Koch, "Schlussbericht über die Tätigkeit der deutschen Expedition," 536.
257. In the letter of February cited above, Koch mentions "temporary blindness"; but by the next report of April, it had turned out to be permanent
258. Bornemann "Erblindung nach Atoxylinjektionen." Published in the *Münchener Medizinische Wochenschrift* (*MMW*), this study was reviewed in the *Deutsche Medizinische Wochenschrift*.
259. "Concerning the atoxyl treatment, you will be interested to learn about the experience of a Dr. Krüdener in Riga, who observed temporary blindness after the application of this medication." Bumm to Koch, 12 Jan. 1907 ( BA, R 86, No. 2616). As Koch's letter of 18 Feb. (see above) indicates, he had by then received this warning of Bumm's.
260. In view of the unclear situation, Koch asked "to be excused if the report that is due will be a little late this time" ( BA, R 86, KGA, No. 3515).

261. In the *Deutsche Medizinische Wochenschrift* (1907), no. 46. Cf. a letter of the Reich Chancellery to the Prussian Ministry of Culture of 10 June 1907 (GStAPK, I. HA Rep. 76 VIII B, No. 4118), which expressly states that Koch did not wish for the newer reports (appended to the letter) to be published.

262. Koch, "Schlussbericht über die Tätigkeit der deutschen Expedition," 537.

263. Koch to Bumm, 3 Aug. 1907 (BA, R 86, KGA, No. 2631).

264. Koch, "Schlussbericht über die Tätigkeit der deutschen Expedition," 542–43.

265. Kudicke before the Imperial Health Board, BA, R 86, KGA, No. 3515.

266. Mortality figures on Sese in Koch, "Schlussbericht über die Tätigkeit der deutschen Expedition," 44. This remark of Kudicke's referred to the highly varying mortality figures in different camps.

267. Koch to Gaffky, RKI/Koch papers, as/b2.

268. Koch, "Schlussbericht über die Tätigkeit der deutschen Expedition," 44, 543.

269. Ibid, 544–46. Minutes of the meeting of the Imperial Board of Health, 18 Nov. 1907, GStAPK I, HA, Rep. 76 VIII B No. 4118.

270. Koch, "Schlussbericht über die Tätigkeit der deutschen Expedition." Reports on the establishment of the camps in BA, R 86, KGA, No. 2622.

271. Cf. Lyons, "Sleeping Sickness," 100–101 and *passim*, which shows that the British authorities employed the same measures to combat the disease in the Congo. For the further development of the anti-sleeping sickness campaign after Koch's departure: Eckart, *Medizin und Kolonialimperialismus*, 346–49; cf. Beck, *Medicine and Society in Tanganyika*, 17–22.

272. Koch, "Erforschung der Schlafkrankheit," 590.

273. See chapter IV.3.

274. For the colonial human economy, see Eckart, *Medizin und Kolonialimperialismus*, 56–72.

275. Müller-Jahnke and Friedrich, *Geschichte der Arzneimitteltherapie*, 223–26.

276. Koch, "Schlussbericht über die Tätigkeit der deutschen Expedition," 542.

277. A colorful description appears in a letter to Libbertz, 27 Nov. 1907, cited in Möllers, *Robert Koch*, 320–21.

278. Koch to Gertrud Pfuhl, 3 March 1907, cited in *Robert Koch*, 325.

279. Koch to Libberts, 27 Nov. 1907, cited in Möllers, *Robert Koch*, 321.

280. This can be inferred from the letters of the office secretary at the Institute for Infectious Diseases, Pohnert. In this case from the letter of March 12, 1907.

281. Koch to Bumm, 6 Nov. 1906 , BA, R 86, KGA, No. 2616.

282. This was Adolf Friedrich, Duke of Mecklenburg. Koch recorded all these visits in his diary: Kleine, "Ein Tagebuch von Robert Koch" 1924. Cf. Adolf Friedrich, *Ins innerste Afrika*, 25. However, the duke places the meeting at Bukoba, not Sese.

283. Koch describes this problem in a letter to his daughter, 27 June 1907 (Möllers, *Robert Koch*, 336).

284. Koch to Gaffky, 28 Apr. 1907 (Möllers, *Robert Koch*, 332).

285. In SPBK Slg. Darmst. 3 b 1882 Koch, Robert. The text is not signed but clearly recognizable as Koch's handwriting. Reprinted in Heymann, *Robert Koch*, part 2, 166–81. On p. 160–61 Heymann also gives his view about the authorship.

286. If Koch happened to meet any of them, he would give them a cursory examination.

287. Möllers, *Robert Koch*, 317. Cf. For instance Koch's letter to Kolle, 7 Sept. 1897:

"It would never have occurred to you, would it, that one day I would turn into a hunter. But here I have gone along a few times and have proven such an adequate shot that everyone thinks I am a good hunter." Kolle, *Robert Koch, Briefe an Wilhelm Kolle*, 37. Koch apparently had gone hunting for the first time at Gaffky's invitation in 1884 (Möllers, *Robert Koch*, 225–26).

288. Koch, "Schlussbericht über die Tätigkeit der deutschen Expedition"; cf. Minutes of the meeting of the Imperial Board of Health, 18 Nov. 1907, GStAPK I, HA, Rep. 76 VIII B No. 4118.

289. MacKenzie, *Imperialism and the Natural World*, 193, 198.

290. Möllers, *Robert Koch*, 312.

291. Ibid.

292. Koch, "Schlussbericht über die Tätigkeit der deutschen Expedition," 545.

293. Fleck, *Genesis and Development*, 154–55.

294. Koch to Bumm, 3 March 1907, BA, R 86, KGA, No. 2631.

295. Kleine, *Ein deutscher Tropenarzt*, 34.

296. Cited in Möllers, *Robert Koch*, 341–42.

297. Cited in Möllers, *Robert Koch*, 343.

298. Bruno Latour has pointed out that these very conditions of a technological, political, or social nature ensure the functioning of laboratory science. Latour, *The Pasteurisation of France*, 90–93.

A Perspective

1. For introductory reading, see Brock, *Bacterial Genetics*, and Foster, *History of Medical Bacteriology*.

2. For the case of the United States, the latest study is Tomes, *Gospel of Germs*. For Imperial Germany, see my studies in "Invisible Enemies." For the most interesting analogous case of entomology, see Jansen, *"Schädlinge."*

3. A classic study is Bein, "'Der jüdische Parasit.'"

4. Sarasin, *"Anthrax."*

5. Mendelsohn, in "Von der 'Ausrottung' zum Gleichgewicht," has impressively drawn attention to such relativising efforts within the context of a holistic epidemiology in the interwar period.

6. Martin, *Flexible Bodies*.

7. Introduction to recent research: Sarasin, "Die Visualisierung des Feindes."

8. See chapter V.2.4.

9. Koch, "Die Bekämpfung der Infektionskrankheiten," 277.

10. Canetti, *Masse und Macht*, 47–48. Italics in the original.

# Bibliography

ARCHIVES CONSULTED

Archiv der Humboldt-Universität, Berlin (HUA)
    Akten der Charité Direktion

Archiv der Nobel Stiftung, Stockholm

Berlin-Brandenburgische Akademie der Wissenschaften, Berlin
    Akademiearchiv
    NL Rudolf Virchow 1115

Bundesarchiv/Berlin (BA)
    R 09.01 Auswärtiges Amt (AA)
    R 15.01 Reichsamt des Inneren (RMDI)

Geheimes Staatsarchiv Preußischer Kulturbesitz, Berlin (GStAPK)
    I. HA, Rep. 92 Althoff
    I. HA, Rep. 76 Kultusministerium

Hoechst-Archiv (Histocom), Frankfurt am Main

Robert Koch correspondence, Henry Barton Jacobs Collection, Johns Hopkins Institute of the History of Medicine, Baltimore

Robert Koch Museum, Institut für Mikrobiologie und Hygiene der Humboldt Universität, Berlin (RKM)

Robert Koch papers, Robert Koch Institute Berlin (RKI/Koch papers)

Rockefeller Archive Center, Sleepy Hollow NY
    RU 650, EH 89 Paul Ehrlich
    RU 650, EH 89 Martha Marquard collection

Staatsbibliothek Preußischer Kulturbesitz–Handschriftenabteilung, Berlin (SBPK)
    Sammlung Darmstädter Stadtarchiv Frankfurt, Frankfurt am Main
    NL Paul Ehrlich

## PRIMARY SOURCES

Adolf Friedrich, Herzog zu Mecklenburg. *Ins innerste Afrika. Bericht über den Verlauf der deutschen wissenschaftlichen Zentral-Afrika-Expedition, 1907–1908.* Leipzig: Klinkhardt & Biermann, 1909.

"Aus der Welt der unendlich Kleinen." *Kladderadatsch* 23 Nov. 1890.

Bary, Anton de. "Referat zu verschiedenen Arbeiten Halliers u.a zur Cholera." *Jahresbericht über die Leistungen und Fortschritte in der gesammten Medicin* 2 (1867): 243–51.

Baumgarten, Paul. "Neuere experimentell-pathologische Arbeiten über Tuberculinwirkung." *Berliner klinische Wochenschrift* 28 (1891): 1206–8, 1218–19, 1233–34.

———. "Tuberkelbakterien." *Centralblatt für die medizinischen Wissenschaften* 20 (1882): 257–59.

———. "Ueber die Einwirkung des Koch'schen Mittels auf Impftuberculose der Kaninchen." *Berliner klinische Wochenschrift* 28 (1891): 464–66.

Bergmann, Ernst von. "Mittheilungen über die mit dem Koch'schen Heilverfahren gewonnenen Ergebnisse." In *Robert Koch's Heilmittel gegen die Tuberkulose*, published by the editors of the *DMW*, vol. 1, 63–82. Berlin and Leipzig, 1890.

Biewend, Robert. "Aus der Familienchronik von Robert Koch. Biographische Mitteilungen." *Deutsche Revue* 16/17 (1891/1892): 179–86, 296–318.

Billroth, Theodor. *Untersuchungen über die Vegetationsformen von Coccobacteria septica und den Anteil welchen sie an der Entstehung und Verbreitung der accidentiellen Wundkrankheiten haben.* Berlin: Reimer, 1874.

Birch-Hirschfeld, Felix Victor. *Lehrbuch der pathologischen Anatomie.* Leipzig: F. C. W. Vogel, 1877.

———. "Die neueren pathologisch-anatomischen Untersuchungen über krankmachende Schmarotzerpilze." *Schmidt's Jahrbücher der gesammten in-und-ausländischen Medicin* 155 (1872): 97–109.

———. "Die neueren pathologisch-anatomischen Untersuchungen über Vorkommen und Bedeutung niederer Pilzformen (Bakterien) bei Infektionskrankheiten." *Schmidt's Jahrbücher der gesammten in-und-ausländischen Medicin* 166 (1875): 169–223.

Bollinger, Otto. "Infectionen durch thierische Gifte (Zoonosen)." In *Handbuch der speciellen Pathologie und Therapie*, Hugo von Ziemssen, ed., vol. 3, 400–607. Leipzig: F. C. W. Vogel, 1874.

———. "Thierkrankheiten." *Jahresbericht über die Leistungen und Fortschritte in der gesammten Medicin* 12 (1877): 583–606.

Bornemann, W. "Ein Fall von Erblindung nach Atoxylinjektionen bei Lichen ruber planus." *Münchener Medizinische Wochenschrift* 52 (1905): 1043–45.

Buchner, Hans. "Tuberculinreaction durch Proteine nicht spezifischer Bacterien." *Münchener Medicinische Wochenschrift* 39 (1891): 841–43.

Cohn, Ferdinand Julius. "Ein Brief über Koch." *Deutsche Revue* 16 (1891): 30–31.

———. "Untersuchungen über Bacterien." *Beiträge zur Biologie der Pflanzen* 1, no. 2 (1875): 127–222.

———. "Untersuchungen über Bacterien II." *Beiträge zur Biologie der Pflanzen* 1, no.2 (1875): 141–207.

Cohnheim, Julius. *Die Tuberkulose vom Standpunkte der Infektionslehre.* 2nd. ed. Leipzig: Edelmann, 1881.

Cornet, Georg. *Ein Rückblick auf die Zeit der Kochschen Publikation*. Printed in manuscript form. [Berlin], 1891.
———. "Die Tuberkulose," In *Spezielle Pathologie und Therapie*, Hermann Nothnagel, ed., vol. 14, 1–1441. Vienna: Alfred Hölder 1907.
———. "Über das Verhalten der Tuberkelbacillen im thierischen Organismus unter dem Einfuß entwicklungshemmender Stoffe." *Zeitschrift für Hygiene* 5 (1889): 98–133.
Cornet, Georg, and Arthur Meyer. "Tuberkulose." In *Handbuch der pathogenen Mikroorganismen*, Wilhelm Kolle and August Wassermann, eds., vol. 2, 78–177. Jena: Gustav Fischer, 1903.
Czerny, Vincenz. "Erster Bericht über die Koch'schen Impfungen." In *Robert Koch's Heilmittel gegen die Tuberkulose*, published by the editors of the *DMW*, 61–68. Berlin and Leipzig, 1891.
Davaine, Casimir-Joseph. "Recherches sur quelques questions relatives à la septicémie." *Bulletin de Académie de Médicine* 1 (1872): 907 ff., 976 ff.
Dehmel, Richard. "Ein Dankopfer. Robert Koch, dem Forscher, dem Menschen." *Freie Bühne für modernes Leben* 1 (1890): 1132–33.
*DMW*, editors of. *Robert Koch's Heilmittel gegen die Tuberkulose*. 12 vols. Berlin and Leipzig: Thieme, 1890/1891.
Ehrlich, Paul. "Chemotherapeutische Trypanosomenstudien." *Berliner Klinische Wochenschrift* 44 (1907): 233–36, 280–83, 310–14, 341–44, 349–50.
"Ehrung Robert Kochs." *Deutsche Medizinische Wochenschrift* 30 (1903): 1246–48.
Eidam, Eduard. *Der gegenwärtige Standpunkt der Mycologie mit Rücksicht auf die Lehre von den Infectionskrankheiten*, 2d ed. Berlin: H. E. Oliven, 1872.
F., Dr. "Der Entdecker des Cholerapilzes." *Die Gartenlaube* (1884): 433.
Flügge, Carl. *Grundriss der Hygiene: für Studierende und praktische Ärzte, Medizinal- und Verwaltungsbeamte*. Leipzig: Veit, 1889.
Fränkel, Carl. *Grundriß der Bakterienkunde*. 3rd ed. Berlin: Hirschwald, 1890.
Fräntzel, Oskar. "Bemerkungen zur Anwendung des Koch'schen Heilverfahrens." *Berliner Klinische Wochenschrift* 27 (1890): 1136–37.
Fräntzel, Oskar, and Runkwitz. "Systematische Anwendung des Koch'schen Specificums gegen Tuberculose bei inneren Krankheiten." In *Robert Koch's Heilmittel gegen die Tuberkulose*, published by the editors of the *DMW*, vol. 1, 27–29. Berlin and Leipzig: Thieme, 1890.
Fürbringer, Paul. "Vierwöchige Koch'sche Behandlung in ihrer Bedeutung für die Abweichung vom Schema." In *Robert Koch's Heilmittel gegen die Tuberkulose*, published by the editors of the *DMW*, vol. 3, 96–109. Berlin and Leipzig: Thieme, 1891.
Gaffky, Georg. *Bericht über die Thätigkeit der zur Erforschung der Cholera im Jahre 1883 nach Ägypten und Indien entsandten Kommission*. Vol. 3. Berlin: Julius Springer, 1887.
"Gratuliere zum neuen Jahre." *Der wahre Jacob* 3 Jan. 1891, 933.
Grawitz, E. "Über Versuche mit dem Koch'schen Mittel bei Affen." In *Robert Koch's Heilmittel gegen die Tuberkulose*, published by the editors of the *DMW*, vol. 11, 91–103. Berlin and Leipzig: Thieme, 1891.
Grünwald. "Bericht über die ersten vier auf der Klinik des Geheimrath von Bergmann nach Koch'scher Methode behandelten Fälle von Kehlkopftuberkulose." *Münchener Medizinische Wochenschrift* 37 (1890).
Guttmann, Paul, and Paul Ehrlich. "Ueber Anfangs-Behandlung der Lungen- und Kehlkopf- Tuberculose mit Koch'schem Tuberkulin." In *Robert Koch's Heilmittel gegen*

*die Tuberkulose*, published by the editors of the *DMW*, vol. 9, 5–7. Berlin and Leipzig: Thieme, 1891.

Hahn, Martin. "Über die chemische Natur des wirksamen Stoffes im Koch'schen Tuberculin." *Berliner Klinische Wochenschrift* 28 (1891): 741–44.

Hallier, Ernst. *Das Cholera-Contagium. Botanische Untersuchungen, Ärzten und Naturforschern mitgeteilt.* Leipzig: Engelmann, 1867.

———. *Gärungserscheinungen. Untersuchungen über Gärung, Fäulnis und Verwesung. Mit Berücksichtigung der Miasmen und Contagien sowie der Desinfection für Ärzte, Naturforscher, Landwirthe und Techniker.* Leipzig: Engelmann, 1867.

Hansemann, David. "Pathologisch-anatomische und histologische Erfahrungen über die Koch'sche Injectionsmethode." *Therapeutische Monatshefte* 5 (1891): 77–80.

Helfrich. "Über die Erfolge, welche mit dem Koch'schen Heilmittel bei Kranken der chirurgischen Klinik bislang erzielt worden sind." In *Robert Koch's Heilmittel gegen die Tuberkulose*, published by the editors of the *DMW*, vol. 2, 40–51. Berlin and Leipzig: Thieme, 1890.

Henle, Jacob. *Pathologische Untersuchungen.* Berlin: Hirschwald, 1840.

———. *Von den Miasmen und Contagien und von den miasmatisch-kontagiösen Krankheiten.* 1840. Reprinted, Leipzig: Barth, 1910.

Henoch, Eduard Heinrich. "Mittheilungen über das Koch'sche Heilverfahren gegen Tuberkulose." *Berliner Klinische Wochenschrift* 2 (1890): 1169–71.

Hertel. "Mittheilungen über die Einwirkungen des Kochschen Mittels auf Kehlkopftuberculose." In *Robert Koch's Heilmittel gegen die Tuberkulose*, published by the editors of the *DMW*, vol. 1, 83–93. Berlin and Leipzig: Thieme, 1890.

Hertwig, Oskar. *Ueber die physiologische Grundlage der Tuberculinwirkung. Eine Theorie der Wirkungsweise bacillärer Stoffwechselprodukte.* Jena: Gustav Fischer, 1891.

Hetsch, Heinrich. "Tuberkulose." In *Spezielle Pathologie und Therapie innerer Krankheiten*, Friedrich Kraus and Theodor Brugsch, eds., vol. II:1, 777–857. Berlin: Urban und Schwarzenberg, 1919.

Hiller, Arnold. *Die Lehre von der Fäulnis auf physiologischer Grundlage.* Berlin: Hirschwald, 1879.

Hueppe, Ferdinand. *Die Methoden der Bakterien-Forschung.* Wiesbaden: Kreidel, 1885.

———. "Robert Koch's Mittheilungen über Tuberkulin. Kritisch beleuchtet." *Berliner klinische Wochenschrift* 28 (1891): 1121–22.

———. "Über Erforschung der Krankheitsursachen und sich daraus ergebende Gesichtspunkte für Behandlung und Heilung von Infektionskrankheiten." *Berliner klinische Wochenschrift* 28 (1891): 279–83, 305–10, 333–36.

———. "Über die Heilung der Tuberculose mit specieller Berücksichtigung der neuen Methode von R. Koch." *Wiener Medizinische Presse* 31 (1890): 1888–92.

Hueppe, Ferdinand, and Hermann Scholl. "Über die Natur der Koch'schen Lymphe." *Berliner klinische Wochenschrift* 28 (1891): 88–89.

Hueter, C[arl]. *Die allgemeine Chirurgie. Eine Einleitung in das Studium der chirurgischen Wissenschaft.* Leipzig: F. C. W. Vogel, 1873.

———. "Die Interpellation über das Koch'sche Heilverfahren." *Vossische Zeitung* 29 Nov. 1890.

Israel, O. "Bericht über die anatomischen Befunde an zwei mit dem Koch'schen Heilmittel behandelten tuberkulösen Lokalerkrankungen." *Berliner Klinische Wochenschrift* 2 (1890): 1025–27.

Johne, Albert. *Die Geschichte der Tuberkulose mit besonderer Berücksichtigung der Tuberkulose des Rindes und die sich daraus ergebenden medicinal-u. veterinärpolizeilichen Consequenzen.* Leipzig: F. C. W. Vogel, 1883.

Jürgens, Dr., Ernst von Leyden, and Dr. Goldscheider. "Aus dem Verein für innere Medicin. Mittheilung über das Koch'sche Heilverfahren." In *Robert Koch's Heilmittel gegen die Tuberkulose*, published by the editors of the *DMW*, vol. 3, 122–29. Berlin and Leipzig: Thieme, 1890.

Klebs, Edwin. *Beiträge zur pathologischen Anatomie der Schusswunden. Nach Beobachtungen in den Kriegslazaretten in Carlsruhe 1870 und 1871.* Leipzig: F. C. W. Vogel, 1872.

———. "Sepsis, septische Infektion." In *Real-Enzyclopädie der gesammten Heilkunde*, Albert Eulenburg, ed., vol. 18, 254–81. Vienna and Leipzig: Urban und Schwarzenberg, 1889.

———. "Tuberkulose." In *Real-Enzyclopädie der gesammten Heilkunde*, Albert Eulenburg, ed., vol. 20, 285–316. Vienna and Leipzig: : Urban und Schwarzenberg, 1890.

———. "Über Cellularpathologie und Infectionskrankheiten." *Tageblatt der 51. Versammlung Deutscher Naturforscher und Aerzte in Cassel 1878.* (1878): 127–34.

———. *Über die Umgestaltung der medicinischen Anschauungen in den letzten drei Jahrzehnten.* Leipzig, 1878.

———. "Über die Wirkung des Koch'schen Mittels auf die Tuberkulose der Thiere, nebst Vorschlägen zur Herstellung eines unschädlichen Tuberkulines." In *Verhandlungen des Congresses für Innere Medizin, 10. Congress*, Ernst von Leyden and Emil Pfeiffer, eds., 191–98. Wiesbaden: Bergmann, 1891.

———. "Die Zusammensetzung des Tuberkulin." *Deutsche Medizinische Wochenschrift* 1 (1891): 1233–34.

Kleine, Friedrich Karl. "Ein Tagebuch von Robert Koch während seiner deutsch-ostafrikanischen Schlafkrankheitsexpedition i.J. 1906/07." Deutsche *Medizinische Wochenschrift* 50 (1924): 21–24, 55–56, 88–89, 121–22, 152–53, 184–85, 216–17, 248–49.

Koch, Robert. "Anthropologische Betrachtungen gelegentlich einer Expedition am Victoria-Nyanza." 1908. In *Gesammelte Werke von Robert Koch*, vol. 2.1, 57–62. Leipzig: Thieme, 1912.

———. "Antrittsrede in der Akademie der Wissenschaften am 1. Juli 1909." 1909. In *Gesammelte Werke von Robert Koch*, vol. 1, 1–4. Leipzig: Thieme, 1912.

———. "Die Ätiologie der Milzbrand-Krankheit, begründet auf die Entwicklungsgeschichte des Bacillus Anthracis." 1876. In *Gesammelte Werke von Robert Koch*, vol. 1, 5–25. Leipzig: Thieme, 1912.

———. "Die Ätiologie der Tuberkulose." 1882. In *Gesammelte Werke von Robert Koch*, vol. 1, 428–45. Leipzig: Thieme, 1912.

———. "Die Ätiologie der Tuberkulose." 1884. In *Gesammelte Werke von Robert Koch*, vol. 1, 467–565. Leipzig: Thieme, 1912.

———. "Die Bekämpfung der Infektionskrankheiten, insbesondere der Kriegsseuchen." 1888. In *Gesammelte Werke von Robert Koch*, vol. 2.1, 276–89. Leipzig: Thieme, 1912.

———. "Bericht über die Tätigkeit der zur Erforschung der Schlafkrankheit im Jahre 1906/07 nach Ostafrika entsandten Kommission." 1909. In *Gesammelte Werke von Robert Koch*, vol. 2.1, 582–645. Leipzig: Thieme, 1912.

———. "Berichte über die Tätigkeit der zur Erforschung der Cholera im Jahre 1883 nach Ägypten und Indien entsandten Kommission." 1883/1884. In *Gesammelte Werke von Robert Koch*, vol. 2.1, 1–19. Leipzig: Thieme, 1912.

———. "Erste Konferenz zur Erörterung der Cholerafrage." 1884. In *Gesammelte Werke von Robert Koch*, vol. 2.1, 20–60. Leipzig: Thieme, 1912.

———. "Experimentelle Studien über die künstliche Abschwächung der Milzbrandbazillen und Milzbrandinfektion durch Fütterung." 1884. In *Gesammelte Werke von Robert Koch*, vol. 1, 232–70. Leipzig: Thieme, 1912.

———. "Fortsetzung der Mitteilungen über ein Heilmittel gegen Tuberkulose." 1891. In *Gesammelte Werke von Robert Koch*, vol. 1, 669–72: Thieme, Leipzig, 1912.

———. *Gesammelte Werke von Robert Koch*. 2 vols. Julius Schwalbe, ed. Leipzig: Thieme, 1912.

———. "Kritische Besprechung der gegen die Bedeutung der Tuberkelbazillen gerichteten Publikationen." 1883. In *Gesammelte Werke von Robert Koch*, vol. 1, 454–66. Leipzig: Thieme, 1912.

———. "Neue Untersuchungen über die Mikroorganismen bei infektiösen Wundkrankheiten." 1878. In *Gesammelte Werke von Robert Koch*, vol. 1, 57–60. Leipzig: Thieme, 1912.

———. Review of *Die niederen Pilze in ihren Beziehungen zu den Infektionskrankheiten und der Gesundheitspflege*, by C. v. Naegeli (Munich, 1877) and *Die Naegelische Theorie der Infektionskrankheiten in ihren Beziehungen zu medizinischen Erfahrungen*, by Hans Buchner (Leipzig, 1871). 1878. In *Gesammelte Werke von Robert Koch*, vol. 1, 51–56. Leipzig: Thieme, 1912.

———. "Schlussbericht über die Tätigkeit der deutschen Expedition zur Erforschung der Schlafkrankheit." 1907. In *Gesammelte Werke von Robert Koch*, vol. 2.1, 534–46. Leipzig: Thieme, 1912.

———. "Über die Ätiologie der Tuberkulose." 1882. In *Gesammelte Werke von Robert Koch*, vol. 1, 446–453. Leipzig: Thieme, 1912.

———. "Über bakteriologische Forschung." 1890. In *Gesammelte Werke von Robert Koch*, vol. 1, 650–660. Leipzig: Thieme, 1912.

———. "Über die Behandlung der Lungentuberkulose mit Tuberkulin." 1901. In *Gesammelte Werke von Robert Koch*, vol. 1, 693. Leipzig: Thieme, 1912.

———. "Über den bisherigen Verlauf der deutschen Expedition zur Erforschung der Schlafkrankheit in Ostafrika." 1906. In *Gesammelte Werke von Robert Koch*, vol. 2.1, 509–24. Leipzig: Thieme, 1912.

———. "Über die Cholerabakterien." 1884. In *Gesammelte Werke von Robert Koch*, vol. 2.1, 61–68. Leipzig: Thieme, 1912.

———. "Über das Entstehen der Bernsteinsäure im menschlichen Organismus." 1865. In *Gesammelte Werke von Robert Koch*, vol.2.2, 814–20. Leipzig: Thieme, 1912.

———. "Über neue Tuberculinpräparate." 1897. In *Gesammelte Werke von Robert Koch*, vol. 1, 683–93. Leipzig: Thieme, 1912.

———. "Über die Trypanosomenkrankheiten. Vortrag gehalten in der Berliner medizinischen Gesellschaft am 26.10.1904." 1904. In *Gesammelte Werke von Robert Koch*, vol. 2.1, 459–72. Leipzig: Thieme, 1912.

———. "Über die Unterscheidung von Trypanosomenarten." 1904. In *Gesammelte Werke von Robert Koch*, vol. 2.1, 473–76. Leipzig: Thieme, 1912.

———. "Untersuchungen über die Ätiologie der Wundinfektionskrankheiten." 1878. In *Gesammelte Werke von Robert Koch*, vol. 1, 61–108. Leipzig: Thieme, 1912.

———. "Verfahren zur Untersuchung, zum Konservieren und Photographieren der

Bakterien." 1877. In *Gesammelte Werke von Robert Koch*, vol. 1, 27–50. Leipzig: Thieme, 1912.

———. "Vorläufige Mitteilungen über die Ergebnisse einer Forschungsexpedition nach Ostafrika." 1905. In *Gesammelte Werke von Robert Koch*, vol. 2.1, 477–86. Leipzig: Thieme, 1912.

———. "Weitere Mitteilungen über ein Heilmittel gegen Tuberkulose." 1890. In *Gesammelte Werke von Robert Koch*, vol. 1, 661–68. Leipzig: Thieme, 1912.

———. "Weitere Mitteilung über das Tuberkulin." 1891. In *Gesammelte Werke von Robert Koch*, vol. 1, 673–682. Leipzig: Thieme, 1912.

———. "Zur Untersuchung von pathogenen Mikroorganismen." 1881. In *Gesammelte Werke von Robert Koch*, vol. 1, 112–163. Leipzig: Thieme, 1912.

———. "Zweite Konferenz zur Erörterung der Cholerafrage." 1885. In *Gesammelte Werke von Robert Koch*, vol. 2.1, 69–166. Leipzig: Thieme, 1912.

———. "Zweiter Bericht über die Tätigkeit der deutschen Expedition zur Erforschung der Schlafkrankheit." 1907. In *Gesammelte Werke von Robert Koch*, vol. 2.1, 525–30. Leipzig: Thieme, 1912.

Köhler, Rudolf. "Mittheilungen über das Koch'sche Heilverfahren der Tuberkulose bei chirurgisch Kranken." *Berliner Klinische Wochenschrift* 2 (1890): 1127–30.

Köhler, Rudolf, and Westphal. "Über die Versuche mit dem von Herrn Geheimrath Koch gegen Tuberkulose empfohlenen Mittel." In *Robert Koch's Heilmittel gegen die Tuberkulose*, published by the editors of the *DMW*, vol. 1, 37–62. Berlin and Leipzig: Thieme, 1890.

———. "Eine neue Theorie zur Erklärung der Wirkung des Koch'schen Heilmittels auf den tuberculösen Menschen nebst therapeutischen Bemerkungen." In *Robert Koch's Heilmittel gegen die Tuberkulose*, published by the editors of the *DMW*, vol. 12, 141–59. Berlin and Leipzig: Thieme, 1891.

Kolle, Wilhelm. "Volksvertretung und Medicin." *Deutsche Medizinische Wochenschrift* 26 (1900): 187–88.

Korach, Dr. "Aus dem Israelitischen Krankenhaus in Hamburg. Über die mit dem Koch'schen Heilmittel auf der medicinischen Abteilung erzielten Resultate." In *Robert Koch's Heilmittel gegen die Tuberkulose*, published by the editors of the *DMW*, vol. 4, 95–100. Berlin and Leipzig: Thieme, 1891.

Korb-Döbeln, Dr. med. *Liederbuch für Deutsche Ärzte und Naturforscher. Zweiter Abschnitt. Ambrosia und Nektar! Enthaltend: 200 ernste und heitere Fest- und Tafellieder, Reden, Aufsätze etc. medicinischen und naturwissenschaftlichen Inhalts, mit mancherlei Illustrationen.* Hamburg: Gebrüder Lüdeking, 1892.

L., Dr. "Der letzte Arzt." *Ulk* 12 Dec. 1890, 2.

L. G. "Das Aerzte-Bankett zu Ehren der Mitglieder der deutschen Cholera-Kommission." *Berliner Tageblatt* 14 May 1884.

Lahmann, Heinrich. *Koch und die Kochianer. Eine Kritik der Koch'schen Entdeckung und der Koch'schen Richtung in der Heilkunde.* Stuttgart: Zimmer, 1890.

Leichtenstern, Otto. "Mittheilungen über das Koch'sche Heilverfahren gegen Tuberkulose." In *Robert Koch's Heilmittel gegen die Tuberkulose*, published by the editors of the *DMW*, vol. 4, 39–56. Berlin and Leipzig: Thieme, 1891.

Levy, William. "Bericht über die ersten nach der Methode des Herrn Geheimrath Dr. Koch behandelten Fälle von chirurgischer Tuberkulose." In *Robert Koch's Heilmittel gegen die Tuberkulose*, published by the editors of the *DMW*, vol. 1, 30–36. Berlin and Leipzig: Thieme, 1890.

Lewin, Leo. "Die Nebenwirkungen der Arzneimittel." *Pharmakologisch-klinisches Handbuch*. 1881, 1893. Reprinted, Berlin: : Hirschwald, 1899.

Leyden, Ernst von. "Bericht über die Anwendung des Koch'schen Heilverfahrens auf der I. medicinischen Klinik vom 20. bis 27. November 1890." *Berliner Klinische Wochenschrift* 2 (1890): 1145–50.

Leyden, Ernst von, and Emil Pfeiffer, eds. *Verhandlungen des Congresses für innere Medicin*. Zweiter Congress, Wiesbaden: Bergmann 1883.

Lichtheim, Dr. "Das Koch'sche Heilverfahren." In *Robert Koch's Heilmittel gegen die Tuberkulose*. Published by the editors of the *DMW*, vol. 7, 65–84. Berlin and Leipzig: Thieme, 1891.

Loeffler, Friedrich. *Vorlesungen über die Entwicklung der Lehre von den Bakterien*. Leipzig: Thieme, 1887.

———. "Zum 25jährigen Gedenktage der Entdeckung des Tuberkelbazillus." *Deutsche Medizinische Wochenschrift* 33 (1907): 449–51, 489–95.

Martius, Friedrich. "Krankheitsursachen und Krankheitsanlage." *Verhandlungen der Gesellschaft deutscher Naturforscher und Ärzte* 70 (1898): 90–110.

*Merck's Index*. Darmstadt, 1902.

Moll, Albert. *Ärztliche Ethik. Die Pflichten des Arztes in allen Beziehungen seiner Tätigkeit*. Stuttgart: Enke, 1902.

"Der neue Ritter St. Georg." *Ulk* 14 Nov. 1890, 8.

Niemeyer, Felix. *Lehrbuch der speziellen Pathologie und Therapie mit besonderer Rücksicht auf Physiologie und pathologische Anatomie*. Vol. 1. Berlin: : Hirschwald, 1863.

Oppenheimer. "Aus dem St. Josephshaus in Heidelberg. Ein Fall von Larynxtuberkulose, rasche Heilung." In *Robert Koch's Heilmittel gegen die Tuberkulose*, published by the editors of the *DMW*, vol. 2, 38–39. Berlin and Leipzig: Thieme, 1890.

Orth, Johannes. *Ätiologisches und Anatomisches über Lungenschwindsucht*. Berlin: Hirschwald, 1887.

Panum, Peter Ludwig. "Das putride Gift, die Bacterien, die putride Intoxikation und die Septicämie." *Archiv für pathologische Anatomie und Physiologie und für klinische Medizin (Virchows Archiv)* 60 (1874): 301–52.

Peiper, Erich. "Über die Wirkung des Koch'schen Mittels auf gesunde und nicht-tuberkulöse Individuen." In *Robert Koch's Heilmittel gegen die Tuberkulose*, published by the editors of the *DMW*, vol. 6, 16–30. Berlin and Leipzig: Thieme, 1891.

Pfuhl, Eduard. "Beitrag zur Behandlung tuberculöser Meerschweinchen mit Tuberculinium Kochii." *Zeitschrift für Hygiene und Infektionskrankheiten* 11 (1892): 241–58.

———. "Privatbriefe von Robert Koch." *Deutsche Revue* (1911): 182–99.

"Politische Tagesübersicht. Die deutsche Cholera-Kommission in Ägypten." *Berliner Tageblatt* 26 Sept. 1883.

Ponfick, Emil. "Pflanzliche und thierische Parasiten." *Jahresbericht über die Leistungen und Fortschritte in der gesammten Medicin* 1 (1878): 288–97.

Predöhl, August. *Die Geschichte der Tuberkulose*. 1888. Reprinted, Wiesbaden: Dr. Martin Sandig OHG, 1966.

"Redaktionelle Mitteilung." *Berliner klinische Wochenschrift* 28 (1891): 86.

Richter, Hermann Eberhard. "Die neueren Kenntnisse von den krankmachenden Schmarotzerpilzen nebst phytophysiologischen Grundbegriffen." *Schmidt's Jahrbücher der gesammten in-und ausländischen Medicin* 3, no. 5 (1867): 81–98.

———. "Die neueren Kenntnisse von den krankmachenden Schmarotzerpilzen nebst phytophysiologischen Grundbegriffen." *Schmidt's Jahrbücher der gesammten in-und ausländischen Medicin* 140 (1868): 101–28.

———. "Die neueren Kenntnisse von den krankmachenden Schmarotzerpilzen nebst phyto-physiologischen Vorbegriffen." *Schmidt's Jahrbücher der gesammten in-und ausländischen Medicin* 151 (1871): 313–53.

———. "Die neueren Kenntnisse von den krankmachenden Schmarotzerpilzen in phyto-physiologischer, pathologischer und sanitätlicher Hinsicht." *Schmidt's Jahrbücher der gesammten in-und ausländischen Medicin* 159 (1873): 169–218.

———. "Neueres über die krankmachenden Schmarotzerpilze." *Schmidt's Jahrbücher der gesammten in-und ausländischen Medicin* 168 (1875): 57–81.

"Robert Koch." *Vossische Zeitung* 16 Nov. 1890, 1.

Rosenbach, Ottomar. *Arzt contra Bakteriologe*. Berlin and Vienna: Urban und Schwarzenberg, 1893.

———. "Beobachtungen über die nach Anwendung des Koch'schen Mittels auftretenden Reactionserscheinungen." In *Robert Koch's Heilmittel gegen die Tuberkulose*, published by the editors of the *DMW*, vol. 2, 1–18. Berlin and Leipzig: Thieme, 1890.

———. *Grundlagen, Aufgaben und Grenzen der Therapie: nebst einem Anhange: Kritik des Koch'schen Verfahrens*. Berlin and Vienna: Urban und Schwarzenberg, 1891.

———. "Über das Verhalten der Körpertemperatur bei Anwendung des Kochschen Verfahrens." In *Robert Koch's Heilmittel gegen die Tuberkulose*, published by the editors of the *DMW*, vol. 5, 12–30. Berlin and Leipzig: Thieme, 1891.

Sander, Friedrich. "Die Bakterienfrage zu London und Berlin." *Deutsche Medizinische Wochenschrift* 1 (1875): 8–10.

Scheube, B. *Die Krankheiten der warmen Länder. Ein Handbuch für Ärzte*. 1896. 3rd ed. Jena: Gustav Fischer, 1910.

Schill, Ernst, and Bernhard Fischer. "Über die Desinfektion des Auswurfs der Phthisiker." *Mittheilungen aus dem Kaiserlichen Gesundheitsamte* 2 (1884): 1312–46.

Schreiber, Julius. "Über das Koch'sche Heilverfahren." In *Robert Koch's Heilmittel gegen die Tuberkulose*, published by the editors of the *DMW*, vol. 8, 11–21. Berlin and Leipzig: Thieme, 1891.

Schroeter, Julius. "Ueber einige durch Bakterien gebildete Pigmente." *Beiträge zur Biologie der Pflanzen* 1 (1870): 109–26.

Schultze, Friedrich. "Weitere Mittheilungen über den diagnostischen und therapeutischen Wert des Koch'schen Mittels." In *Robert Koch's Heilmittel gegen die Tuberkulose*, published by the editors of the *DMW*, vol. 10, 16–21. Berlin and Leipzig: Thieme, 1891.

Schulz, H. "Quecksilber." In *Real-Enzyclopädie der gesammtem Heilkunde*, Albert Eulenburg, ed., vol. 20, 3rd ed., 112–41. Berlin and Vienna: Urban und Schwarzenberg, 1899.

Schwimmer, Ernst. "Die Behandlung mit Koch'scher Lymphe vom dermatologischen Standpunkt aus beurteilt." In *Robert Koch's Heilmittel gegen die Tuberkulose*, published by the editors of the *DMW*, vol. 4, 64–70. Berlin and Leipzig: Thieme, 1891.

Seiffert, G. "Die Tuberkulose als übertragbare Krankheit und ihre Bekämpfung vor Robert Koch." *Münchener Medizinische Wochenschrift* 79 (1932): 501–6.

Senator, Hermann. "Mittheilungen über das Koch'sche Heilverfahren gegen Tuberkulose." *Berliner klinische Wochenschrift* 27 (1890): 1167.
Siegmund, G. "Die Stellung des Arztes zur Tuberculinbehandlung." *Therapeutische Monatshefte* 5 (1891): 415–21.
Sonnenburg, Dr. "Aus der chirurgischen Abteilung des städtischen Krankenhauses Moabit in Berlin: Das Koch'sche Verfahren combiniert mit chirurgischen Eingriffen." In *Robert Koch's Heilmittel gegen die Tuberkulose*, published by the editors of the *DMW*, vol. 5, 102–9. Berlin and Leipzig: Thieme, 1891.
*Stenographische Berichte über die Verhandlungen [...] beider Häuser des Landtages. Haus der Abgeordneten.* Berlin: W. Moeser Hofdruckerei, 1890 and 1891.
Stuhlmann, Franz. "Beiträge zur Kenntnis der Tsetse-Fliege." *Arbeiten aus dem Kaiserlichen Gesundheitsamt* 26 (1907): 301–83.
Valerius. "Die Cholera-Gefahr." *Die Gartenlaube* (1884).
———. "Der Kommabacillus." *Die Gartenlaube* (1884): 598–99.
Virchow, Rudolf. *Die Fortschritte der Kriegsheilkunde, besonders im Gebiete der Infectionskrankheiten.* Berlin: Hirschwald, 1874.
———. "Der Kampf der Zellen und Bakterien." *Archiv für pathologische Anatomie und Physiologie* 101 (1885): 1–13.
———. "Krankheitsursachen." *Archiv für pathologische Anatomie, Physiologie und klinische Medizin* 79 (1880): 1–19, 185–228.
———. "Über die Wirkung des Koch'schen Mittels auf innere Organe Tuberkulöser." *Berliner klinische Wochenschrift* 28 (1891): 49–52.
Weigert, Carl. "Zur Lehre von der Tuberculose und von verwandten Erkrankungen." *Archiv für pathologische Anatomie und Physiologie und für klinische Medizin* (Virchows Archiv) 77 (1879): 269–98.
"Willkommen, Ihr Sieger." *Berliner Tageblatt* 3 May 1884.
"Die Wirksamkeit des Kochschen Heilmittels gegen Tuberkulose. Amtliche Berichte der Kliniken, Polikliniken und pathologisch-anatomischen Institute der preussischen Universitäten." *Klinisches Jahrbuch*, Ergänzungsband, Berlin, 1891.
Wolff, F. "Die Reaktion der Lungenkranken bei den Koch'schen Impfungen." In *Robert Koch's Heilmittel gegen die Tuberkulose*, published by the editors of the *DMW*, vol. 2, 53–70. Berlin and Leipzig: Thieme, 1890.
———. "Über die Anwendung des Tuberkulins bei Lungenkranken." In *Robert Koch's Heilmittel gegen die Tuberkulose*, published by the editors of the *DMW*, vol. 9, 95–110. Berlin and Leipzig, 1891.

## SECONDARY SOURCES

Ackerknecht, Erwin H. *Medicine at the Paris Hospital 1794–1848.* Baltimore: Johns Hopkins University Press, 1966.
———. *Rudolf Virchow. Arzt, Politiker, Anthropologe.* Stuttgart: Enke, 1957.
Altman, Lawrence K. *Who Goes First? The Story of Self-Experimentation in Medicine.* Berkeley, Los Angeles, London: University of California Press, 1987.
Amsterdamska, Olga. "Medical and Biological Contraints: Early Research on Variation in Biology." *Social Studies of Science* 1 (1987): 657–87.
Arnold, David, ed. *Warm Climates and Western Medicine: The Emergence of Tropical Medicine, 1500–1900.* Amsterdam: Rodopi, 1996.

Bald, Detlef, and Gerhild Bald. *Das Forschungsinstitut Amani. Wirtschaft und Wissenschaft in der deutschen Kolonialpolitik. Ostafrika 1900–1918*. IFO-Institut, 1972.
Barnes, David S. *The Making of a Social Disease: Tuberculosis in 19th-Century France*. Berkeley: University of California Press, 1995.
Bauer, Axel. *Die Krankheitslehre auf dem Weg zur naturwissenschaftlichen Morphologie. Pathologie auf den Versammlungen Deutscher Naturforscher und Ärzte*. Stuttgart: Wissenschaftliche Verlagsgesellschaft, 1989.
Bäumler, Ernst. *Farben, Formeln, Forscher. Hoechst und die Geschichte der industriellen Chemie in Deutschland*. Munich and Zürich: Pieper, 1989.
———. *Paul Ehrlich. Forscher für das Leben*. 1979. Reprinted, Frankfurt/Main: Societaets-Verlag, 1989.
Beck, Ann. *Medicine and Society in Tanganjika 1890–1930. A Historical Inquiry*. Philadelphia: American Philosophical Society, 1977.
———. "The Role of Medicine in German East Africa." *Bulletin of the History of Medicine* 6 (1971) pp. 170–71.
Bein, Alexander. "'Der jüdische Parasit.' Bemerkungen zur Semantik der Judenfrage." *Vierteljahrshefte für Zeitgeschichte* 1 (1965): 121–49.
Benaroyo, Lazare. "La contribution de Friedrich Wilhelm Zahn (1845–1904) à l'étude de l'inflammation." *Gesnerus* 8 (1991): 395–408.
Berghoff, Emanuel. *Entwicklungsgeschichte des Krankheitsbegriffes*. Vienna: Verlag Wilhelm Handrich, 1947.
Bleker, Johanna. *Die naturhistorische Schule 1825–1845*. Stuttgart and New York: Gustav Fischer, 1981.
Bochalli, Richard. *Die Entwicklung der Tuberkuloseforschung in der Zeit von 1878–1958. Rückblick eines deutschen Tuberkulosearztes*. Stuttgart: Georg Thieme, 1958.
Bourdelais, Patrice, and André Dodin, eds. *Visages du Cholera*. Paris: Belin 1987.
Bourguet, Marie-Noelle, Christian Licoppe, and H. Otto Sibum, eds., *Instruments, Travel, and Science: Itineraries of Precision from the Seventeenth to the Twentieth Century*. London and New York: Routledge, 2002.
Boyd, John. "Sleeping Sickness: The Castellani-Bruce Controversy." *Notes and Records of the Royal Society of London* 28 (1973): 93–110.
Bracegirdle, Brian. *A History of Microtechnique*. New York: Cornell University Press, 1978.
Brecht, Christine. "Das Publikum belehren—Wissenschaft zelebrieren. Bakterien in der Ausstellung 'Volkskrankheiten und ihre Bekämpfung' von 1903." In *Strategien der Kausalität. Konzepte der Krankheitsverursachung im 19. und 20. Jahrhundert*, Christoph Gradmann and Thomas Schlich, eds., 53–76. Pfaffenweiler: Centaurus, 1998.
Brecht, Christine, and Sybilla Nikolow. "Displaying the Invisible: Volkskrankheiten on Exhibition in Imperial Germany." *Studies in History and Philosophy of Biological and Biomedical Sciences* 31 (2000): 511–30.
Brenner, Peter J., ed., "Gefühl und Sachlichkeit. Humboldts Reisewerk zwischen Naturwissenschaft und Naturphilosophie." *Archiv für Kulturgeschichte* 73(1991): 135–68.
———. "Der Reisebericht in der deutschen Literatur. Ein Forschungsüberblick als Vorstudie zu einer Gattungsgeschichte." *IASL Sonderheft* 2. Tübingen, 1990.
———. *Der Reisebericht. Die Entwicklung einer Gattung in der deutschen Literatur*. Frankfurt/Main: Suhrkamp, 1989.
Briese, Olaf. *Angst in den Zeiten der Cholera*. 4 vols. Berlin: Akademie Verlag, 2003.

———. "Defensive, Offensive, Strassenkampf. Die Rolle von Medizin und Militär am Beispiel der Cholera in Preussen." *MedGG* 16 (1997): 9–32.

Brock, Thomas D. *The Emergence of Bacterial Genetics*. Cold Spring Harbour: Cold Spring Harbour Laboratory Press, 1990.

———. *Robert Koch: A Life in Medicine and Bacteriology*. Madison, Wisc.: Science Tech Publishers, 1988.

Brock, Thomas D., ed. *Milestones in Microbiology*. Washington, D.C.: American Society for Microbiology, 1975.

Brugsch, Theodor. *Arzt seit fünf Jahrzehnten*. Berlin: Rütten und Loening, 1957.

Bulloch, William. *The History of Bacteriology*. 1938. London: Oxford University Press, 1960.

Burke, Donald S. "Of Postulates and Peccadilloes: Robert Koch and Vaccine (Tuberculin) Therapy for Tuberculosis." *Vaccine* 11 (1993): 795–804.

Bynum, W. F., and Caroline Overy, eds. *The Beast in the Mosquito: The Correspondence of Ronald Ross and Patrick Manson*. Amsterdam and Atlanta: Rodopi, 1998.

Bynum, William F. "C'est un malade: Animal Models and Concepts of Human Diseases." *Journal of the History of Medicine and Allied Sciences* 5 (1990): 397–413.

———. *Science and the Practice of Medicine in the Nineteeth Century*. Cambridge: Cambridge University Press, 1994.

Canetti, Elias. *Masse und Macht*. 1960. Reprinted, Frankfurt/Main: Fischer, 1993. English translation, *Crowds and Power*, 1960.

Canguilhem, Georges. "Der Beitrag der Bakteriologie zum Untergang der 'medizinischen Theorien' im 19. Jahrhundert." In *Wissenschaftsgeschichte und Epistemologie*, Georges Canguilhem, ed., 110–32. Frankfurt/Main: Suhrkamp, 1979.

———. *Das Experimentieren in der Tierbiologie*. Henning Schmidgen, translator, MPI Preprint No. 189. Berlin: MPI für Wissenschaftsgeschichte, 2001.

———. *Das Normale und das Pathologische*. Munich: Hauser, 1974.

Carter, K. Codell. "Edwin Klebs' Criteria for Disease Causality." *Medical History* 22 (1987): 80–89.

———. "Edwin Klebs's *Grundversuche*." *Bulletin of the History of Medicine* 5 (2001): 771–81.

———. "Jacob Henle's Views on Disease Causation." *NTM: Zeitschrift fur Geschichte der Naturwissenschaft, Technik und Medizin* 28 (1991): 259–64.

———. "The Koch-Pasteur Dispute on Establishing the Cause of Anthrax." *Bulletin of the History of Medicine* 62 (1988): 42–57.

———. "Koch's Postulates in Relation to the Work of Jacob Henle and Edwin Klebs." *Medical History* 29 (1985): 353–75.

———. "The Rise of Causal Concepts of Disease" *Case Histories*. Aldershot: Ashgate, 2003.

Chauvet, Michel. "Une Cenetaire qui n'a pas tenu toutes ses promesses." *Revue médicale de la suisse romande* 110 (1990): 1067–70.

Churchill, Frederick B. "Life Before Model Systems: General Zoology at August Weismann's Institute." *American Zoologist* 37 (1997): 260–68.

Cittadino, Eugene. *Nature as Laboratory: Darwinian Plant Ecology in the German Empire 1880–1900*. Cambridge: Cambridge University Press, 1990.

Clark, George, and Frederick H. Kasten. *History of Staining*. 3rd ed. Baltimore: Williams & Wilkins, 1983.

Coleman, William. "Koch's Comma Bacillus: The First Year." *Bulletin of the History of Medicine* 61 (1987): 315–42.

Coleman, William, ed. *The Investigative Enterprise: Experimental Physiology in Nineteenth-Century Medicine.* Berkeley: University of California Press, 1988.

Condrau, Flurin. *Lungenheilanstalt und Patientenschicksal. Sozialgeschichte der Tuberkulose in Deutschland und England im späten 19. und frühen 20. Jahrhundert.* Göttingen: Vandenhoek and Ruprecht, 2000.

Conrad, Lawrence, and Dominik Wujastyk, eds. *Contagion: Perspectives from Premodern Societies.* Aldershot: Ashgate, 2000.

Cooter, Roger, and Steven Pumfrey. "Separate Spheres and Public Places: Reflections on the History of Science Popularisation and Science in popular Culture." *History of Science* 32 (1994): 237–67.

Craeger, Angela N. H. *The Life of a Virus: Tobacco Mosaic Virus as an Experimental Model 1930–1965.* Chicago: University of Chicago Press, 2002.

Cranefield, Paul F. *Science and Empire: East Coast Fever in Rhodesia and the Transvaal.* Cambridge: Cambridge University Press, 1991.

Cunningham, Andrew. "Transforming Plague: The Laboratory and the Identity of Infectious Disease." In *The Laboratory Revolution in Medicine*, Andrew Cunningham and Perry Williams, eds., 209–24. Cambridge: Cambridge University Press, 1992.

Cunningham, Andrew, and Perry Williams, eds. *The Laboratory Revolution in Medicine* Cambridge: Cambridge University Press, 1992.

Daum, Andreas. *Wissenschaftspopularisierung im 19. Jahrhundert. Bürgerliche Kultur, naturwissenschaftliche Bildung und die deutsche Öffentlichkeit 1848–1914.* Munich: Oldenbourg, 1998.

David, Heinz. *Rudolf Virchow und die Medizin des 20. Jahrhunderts.* Werner Selberg and Hans Hamm, eds. Munich: Quintessenz-Verlag, 1993.

De Kruif, Paul. *Microbe Hunters.* 1926. Reprinted, Zürich: Füssli 1927.

Diepgen, Paul. "Krankheitswesen und Krankheitsursache in der spekulativen Pathologie des 19. Jahrhunderts." *Sudhoffs Archiv* 18 (1926): 302–27.

———. "Die Lehre von der Entzündung. Von der Begründung der Zelluarpathologie bis zum Aufkommen der Bakteriologie." *Akademie der Wissenschaften und der Literatur. Abhandlungen der mathematisch-naturwissenschaftlichen Klasse* 3 (1953): 67–85.

Dierig, Sven. "Feinere Messungen in der Mitte einer belebten Stadt—Berliner Großstadtverkehr und die apparativen Hilfsmittel der Elektrophysiologie 1845–1910." *NTM: Zeitschrift für Geschichte der Naturwissenschaft, Technik und Medizin* 6 (1998): 148–69.

Dinges, Martin, ed. *Medizinkritische Bewegungen im Deutschen Reich.* Stuttgart: Steiner, 1996.

Doerr, Wilhelm. "Cohnheims Entzündungslehre und die aktuelle Debatte." *Zentralblatt für allgemeine Pathologie und pathologische Anatomie* 130 (1985): 299–306.

Dolman, Claude E. "Brefeld, Julius Oscar." In *Dictionary of Scientific Biography*, vol. 2, 436–38. New York: Charles Scribner's Sons, 1973.

———. "Ehrlich, Paul." In *Dictionary of Scientific Biography*, vol. 4, 295–305. New York: Charles Scribner's Sons, 1971.

———. "Koch, Heinrich Hermann Robert." In *Dictionary of Scientific Biography*, vol. 7, 420–35. New York: Charles Scribner's Sons, 1973.

D'Orazio, Ugo. "Scienza Tedesca e Università Italiana: Recezione di Modelli Esteri nell'Istituzionalizzazione delle Discipline Igieniche in Italia (1885–1900)." *Medizinhistorisches Journal* 33 (1998): 293–321.

Drews, Ferdinand. "Ferdinand Cohn, ein Wegbereiter der modernen Mikrobiologie und Pflanzenbiologie." *Freiburger Universitätsblätter* 1, no. 2 (1998): 29–84.

Dubos, René Jules, Jean Dubos, Rosenkrantz Dubos, and Barbara Gutmann, eds. *The White Plague: Tuberculosis, Man, and Society*. New Brunswick: Rutgers University Press, 1987.

Duffin, Jacalyn. *To See with a Better Eye. A Life of R. T. H. Laënnec*. Princeton, N.J.: Princeton University Press, 1998.

Dwork, Deborah. "Koch and the Colonial Office, 1902–1904: The Second South Africa Expedition." *NTM: Zeitschrift für Geschichte der Naturwissenschaft, Technik und Medizin* 20 (1983): 67–74.

Eckart, Wolfgang U. "Malaria and Colonialism in the German Colonies New Guinea and the Cameroons. Research, Control, Thoughts of Eradication." *Parasitologia* 40 (1998): 83–90.

———. *Medizin und Kolonialimperialismus in Deutschland 1884–1945*. Paderborn: Schöningh, 1997.

———. "Von der Idee eines 'Reichsinstituts' zur unabhängigen Forschungsinstitution—Vorgeschichte und Gründung des Hamburger Instituts für Schiffs-und Tropenkrankheiten." In *Formen außerstaatlicher Wissenschaftsförderung. Deutschland im europäischen Vergleich*, Rüdiger vom Bruch and Rainer A. Müller, eds., 31–52. Stuttgart: Steiner, 1990.

———. "Wissenschaft und Reisen—Einleitung und Bericht." *Berichte zur Wissenschaftsgeschichte* 22 (1999): 75–80.

Eckart, Wolfgang U., and Meike Cordes. "People Too Wild? Pocken, Schlafkrankheit und koloniale Gesundheitskontrolle im Kaiserlichen 'Schutzgebiet' Togo." In *Neue Wege in der Seuchengeschichte*, Martin Dinges and Thomas Schlich, eds., 175–206. Stuttgart: Steiner, 1995.

Elkeles, Barbara. *Der moralische Diskurs über das medizinische Menschenexperiment im 19. Jahrhundert*. Stuttgart, Jena, New York: Gustav Fischer, 1996.

———. "Robert Koch (1843–1910)." In *Klassiker der Medizin*, Dietrich von Engelhardt and Fritz Hartmann, eds., 247–71. Munich: Beck, 1991.

———. "Der 'Tuberkulinrausch' von 1890." *Deutsche Medizinische Wochenschrift* 115 (1990): 1729–32.

Engelhardt, Dietrich von. "Kausalität und Konditionalität in der modernen Medizin." In *Pathogenese. Grundzüge und Perspektiven einer theoretischen Pathologie*, Heinrich Schipperges, eds., 32–58. Berlin, Heidelberg, New York, Tokyo: Springer-Verlag, 1985.

———. *Medizin in der Literatur der Neuzeit, Vol. I. Darstellung und Deutung*. Hürtgenwald: Guido Pressler, 1991.

Epstein, Steven. *Impure Science: Aids, Activism, and the Politics of Knowledge*. Berkeley, Los Angeles, London: University of California Press, 1996.

Eschenhagen, Gerhard. *Das Hygiene-Institut der Berliner Universität unter der Leitung Robert Kochs 1883–1891*. Diss. med., Berlin (HU), 1983.

Eulner, Hans-Heinz. *Die Entwicklung der medizinischen Spezialfächer an den Universitäten des deutschen Sprachgebietes*. Stuttgart: Enke, 1970.

Evans, Alfred S. *Causation and Disease: A Chronological Journey*. New York, London: Plenam Medical Book Company, 1993.

Evans, Richard J. *Death in Hamburg: Society and Politics in the Cholera Years, 1830–1910*. Oxford: Clarendon, 1987.

Faber, Knud. *Nosography: The Evolution of Clinical Medicine in Modern Times*. New York: Paul B. Hoeber, 1930.

Fabian, Johannes. *Im Tropenfieber: Wissenschaft und Wahn in der Erforschung Zentralafrikas*. Munich: Beck, 2001.

Farley, John. *The Spontaneous Generation Controversy from Descartes to Opain*. Baltimore and London: Johns Hopkins University Press, 1967.

Fisch, Stefan. "Forschungsreisen im 19. Jahrhundert." In *Der Reisebericht. Die Entwicklung einer Gattung in der deutschen Literatur*, Peter J. Brenner, eds., 383–401. Frankfurt/Main: Suhrkamp 1989.

Fischer, Isidor. *Biographisches Lexikon der hervorragenden Ärzte der letzten fünfzig Jahre*. Munich and Berlin: Urban und Schwarzenberg, 1932/1933.

Fleck, Ludwik. *Genesis and Development of a Scientific Fact*. 1935. Reprinted, Chicago: University of Chicago Press, 1988.

Fleischer, Arndt. *Patentgesetzgebung und chemisch-pharmazeutische Industrie im deutschen Kaiserreich (1871–1918)*. Stuttgart: Deutscher Apotheker-Verlag, 1984.

Foster, William. *A History of Medical Bacteriology and Microbiology*. London: William Heinemann Medical Books, 1970.

Franceschini, Pietro. "Pacini, Antonio." In *Dictionary of Scientific Biography*, vol. 10, 266–68. New York: Scribner, 1974.

Frevert, Ute. *Krankheit als politisches Problem 1770–1880. Soziale Unterschichten in Preußen zwischen medizinischer Polizei und Sozialversicherung*. Göttingen: Vandenhoek and Ruprecht, 1984.

Geison, Gerald. *The Private Science of Louis Pasteur*. Princeton, N.J.: Princeton University Press, 1995.

Genschorek, Wolfgang. *Wegbereiter der Chirurgie. Joseph Lister, Ernst von Bergmann*. Leipzig: Hirtzel, 1984.

Göckenjahn, Gerd. *Kurieren und Staat machen. Gesundheit und Medizin in der bürgerlichen Welt*. Frankfurt/Main: Suhrkamp, 1985.

Goltz, Dietlinde. "'Das ist eine fatale Geschichte für unseren medizinischen Verstand.' Pathogenese und Therapie der Cholera um 1830." *Medizinhistorisches Journal* 33 (1998): 211–44.

Gorsboth, Thomas, and Bernd Wagner. "Die Unmöglichkeit der Therapie. Am Beispiel der Tuberkulose." *Kursbuch* 94 (Die Seuche) (1988): 123–45.

Goschler, Constantin. *Rudolf Virchow. Mediziner, Anthropologe, Politiker*. Köln: Böhlau 2002.

Gossel, Patricia Peck. "A Need for Standard Methods: The Case of American Bacteriology." In *The Right Tools for the Job: at Work in Twentieth Century Life Sciences*, Adele Clarke and Joan Fujimura, eds., 287–311. Princeton, N.J.: Princeton University Press, 1992.

Gradmann, Christoph. "'Auf Kollegen zum fröhlichen Krieg.' Popularisierte Bakteriologie im Wilhelminischen Zeitalter." *Medizin, Gesellschaft und Geschichte* 1 (1995): 35–54.

———. "Bazillen, Krankheit und Krieg. Bakteriologie und politische Sprache im deutschen Kaiserreich." *Berichte zur Wissenschaftsgeschichte* 19 (1996): 81–94.

———. "Die Entdeckung der Cholera in Indien—Robert Koch und die *DMW*." *Deutsche medizinische Wochenschrift* 124 (1999): 1253–56.

———. "Ein Fehlschlag und seine Folgen: Robert Kochs Tuberkulin und die Gründung des Instituts für Infektionskrankheiten in Berlin 1891." In *Strategien der Kausalität. Konzepte der Krankheitsverursachung im 19. und 20. Jahrhundert*, Christoph Gradmann and Thomas Schlich, 29–52. Pfaffenweiler: Centaurus, 1999.

———. "Geschichte als Naturwissenschaft: Ernst Hallier und Emil du Bois-Reymond als Kulturhistoriker." *Medizinhistorisches Journal* 5 (2000): 31–54.

———. "Invisible Enemies: Bacteriology and the Language of Politics in Imperial Germany." *Science in Context* 13 (2000): 9–30.

———. "Hermann von Helmholtz als Physiologe und Mediziner." In *Hermann von Helmholtz. Klassiker an der Epochenwende*, Helmut Klages and Heinz Lübbig, eds., 127–51. Braunschweig, 1998.

———. *Money, Microbes, and More: Robert Koch, Tuberkulin and the Foundation of the Institute for Infectious Diseases in Berlin 1891*. MPI Preprint No. 69. Berlin, 1997.

———. "Nur Helden in weissen Kitteln? Anmerkungen zur medizinhistorischen Biographik in Deutschland." In *Biographie schreiben*, Hans Erich Boedeker, ed., 243–84. Göttingen: Wallstein-Verlag, 2003.

———. "Robert Koch." In *Encyclopaedia of the Life Sciences*. London, 2001. www.els.net.

———. "Robert Koch and the Pressures of Scientific Research: Tuberculosis and Tuberculin." *Medical History* 5 (2001): 1–32.

———. "Robert Koch und das Tuberkulin—Anatomie eines Fehlschlags." *Deutsche medizinische Wochenschrift* 124 (1999): 1253–56.

Grafe, Alfred. "Die sogenannten Kochschen Postulate." *Gesnerus* 42 (1988): 411–18.

Gründer, Karlfried. *Geschichte der deutschen Kolonien*. 1985. Paderborn: Schöningh, 1995.

Haddad, George E. "Medicine and the Culture of Commemoration: Representing Robert Koch's Discovery of the Tubercle Bacillus." *Osiris* 1 (2000): 118–37.

Hagner, Michael. "Mikro-Anthropologie und Fotografie. Gustav Fritschs Haarspaltereien und die Klassifizierung der Rassen." In *Ordnungen der Sichtbarkeit. Fotografie in Wissenschaft, Kunst und Technologie*, Peter Geimer, ed., 252–84. Frankfurt: Suhrkamp, 2002.

Hansen, Bert. "New Images of a New Medicine: Visual Evidence for the Widespread Popularity of Therapeutic Discoveries in America after 1885." *Bulletin of the History of Medicine* 73 (1999): 629–78.

Harden, Victoria A. "Koch's Postulates and the Etiology of AIDS: A Historical Perspective." *History and Philosophy of the Life Science* 14 (1992): 249–69.

Hardy, Anne. "Relapsing Fever." In *The Cambridge World History of Human Disease*, Kenneth F. Kiple, ed., 967–70. Cambridge: Cambridge University Press, 1994.

Harrison, Mark. "A Question of Locality: The Identity of Cholera in British India, 1860–1890." In *Warm Climates and Western Medicine: The Emergence of Tropical Medicine, 1500–1900*, David Arnold, ed., 133–59. Atlanta: Rodopi, 1996.

Heinicke, Petra-Heike, and Klaus Heinicke. *Zur Geschichte des Lehrstuhls für Hygiene an der Universität zu Berlin von der Gründung bis zur Berufung Max Rubners*. Diss. med., HU Berlin, 1979.

Hellmuth, Edith, and Wolfgang Mühlfriedel. *Zeiss 1846–1905. Vom Atelier für Mechanik* zum führenden Unternehmen des optischen Gerätebaus. Weimar, Köln, Vienna: Böhlau, 1996.

Helvoort, Ton van. "A Bacteriological Paradigm in Influenza Research in the First Half of the Twentieth Century." *History and Philosophy of the Life Sciences* 15 (1993): 3–21.

———. "Bacteriological and Physiological Research Styles in the Early Controversy on the Nature of the Bacteriophage Phenomenon." *Medical History* 26 (1992): 243–70.

Heninger, Johannes. "Leeuwenhoek, Antoni." In *Dictionary of Scientific Biography*, vol. 8, 126–30. New York: Charles Scribner's Sons, 1973.

Hess, Volker, ed. *Der wohltemperierte Mensch. Wissenschaft und Alltag des Fiebermessens (1850–1900)*. Frankfurt/Main and New York: Campus, 2000.

———. "Disease as Parasite: The Discovery of Time for a Theory of Parasites." In *Pathology in the 19th and 20th Centuries: The Relationship between Theory and Practice*, Cay-Rüdiger Prüll, ed., 11–30. Sheffield: EAHMH Publications, 1998.

———. *Normierung der Gesundheit. Messende Verfahren der Medizin als kulturelle Praktik um 1900*. Husum: Matthieson, 1997.

Heymann, Bruno. "Robert Koch als Patient in seinen letzten Lebenstagen. Zum Gedenken an Bruno Heymann aus seinen nachgelassenen Papieren herausgegeben von Walter Artelt." *Medizinhistorisches Journal* 1 (1966): 156–60.

———. *Robert Koch Biographie*. Part 2. Georg Henneberg, Klaus Janitschke, Klaus Stürzbecher, und Rolf Winau, eds. Berlin: Robert Koch–Institute, 1997.

———. *Robert Koch*. Part 1, 1843–82. Leipzig: Akademische Verlagsanstalt, 1932.

Hintzsche, Erich. "Henle, Friedrich Gustav Jacob." In *Dictionary of Scientific Biography*, vol. 6, 268–70. New York: Charles Scribner's Sons, 1972.

Holmes, Frederic L. "The Fine Structure of Scientific Creativity." *History of Science* 19 (1981): 60–70.

———. "The Old Martyr of Science: The Frog in Experimental Physiology." *Journal of the History of Biology* 26 (1993): 311–28.

———. "Scientific Writing and Scientific Discovery." *Isis* 78 (1987): 220–35.

Holz, Ehrentraut. *Robert Koch (1841–1910). Auswahlbibliographie*. Berlin: HU Bibliothek, 1981.

Hoppe, Brigitte. "Die Biologie der Mikroorganismen von F. J. Cohn." *Sudhoffs Archiv* 67 (1983): 158–89.

Hubenstorf, Michael. "Die Genese der Sozialen Medizin als universitäres Lehrfach in Österreich bis 1914." Freie Universität Berlin, diss., 1999.

Huentelmann, Axel Cäsar. *Die Medizin "ist bloß eine Fortsetzung der Politik mit anderen Mitteln." Das Kaiserliche Gesundheitsamt/Reichsgesundheitsamt von seiner Gründung 1876 bis zum Ende der Weimarer Republik 1926/1933*. PhD thesis, Humboldt-Universität, Berlin, 2003.

Huerkamp, Claudia. *Der Aufstieg der Ärzte im 19. Jahrhundert. Vom gelehrten Stand zum professionellen Experten*. Göttingen: Vandenhoek and Ruprecht, 1985.

Jahn, Ellen. *Die Cholera in Medizin und Pharmazie im Zeitalter des Hygienikers Max von Pettenkofer*. Stuttgart: Steiner, 1994.

Jahn, Ilse. *Grundzüge der Biologiegeschichte*. Jena: Gustav Fischer Verlag, 1990.

———. "Hallier, Ernst." In *Neue Deutsche Biographie*, Historische Kommission der Bayrischen Akademie der Wissenschaften, ed., vol. 7, 563–64. Berlin: Duncker & Humblot, 1966.

Jansen, Sarah. *"Schädlinge." Geschichte eines wissenschaftlichen und politischen Konstrukts; 1840–1920.* Frankfurt: Campus, 2003.
Johnston, William D. "Tuberculosis." In *The Cambridge World History of Human Disease*, Kenneth F. Kiple, ed., 1059–68. Cambridge: Cambridge University Press, 1994.
Jusatz, H. J. "Robert Kochs Bedeutung für die Wandlung der Tropenmedizin am Ende des 19. Jahrhunderts." *Deutsche Medizinische Wochenschrift* 100 (1975): 1933–36.
Kaufmann, Stefan. "Robert Koch: Höhen und Tiefen auf der Suche nach einem Heilmittel gegen Tuberkulose." Robert-Koch-Stiftung e.V., *Beiträge und Mitteilungen* 26 (2002): 5–14.
King, Lester. *Medical Thinking: A Historical Preface.* Princeton, N.J.: Princeton University Press, 1982.
Kirchberger, Ulrike. "Deutsche Naturwissenschaftler im britischen Empire. Die Erforschung der außereuropäischen Welt im Spannungsfeld zwischen deutschem und britischem Imperialismus." *Historische Zeitschrift* 271 (2000): 621–60.
Kleine, Friedrich Karl. *Ein deutscher Tropenarzt.* Hannover: Schmorl und von Seefeld, 1949.
Köhler, Robert E. "Bacterial Physiology: The Medical Context." *Bulletin of the History of Medicine* 59 (1985): 54–74.
Köhler, Werner, and Hanspeter Mochmann. "Edwin Klebs (1834–1913), Pathologe und Wegbereiter der Bakteriologie. Zum Gedenken an seinen 75. Todestag." *Zeitschrift für ärztliche Fortbildung* 82 (1988): 1037–42.
Kolle, Kurt, ed. *Robert Koch, Briefe an Wilhelm Kolle. Mit einem Geleitwort von Georg B. Gruber.* Stuttgart: Georg B. Gruber, 1959.
Kollenbrock-Netz, Jutta. "Wissenschaft als nationaler Mythos. Anmerkungen zur Haeckel-Virchow-Kontroverse auf der 50. Jahresversammlung der deutschen Naturforscher und Ärzte in München (1877)." In *Nationale Mythen und Symbole in der zweiten Hälfte des 19. Jahrhunderts*, Jürgen Link und Wulf Wülfling, ed., 212–36. Stuttgart: Klett-Cotta 1991.
Kühn, Bettina. *Robert Kochs Bedeutung für die Tropenmedizin anhand seiner Protozoenforschung, insbesondere des Studiums der Schlafkrankheit.* Diss. med., Martin-Luther-Universität Halle-Wittenberg, 1994.
Labisch, Alfons. *Homo Hygienicus. Gesundheit und Medizin in der Neuzeit.* Frankfurt/Main: Campus, 1992.
Latour, Bruno. "Give Me a Laboratory and I will Raise the World." In *Science Observed: Perspectives on the Social Study of Science*, Karin Knorr-Cetina und Michael Mulkay, ed., 141–70. London: Sage, 1983.
———. *The Pasteurisation of France.* 1984. Translation, Cambridge, Mass., and London: Harvard University Press, 1988.
———. *Science in Action: How to Follow Scientists and Engineers through Society.* Cambridge, Mass.: Harvard University Press, 1987.
Latour, Bruno, and Steve Woolgar. *Laboratory Life: The Construction of Scientific Facts.* Princeton, N.J.: Princeton University Press, 1979.
Lechevalier, H. A., and M. Solotorovsky. *Three Centuries of Microbiology.* New York: Dover, 1974.
Lederer, Susan E. *Subjected to Science. Human Experimentation in America before the Second World War.* Baltimore and London: Johns Hopkins University Press, 1995.
Lehman, K. B. "Frohe Lebensarbeit." Munich: J. F. Lehmann, 1924.

Leibowitz, David. "Scientific Failure in an Age of Optimism: Public Reaction to Robert Koch's Tuberculin Cure." *New York State Journal of Medicine* 93 (1993): 41–48.

Lennox, John. "Those Deceptively Simple Postulates of Professor Robert Koch." *American Biology Teacher* 47 (1985): 216–21.

Lenoir, Timothy. *Politik im Tempel der Wissenschaft. Forschung und Machtausübung im deutschen Kaiserreich.* Frankfurt/Main: Campus, 1992.

Leven, Karl-Heinz. *Die Geschichte der Infektionskrankheiten. Von der Antike bis ins 20. Jahrhundert.* Landsberg/Lech: Ecomed, 1997.

Liebenau, Jonathan. "Paul Ehrlich as Commercial Scientist and Research Administrator." *Medical History* 34 (1990): 65–78.

Logan, Cheryl A. "Before There Were Standards: The Role of Test Animals in the Production of Empirical Generality in Physiology." *Journal of the History of Biology* 35 (2002): 329–63.

Löwy, Ilana. "On Hybridizations, Networks and New Disciplines: The Pasteur-Institute and the Development of Microbiology in France." *Studies in the History and Philosophy of Science* 25 (1994): 655–88.

Lundgreen, Peter, Bernd Horn, Wolfgang Krohn, Günter Küppers, Rainer Paslack. *Staatliche Forschung in Deutschland 1870–1980.* Frankfurt/Main: Campus, 1986.

Lyons, Maryinez. "African Trypanosomiasis." In *The Cambridge World History of Human Disease*, Kenneth F. Kiple, ed., 552–61. Cambridge: Cambridge University Press, 1993.

———. *The Colonial Disease: A Social History of Sleeping Sickness in Northern Zaire, 1900–1940.* Cambridge: Cambridge University Press, 1992.

———. "Sleeping Sickness and Public Health in the Belgian Congo, 1903–30." *Society for the Social History of Medicine Bulletin* 39 (1986): 44–46.

MacKenzie, John M. *Imperialism and the Natural World.* Manchester and New York: Manchester University Press, 1990.

Maehle, Andreas-Holger. "Assault and Battery, or Legitimate Treatment? German Legal Debates on the Status of Medical Interventions Without Consent, c. 1890–1914." *Gesnerus* 57 (2000): 206–21.

———. *Kritik und Verteidigung des Tierversuchs. Die Anfänge der Diskussion im 17. und 18. Jahrhundert.* Stuttgart: Steiner, 1992.

———. "Zwischen medizinischem Paternalismus und Patientenautonomie: Albert Molls 'Ärztliche Ethik' (1902) im historischen Kontext." In *Medizingeschichte und Medizinethik. Kontroversen und Begründungsansätze 1900–1950*, Andreas Frewer und Josef N. Neumann, eds., 44–56. Frankfurt/Main: Campus, 2001.

Maiwald, Matthias. "Charakterisierung 'neuer' Infektionserreger: eine Herausforderung an die Kochschen Postulate." *Hygiene und Mikrobiologie* (1998): 16–26.

Martin, Emily. *Flexible Bodies: Tracking Immunity in American Culture. From the Days of Polio to the Age of Aids.* Boston: Beacon Press, 1994.

Martin, Michael. "Die Bedeutung und Funktion des medizinischen Messens in geschlossenen Patienten-Kollektiven. Das Beispiel der Lungensanatorien." In *Normierung der Gesundheit. Messende Verfahren der Medizin als kulturelle Praktik um 1900*, Volker Hess, ed., 145–64. Husum: Matthiesen, 1997.

Maulitz, Russell C. "Physician versus Bacteriologist: The Ideology of Science in Clinical Medicine." In *The Therapeutic Revolution. Essays in the Social History of American Medi-*

*cine*, M. J. Vogel and C. E. Rosenberg, eds., 91–107. Philadelphia: University of Pennsylvania Press, 1979.

———. "Robert Koch in the United States of America." *NTM: Zeitschrift für Geschichte der Naturwissenschaft, Technik und Medizin* 20 (1983): 75–84.

———. "Rudolf Virchow, Julius Cohnheim and the Program of Pathology." *Bulletin of the History of Medicine* 52 (1978): 162–82.

Mazumdar, Pauline M. H. *Species and Specificity: An Interpretation of the History of Immunology*. Cambridge: Cambridge University Press, 1995.

Mendelsohn, John Andrew. *Cultures of Bacteriology: Formation and Transformation of a Science in France and Germany, 1870–1914*. Diss. phil., Princeton University, 1996.

———. "Von der 'Ausrottung' zum Gleichgewicht: Wie Epidemien nach dem Ersten Weltkrieg komplex wurden." In *Strategien der Kausalität. Konzepte der Krankheitsverursachung im 19. und 20. Jahrhundert*, Christoph Gradmann and Thomas Schlich, eds., 227–71. Pfaffenweiler: Centaurus, 1999.

Métraux, Alexandre. "Reaching the Invisible: A Case Study of Experimental Work in Microbiology (1880–1900)" In *Social Organisation and Social Process. Essays in Honor of Anselm Strauss*, David R. Maines, ed., 249–60. New York: Aldine de Gruyter, 1991.

Miller, David Philip, and Hans Peter Reill, eds. *Visions of Empire: Voyages, Botany, and Representations of Nature*. Cambridge, 1996.

Möllers, Bernhard. *Robert Koch. Persönlichkeit und Lebenswerk 1843–1910*. Hannover: Schmorl und von Seefeld, 1950.

Müller-Jahnke, Wolf-Dieter, and Christoph Friedrich. *Geschichte der Arzneimitteltherapie*. Stuttgart: Deutscher Apotheker Verlag, 1996.

Münch, Ragnhild. *Robert Koch und sein Nachlass in Berlin*. Berlin and New York: Aldine de Gruyter, 2003.

Münch, Ragnhild, and Stefan S. Biel. "Expedition, Experiment und Expertise im Spiegel des Nachlasses von Robert Koch." *Sudhoff's Archiv* 82 (1998): 1–29.

Myers, J. Arthur. "Development of Knowledge of Unity of Tuberculosis and of the Portals of Entry of Tubercule Bacilli." *Journal of the History of Medicine and Allied Sciences* 29 (1974): 213–28.

Nolte, Oliver. "Autovakzine—Ein Überblick." *Der Mikrobiologe* 11 (2001): 1116.

Ogawa, Mariko. "Uneasy Bedfellows: Science and Politics in the Refutation of Koch's Bacterial Theory of Cholera." *Bulletin of the History of Medicine* 74 (2000): 671–707.

Opitz, Bernhard. "Robert Koch's Ansichten über die zukünftige Gestaltung des Kaiserlichen Gesundheitsamtes." *Medizinhistorisches Journal* 29 (1994): 363–77.

Opitz, Bernhard, and Herwarth Horn. "Die Tuberkulinaffäre. Neue medizinhistorische Untersuchungen zum Kochschen Heilverfahren." *Zeitschrift für die gesamte Hygiene* 30 (1984): 731–34.

Opitz, Gerda. *Tierversuche und Versuchstiere in der Geschichte der Biologie und Medizin*. Diss. rer. nat., Friedrich-Schiller-Universität, Jena, 1968.

Osterhammel, Jürgen. "Forschungsreise und Kolonialprogramm. Ferdinand von Richthofen und die Erschließung Chinas im 19. Jahrhundert." *Archiv für Kulturgeschichte* 69 (1987): 150–95.

Otis, Laura. *Membranes: Metaphors of Invasion in Nineteenth-Century Literature, Science, and Politics*. Baltimore: Johns Hopkins University Press, 1999.

Perleth, Mathias. "The Discovery of Chagas' Disease and the Formation of the Early

Chagas' Disease Concept." *History and Philosophy of the Life Sciences* 19 (1997): 211–36.
Pfohl, Gerhard. "Und Naunyn hat's doch gesagt," *Die medizinische Welt* 38 (1987): 597–600.
Pickstone, John V., ed. "Medical Innovations in Historical Perspective." Houndsmills, Basingstoke: Macmillan, 1992.
Power, Helen J. *Tropical Medicine in the Twentieth Century: A History of the Liverpool School of Tropical Medicine 1898–1990*. London and New York: Keagan Paul International, 1999.
Radday, Helmut, ed. *… und geben unsere Auswanderung bekannt. Ein Beitrag zur Sozialgeschichte des Oberharzes am Beispiel des Familienverbandes Koch*. Clausthal-Zellerfeld: Oberharzer Geschichts–und Museums-Verein, 2000.
Rather, L. J. "Virchow und die Entzündungsfrage im neuzehnten Jahrhundert." In *Verhandlungen des 20. Internationalen Kongresses für Geschichte der Medizin, Berlin, 22–27 August 1966*, Heinz Goerke, ed., 161–77. Hildesheim: Olms, 1968.
*Das Reichsgesundheitsamt 1876–1926. Festschrift hg. vom Reichsgesundheitsamt aus Anlaß seines fünfzigjährigen Bestehens*. Berlin: J. Springer, 1926.
Reim, Ulrike. "Der Robert Koch—Film (1939) von Hans Steinhoff. Kunst oder Propaganda?" In *Medizin im Spielfilm des Nationalsozialismus*, Udo Benzenhöfer, ed., 22–33. Tecklenburg: Burgverlag, 1990.
Revill, George, and Richard Wrigley. *Pathologies of Travel*. Amsterdam: Rodopi, 2000.
Rheinberger, Hans-Jörg. *Experimentalsysteme und epistemische Dinge. Eine Geschichte der Proteinsynthese im Reagenzglas*. Göttingen: Wallstein Verlag, 2001.
Rheinberger, Hans-Jörg, and Michael Hagner, eds. *Die Experimentalisierung des Lebens. Experimentalsysteme in den biologischen Wissenschaften 1850/1950*. Berlin: Akademie Verlag, 1993.
Rheinberger, Hans-Jörg, Michael Hagner, and Bettina Wahrig-Schmidt, eds. *Räume des Wissens. Repräsentation, Codierung, Spur*. Berlin: Akademie Verlag, 1997.
Riethmiller, Steven. "Ehrlich, Bertheim, and Atoxyl." *Bulletin of the History of Chemistry* 23 (1999): 28–33.
Ritter, Gerhard A. *Grossforschung und Staat in Deutschland. Ein historischer Überblick*. Munich: Beck, 1992.
Rosenkrantz, Barbara Gutmann. *From Consumption to Tuberculosis: A Documentary History*. New York and London: Garland Publishers, 1994.
———. "The Trouble with Bovine Tuberculosis." *Bulletin of the History of Medicine* 59 (1985): 155–75.
Roudolf, L. "Die wissenschaftliche Bibliothek Robert Kochs." *Zentralblatt für Bakteriologie, Parasitenkunde, Infektionskrankheiten und Hygiene* 1, no. 5 (1960): 447–72.
Rupke, Nicolas, ed. *Medical Geography in Historical Perspective*. London: Wellcome Trust Centre for the History of Medicine at UCL, 2000.
———. *Vivisection in Historical Perspective*. London and New York: Routledge, 1987.
Sarasin, Philipp. *'Anthrax.' Bioterror als Phantasma*. Frankfurt/Main: Suhrkamp, 2004.
———. "Die Visualisierung des Feindes. Über metaphorische Technologien der frühen Bakteriologie." *Geschichte und Gesellschaft* 30 (2004): 250–76.
Sarasin, Philipp, and Jakob Tanner, eds. *Physiologie und industrielle Gesellschaft*. Frankfurt, 1998.

Sauerteig, Lutz. "Ethische Richtlinien, Patientenrechte und ärztliches Verhalten bei der Arzneimittelerprobung 1892–1931." *Medizinhistorisches Journal* 35 (2000): 303–34.
Schadewaldt, Hans. "Die Entdeckung des Tuberkulins." *Deutsche Medizinische Wochenschrift* 100 (1975): 1925–32.
Schlich, Thomas. "Ein Symbol medizinischer Fortschrittshoffnung. Robert Koch entdeckt den Erreger der Tuberkulose." In *Meilensteine der Medizin*, Heinz Schott, ed., 368–74. Düsseldorf: Harenberg, 1996.
———."Die Kontrolle notwendiger Krankheitsursachen als Strategie der Krankheitsbeherrschung im 19. und 20. Jahrhundert." In *Strategien der Kausalität. Konzepte der Krankheitsverursachung im 19. und 20. Jahrhundert*, Christoph Gradmann and Thomas Schlich, eds., 3–28. Pfaffenweiler: Centaurus, 1999.
———. "Repräsentationen von Krankheitserregern. Wie Robert Koch Bakterien als Krankheitserreger dargestellt hat." In *Räume des Wissens. Repräsentation, Codierung, Spur*, Hans-Jörg Rheinberger, Michael Hagner, und Bettina Wahrig-Schmidt, eds., 165–90. Berlin: Akademie Verlag 1997.
———. "Wichtiger als der Gegenstand selbst—Die Bedeutung des fotografischen Bildes in der Begründung der bakteriologischen Krankheitsauffassung durch Robert Koch." In *Neue Wege in der Seuchengeschichte*, Martin Dinges and Thomas Schlich, eds., 143–74. Stuttgart: Steiner, 1995.
Schmidt, Josef M. "Geschichte der Tuberkulin-Therapie—Ihre Begründung durch Robert Koch, ihre Vorläufer und ihre weitere Entwicklung." *Pneumologie* 45 (1991): 776–84.
Shortland, Michael, and Richard Yeo, eds. *Telling Lives in Science: Essays on Scientific Biography*. Cambridge: Cambridge University Press, 1996.
Silverstein, Arthur. *A History of Immunology*. San Diego: Academic Press, 1989.
Skrobacki, Andrej. "Robert Koch und die Polen. Ein Beitrag zur Genese Robert Kochs wissenschaftlicher Untersuchungen." *Zeitschrift für ärztliche Fortbildung* 85 (1992): 183–85.
———. "Robert Koch in Wolsztyn." In *Robert Koch: (1843–1910)*, Wolfram Kaiser and Hans Hübner, eds., 51–61. Halle: Martin Luther–Universität Halle–Wittenberg, 1983.
Smith, Francis Barrymore. *The Retreat of Tuberculosis 1850–1950*. London: Croom Helm, 1987.
Stürzbecher, Manfred. "Klebs, Edwin." In *Neue Deutsche Biographie, hg. von Bayrische Akademie der Wissenschaften*, vol. 11, 719–20. Berlin: Dunker & Humblot, 1977.
Summers, William C. "Pasteur's 'Private Science.'" *The New York Review of Books* 44 (1997).
Szöllösi-Janze, Margit. *Fritz Haber 1868–1934; eine Biographie*. Munich: Beck, 1998.
Thagard, Paul. *How Scientists Explain Disease*. Princeton, N.J.: Princeton University Press, 1999.
Théodoridès, Jean. "Hallier, Ernst Hans." In *Dictionary of Scientific Biography*, Charles Coulston Gillespie, ed., vol. 6, 72–73. New York: Charles Scribner's Sons, 1972.
Throm, Carola. *Das Diphterieserum. Ein neues Therapieprinzip, seine Entwicklung und Markteinführung*. Stuttgart: Wissenschaftiche Verlags-Gesellschaft, 1995.
Tomes, Nancy. *The Gospel of Germs: Men, Women, and the Microbe in American Life*. Cambridge, Mass.: Harvard University Press, 1998.
Tomes, Nancy J., and John Harley Warner. "Introduction to the Special Issue on Rethink-

ing the Reception of the Germ Theory of Disease: Comparative Perspectives." *Journal of the History of Medicine and Allied Sciences* 52 (1997): 7–16.

Travis, Anthony. "Science as Receptor of Technology: Paul Ehrlich and the Synthetic Dyestuffs." *Science in Context* 3 (1989): 383–408.

Tröhler, Ulrich, and Andreas-Holger Maehle. "Antivivisection in Nineteenth-Century Germany and Switzerland: Motives and Models." In *Vivisection in Historical Perspective*, Nicolas Rupke, ed., 149–87. London and New York: Routledge, 1987.

Vasold, Manfred. *Robert Koch, der Entdecker von Krankheitserregern.* Heidelberg: Spektrum der Wissenschaften-Verlagsgesellschaft, 2002.

vom Brocke, Bernhard, ed. *Wissenschaftsgeschichte und Wissenschaftspolitik im Industriezeitalter. Das 'System Althoff' in historischer Perspektive.* Hildesheim: Lax, 1991.

Wainwright, Milton. *Miracle Cure: The Story of Penicillin and the Golden Age Of Antibiotics.* Cambridge, Mass.: Basil Blackwell, 1990.

Walter, Heinz. "Külz, Eduard." In *Neue Deutsche Biographie*, hg. von Bayrische Akademie der Wissenschaften, vol. 1, 210. Berlin: Duncker & Humblot, 1982.

Warner, John Harley. "The History of Science and the Sciences of Medicine." *Osiris* 10 (1995): 164–93.

Watts, Sheldon. *Epidemics and History: Disease, Power and Imperialism.* New Haven, Conn.: Yale University Press, 1997.

Weindling, Paul. *Epidemics and Genocide in Eastern Europe 1890–1945.* Oxford: Oxford University Press, 2000.

———. "From Medical Research to Clinical Practice: Serum Therapy for Diphtheria in 1890s." In *Medical Innovations in Historical Perspective*, John V. Pickstone, ed., 72–83. Houndmills, Basingstoke: Macmillan, 1992.

———. *Health, Race, and German Politics Between National Unification and Nazism 1870–1945.* Cambridge: Cambridge University Press, 1989.

———. "Scientific Elites and Laboratory Organisation in Fin de Siècle Paris and Berlin. The Pasteur Institute and Robert Koch's Institute for Infectious Diseases Compared." In *The Laboratory Revolution in Medicine*, Andrew Cunningham and Perry Williams, eds., 170–88. Cambridge: Cambridge University Press, 1992.

Wells, Herbert George. "The Stolen Bacillus." In *The Stolen Bacillus and Other Incidents.* London: Methuen, 1912.

Whitrow, Magda. *Julius Wagner-Jauregg (1857–40).* London: Smith-Gordon, 1991.

———. "Wagner-Jauregg and Fever Therapy." *Medical History* 34 (1990): 294–310.

Wilson, Catherine. *The Invisible World: Early Modern Philosophy and the Invention of the Microscope.* Princeton, N.J.: Princeton University Press, 1995.

Wilson, Leonard G. "The Historical Decline of Tuberculosis in Europe and America: Its Causes and Significance." *Journal of the History of Medicine and Allied Sciences* 45 (1990): 366–96.

Wimmer, Wolfgang. *"Wir haben fast immer was Neues": Gesundheitswesen und Innovation der Pharma-Industrie in Deutschland 1880–1935.* Berlin: Duncker & Humblot, 1994.

Winau, Rolf. "Medizin und Menschenversuch. Zur Geschichte des 'informed consent.'" In *Medizin und Ethik im Zeichen von Auschwitz*, Claudia Wiesemann and Andreas Frewer, eds., 13–29. Erlangen und Jena: Palm und Enke, 1996.

———. "Serumtherapie: Die Entdeckung eines bahnbrechenden Therapieprinzips im Jahr 1890." *Deutsche Medizinische Wochenschrift* 115 (1990): 1883–86.

Wohlrab, Frank, and Ulf Henoch. "Zum Leben und Wirken von Carl Weigert (1845–

1904) in Leipzig." *Zentralblatt für allgemeine Pathologie und pathologische Anatomie* 134 (1988): 743–51.

Wolf, Claudia. *Georg Gaffky. Erster Vertreter der Hygiene in Giessen von 1888 bis 1904.* Giessen: Wilhelm Schmitz, 1992.

Worboys, Michael. "The Comparative History of Sleeping Sickness in East and Central Africa, 1900–1914." *History of Science* 32 (1994): 89–102.

———. "The Sanatory Treatment for Consumption in Britain 1890–1914." In *Medical Innovations in Historical Perspective*, John V. Pickstone, ed., 47–71. Houndsmills, Basingstoke: Macmillan, 1992.

———. *Spreading Germs: Disease Theories and Medical Practice in Britain, 1985–1900*. Cambridge: Cambridge University Press, 2000.

Zeiss, Heinz, and Richard Bieling. *Behring. Gestalt und Werk*. Berlin: Bruno Schultz Verlag, 1940.

Zepernik, Bernhard. "Zwischen Wirtschaft und Wissenschaft—die deutsche Schutzgebiets-Botanik." *Berichte zur Wissenschaftsgeschichte* 13 (1990): 207–17.

# Index

*Achorion schönleini*, 23
Alexandria (Egypt), working conditions in, 184–85
Althoff, Friedrich, 102, 105, 111, 126, 146
Amani (East Africa), research station at, 207, 210–11
anatomy, pathological, 8, 24, 30, 44; concept of, 83; in tuberculin research, 133
aniline dyes, 53; trials with, 93
*animalcula*, 23, 28
animal experiments, 13, 48, 49, 58; limited usefulness of, 125; as model, 59; replication of, 145. *See also* experimentation
animal model: of disease process, 66; relevance of, 61
animals, experimental, 64; choice of, 64, 67; as culturing aparatus, 33, 68. *See also* guinea pigs
anthrax, 45–54 passim; first demonstration of Koch's work on, 20; Koch's first study of, 12; vaccine for, 91, 182
antipyresis, 135
antivivisectionism, 66, 118
arsenic, 166
atoxyl, 159–70; dangers of therapy with, 217, 219; dosages of, 161; resistance to, 166; side effects of, 163; testing of on Sese Islands, 150–61, 214, 215–16
auramine, 93
auto-vaccination, 107

bacilli, attenuation of, 91
*bacillus anthracis*, 21; life cycle of, 12; spores of, 51

*bacillus subtilis*, 50
bacillus, tubercle: identification of, 8, 69; spores of, 77–78, 260n220
bacteria, 1; attenuation of, 91; classification of, 58; identification of, 2, 56; monomorphism and polymorphism of, 45; morphological characteristics of, 57; physiology of, 114; spore phase of, 49; staining of, 2, 52; transformation of, 14. *See also* "fraud bacillus"; microbe hunting; *specific species*
bacterial species, constancy of, 31–32, 37, 40, 44, 61, 77; theory of, 49
bacterial specificity, 3
bacteriological knowledge, transfer of, 9
bacteriologists, prestige of, 116
bacteriology, medical: development of, 175, 228, 230; place of, in science, 5, 7, 9, 231; political language of, 231; popularity of, 7
Bassi, Antonio, 23, 28
Baumgarten, Paul, 76, 99, 104
Bayle, Pierre, 81
Behring, Emil von, 3, 17, 112, 149; Koch's resentment of, 176
Bengal, swamps of, 232
Bergmann, Ernst von, 126, 134
Berlin University, Koch's position at, 15, 90
Biewend, Eduard, 26
Billroth, Theodor, 41, 46, 47, 55
Birch-Hirschfeld, Felix Victor, 26, 37, 41, 43
Bois-Raymond, Emil du, 24
Bollinger, Friedrich August, 49, 50, 52
botany, 8
Brefeld, Oscar, 37

Breslau University, 40, 74; Koch's position at, 13
Brieger, Ludwig, 17
Broemel (delegate), 147
Bruce, David, 203, 223
Buchner, Eduard, 19, 61, 110
Bukoba, East Africa, 202
Bumangi, Sese Islands, 213
Bumm, Franz, 207, 219
Burden-Sanderson, John, 40, 184

Calcutta, working conditions in, 188
Canetti, Elias, 233
Canguilhem, Georges, 3, 6, 248n216, 256n116
Caprivi, Georg Leo, 102
carrier: asymptomatic, 177; healthy, 17, 111
Casella Company, 163
Castellani, Aldo, 203
cell, giant, 87
Chaveau, A., 40
chemotherapy, 118, 120, 149; history of, 155; testing of, on Africans, 221
cholera: character of, 182; compared to sleeping sickness, 226–27; pathogen of, 182; popular image of, 198
cholera bacilli, pure cultures of, 189
cholera epidemic, 11, 16, 34, 37, 183–84; political implications of, 16
cholera expedition: cost of, 199; of 1866, 14; of 1883–84, 173; French, 183, 187; scientific yield of, 200
chromatogeny, 47
climate therapy, 93
*Coccobacteria septica*, 21, 47, 63
Cohn, Ferdinand Julius, 12, 20, 31, 37, 49–51
Cohnheim, Julius, 12, 20, 43, 52, 67, 74
Coler, Alwin von, 126
colonial human economy, 205
colonialism, science as a resource for, 181, 201; ideology of, 221
comma bacillus, discovery of, 14, 188
Congo Basin, 20
Congress for Internal Medicine, Tenth, 119
Congress of Hygiene, 182
*contagium vivum*, 60
Cornet, Georg, 92, 124, 129
crocodiles, 223

culture media, solid, 13, 75, 184; human body as, 110; liquefying of, 185, 192
cultures, pure, 37, 75, 80, 249n234; experiments with, 80. See also animals, experimental
culturing: apparatuses for, 33, 65; experimentation with, 43, 55; nutrient, 80; techniques for, 41, 47, 181
Cunningham, J. M., 188
Czerny, Vinzenz, 135, 141

Damiette, Egypt, 185
Davaine, Casimir Joseph, 12, 48, 49, 55
de Bary, Anton, 31, 34, 36
Dehmel, Richard, 266n85
Dettweiler, Peter, 146
diphtheria serum, 118, 149, 154
disease, infectious: animal model of, 61; bacteriological concept of, 68, 109–10; early concepts of, 24, 27; invasion model of, 190, 231; Koch's understanding of, 224; laboratory model of, 60; vector-dependent, 178
disinfection: experiments with, failure of, 95; internal, 99
doctor-patient relationship, 9
Dönitz, Wilhelm, 157, 203
drinking water, Calcutta, 189, 191
dye therapy, 164–65. See also chemotherapy

East Africa, expedition to, 173
east coast fever, 17, 172, 177
Egypt: cholera epidemic in, 183; Koch's flight to, 103
Ehrenberg, Johann Gottfried, 23
Ehrlich, Paul, 3, 17, 69, 76, 112; relations of, with chemical industry, 221; as tuberculosis patient, 124
Eidam, Eduard, 52
Entebbe, British Uganda, laboratories in, 213
epidemics, 7; cholera, 11, 16, 34, 37, 183–84; popular imagery of, 232; tank, 109
epidemiology, 17, 111; Koch's turn toward, 176–77, 188
epizootics, 172, 177
ethics, medical, 120, 152
etiology, concept of, 83
exhibition, hygiene, 181

experimental subjects, human, 122; absence of consent of, 151; consent of, 153
experimental techniques, improvements in, 53
experimentation: animal, 121, 169; discussion of, in House of Representatives, 147–48; human, 118, 120, 130–31, 132, 141, 148, 169; infection, example of, 58–59; secrecy of, 170; self-, 121, 140; therapeutic, 120, 148, 220
extraction methods, 258n17

Feldmann, Karl, 202, 211–12
fermentation, 28, 31
fever: concept of, 152; east coast, 17, 172, 177; horse, 177; measuring of, 24; paroxysm, 144; as reaction to tuberculin, 143–45; relapsing, 48, 246n150; in sleeping sickness, 159–70 passim; typhoid, 191
fever therapy, 149
Fleck, Ludwik, 71, 225
Fracastoro, Girolamo, 22–23
Franco-Prussian war, 11, 41, 195
Fräntzel, Oscar, 91, 122, 130
"fraud bacillus," 152. *See also* "tuberculin affair"; "tuberculin swindle"
Freiberg, Hedwig, 16, 121, 210, 240n58
fuchsin, 164, 168, 273n281. *See also* chemotherapy
Fülleborn, Ernst, 202, 207
fungi, parasitical, 30
Fürbringer, Paul, 146

Gaffky, Georg, 13, 92; on sleeping sickness, 157
*Gartenlaube* (magazine), 198
Geison, Gerald, 4
genus *bacillus*, 246n165
geography, medical, 188
German Mandated Territory, 202
"German science," national solicitude for, 103, 146, 196, 217
*glossina palpalis*, 204, 210. *See also* tsetse fly
Goldschmidt (delegate), 147
Gossler, Gustav von, 126
Göttingen University, 25
grant, government, 104

Grawitz, Ernst, 99
Greifswald, tuberculin celebration in, 119
guinea pigs, 67, 78, 107, 121; susceptibility of, 124

haleridia, 283n195
Hallier, Ernst, 32–37; Koch's reading of, 183
"harmony of illusions," 107–8
Hasse, Karl, 11, 25, 241n10
healing, spontaneous, 86
Henle, Jacob, 8, 11, 25–29
Henoch, Eduard, on tuberculin, 131–32, 144
Héricourt, Jules H., 104
Hiller, Arnold, 55
Hoechst Company, 149, 270n211
Hueppe, Ferdinand, 13, 110, 132, 176, 257n152
Hueter, Carl, 41
human body, as culture medium, 110
human economy, colonial, 205
hunting, Koch's pursuit of, in Africa, 222–24
Huxley, Thomas Henry, 33
hygiene, bacteriological, 82–83; teaching of, 174
hygiene exhibition, 181

immune reaction, 107
immunology, 17, 111, 175
Imperial Health Office (IHO), Koch's position at, 13, 113
India: cholera expedition to, 188–94; Koch's cultural enrichment in, 193–94
industry, chemical-pharmaceutical, 120; products of German, 221; rise of, 153
infection: bacterial, 57–58; traumatic, 8, 54, 60
infection experiment, example of, 58–59
infectious disease: Koch's model of, 83, 85–86; Virchow on, 86
inflammation, views of, 48
Institute for Hygiene, Berlin, 113
Institute for Infectious Diseases: experimentation at, 151; hospital section of, 150; Koch's directorship of, 16, 104; Koch's salary at, 16, 105, 113; sleeping sickness at, 120
Institute for Naval and Tropical Hygiene, Hamburg, 207

Institute for Plant Physiology, Breslau, 47, 51
Institute for Serum Research and Serum Testing, 112
Institute for Tropical Medicine, Hamburg, 155
Institut Pasteur, 15
Internal Medicine, Tenth Congress for, 146
intoxication, putrid, 55
intrigues, institutional, 209
investigation, techniques of, 72

Jauregg, Julius Wagner von, 149

Kitasato, Shibasaburo, 122
Klebs, Edwin, 26, 42–43, 74, 88; tuberculocidin work of, 150–51
Klenke, Friedrich, 74
Koch, Emmy, 11, 214n58; Koch's separation from, 15
Koch, Hedwig. *See* Freiberg, Hedwig
Koch, Robert: academic teachers of, 25–26; attitude of, toward theory, 3; biographical information on, 10–18; career of, 90; disgruntlement of, 172, 175–76; flexibility of, 225; honors and rewards received by, 18, 172, 194, 196, 200, 217, 259n185; innovations of, 3; love of travel of, 172; perception of nature of, 224–25; personality of, 113, 208; personal life of, 90
Koch's postulates, 28, 42, 44, 83; deviation from, 190–91
Kolle, Wilhelm, 154
Krause, Wilhelm, 25
Külz, Eduard, 105

laboratory, culture of the, 276n44
Laënnec, Théophile, 30, 81
Lahmann, Heinrich, 110
Lake Victoria, 203
Latour, Bruno, 83, 255n90, 276n44
Leeuwenhoek, Antoni van, 23
Levy, William, 123
Leyden, Ernst von, 132–33, 141, 146
Libbertz, Arnold, 103, 149
Liebig, Justus, 30
Linnaeus, 23
Lister, Joseph, 8, 33, 41

Livingston, David, 223
Loeffler, Friedrich, 13, 36, 69, 77, 82, 169
Lott, Dr., 102
lupus, 73, 101; description of, 128; tuberculin treatment of, 123, 126

Martini, Fritz, 202
Martius, Friedrich, 110
Mazumdar, Pauline, 31
Medical College Hospital, Calcutta, 188
medicine: bacteriological, passim; clinical vs. experimental, 117; tropical, 17, 175, 177
Meissner, Georg, 11, 25
mercury inunction, 167
Metchnikov, Elias, 106, 251n13
methylene blue, 75. *See also* staining techniques
microbe hunting, 115, 116
micrococci, 39–45
microphotography, 12, 53, 76–77
*microsporon septicum*, 42, 55, 61
Moabit Hospital, 158
Moll, Albert, 152, 154
Möllers, Bernhard, 157–70 passim
morphic cycles, 34
morphs, 33
Muanza, East Africa, 202
Müller, Johannes, students of, 26

Naegeli, Carl von, 19, 20, 31, 61; transformist bacteriology of, 114
nationalism, in science, 13, 181, 187, 194, 217; rhetoric of, 187
Naturalist Society: 1878 meeting of, 46, 56, 61; 1875 meeting of, 52
Naunyn, Bernhard, 6, 146–147
Neisser, Albert, 119, 153
Niemeyer, Felix, 74
Nobel prize, 207, 284n205
Nocht, Bernhard, 176
nosology, 68
notes, laboratory, 3, 4, 253n59

Obermeyer, Otto, 48
optical aids, 22
Orth, Johannes, 137

Panum, Peter Ludwig, 55
pararosanilin, 163
parasitism, 24
parasitology, 175, 178
paroxysm, fever, 144. *See also* fever
Pasteur, Louis, 3, 6; Koch's first controversy with, 13; Koch's reading of, 57; rivalry with, 90, 102, 110, 174, 181–82, 197; work of, on fermentation, 33; work of, on spontaneous generation, 32
patents, 112
pathology, experimental, 8, 14, 54
patients: African, 214; internment of, 220; moribund, 122, 130–31; number of, treated with tuberculin, 122, 139; place of, in Koch's work, 87, 115; terminally ill, 118
Peiper, Erich, 108
permission for treatment, absence of, 151
Pettenkofer, Max von, 16, 110; cholera etiology of, 182, 197; Koch's debate with, 14–15; rivalry with, 172
Pfeiffer, Richard, 17
Pfuhl, Eduard, 99, 103, 122
phthisis, 72, 102
Physiological Society of Berlin, 72
Pirquet, Clemens von, 108
pleomorphism, 32
policy, Prussian science, 111
practices, experimental, 3
Proskauer, Bernhard, 13, 106
protozoa diseases, 207
protozoa, pathogenic, 283n195
protozoology, 206
Prussian House of Representatives, discussion of medical experiments in, 147–48
psychiatry, tuberculin use in, 149
publicity for science, 126–27, 187–88
pus, 42
putrefaction, 28
pyemia, 54, 244n111, 245n119

rabies serum, 15, 117, 118, 119
Recklinghausen, Friedrich von, 41, 43
relapsing fever, 48, 246n150
research, medical, public staging of, 6
Richet, Charles Robert, 104
Richter, Eberhard, 30–31

rinderpest, 17, 172, 177
rivalry, Franco-German, 14, 15. *See also* "German Science"; nationalism; Pasteur, Louis
rosaniline, 164, 165, 166
Rosenbach, Ottomar, 6, 110–11, 134

Salomonsen, Carl, 74
salvarsan, 154
Sander, Friedrich, 46–47
Schaudinn, Friedrich, 206, 207, 208
schizomycetes, 19
Schleiden, Matthias, 31
Schmidt, Berthold: questionable diagnosis for, 167; trypanosoma infection of, 155–170 passim, 272n254
Schönlein, Johann Lucas, 23
Schreiber, Julius, 108
Schroeter, Julius, 40
Schwann, Theodor Ambrose, 18, 23
scrofula, 73, 81
self-delusion, 107, 110
self-experimentation, 121, 140
*seminaria contagionis*, 22, 23,
Senator, Hermann, 122, 130
sepsis, 67
septicemia, 54, 56, 244n111, 245n119; bacterial, 55
serology, 175
Sese Islands, Lake Victoria, 161, 213; Koch's leisure activities on, 222; visitors to, 222
Siegmund, G., 148–49
sleeping sickness (trypanomiasis): absence of, on Lake Victoria, 211–12; in Berlin, 155–70; compared to cholera, 226–27; control measures against, 210; experimentation with, 170; importance of, for colonial empires, 155, 201
sleeping sickness expedition (1906–07), 201–29; financing of, 202–3; intrigues connected with, 209; team participating in, 210
social controls, medically legitimized, 83
specificity, bacterial, 3
*spirochaeta pallida*, 206
spontaneous generation, 23–24, 28, 31, 32
spores: anthrax, 49–50; tubercle bacillus, 77–78, 260n220

staining techniques, 12, 40, 76; failure of, 74; refining of, 75
syphilis, etiology of, 206

taxonomy, 68
testing, bacteriological, 185
therapeutic agents. *See* atoxyl; rosaniline; tuberculin
therapeutic trials, regulation of, 119
therapies: antitubercular, failed, 92–93; experimental, 149–57; inhalation, 263n31; sleeping sickness, 159. *See also* chemotherapy
Thiersch, Carl, 184
transformation, bacterial, 14
travel: role of, in bacteriological hygiene, 179–80; scientific, evolution of, 180
treatment, absence of permission for, 151
tropical medicine, importance of, for colonization, 177
tropics, exotic environment in, 177–78
*trypanosoma gambiense*, 155, 283n195; atoxyl resistance of, 163; experimental work on, 206–7; hosts of, in Africa, 212
tsetse fly, 207, 212
tubercle bacillus: identification of, 8, 69; spores of, 77–78
tubercles, miliar, 136
tuberculin, 9, 71, 91; chemical analysis of, 105–6; composition of, 97, 103–4; diagnostic potential of, 139; dosage of, 124; early critics of, 134–35; euphoria about, 127–28; experimental use of, 140; fatalities from, 133–34; financial aspects of, 102; introduction of, 95–97; marketing campaign for, 126; misleading information about, 97; public demand for, 126; public demonstration of, 126–27; reports on, 128–30, 138–42; sale of, in pharmacies only, 137, 145–46; as secret nostrum, 137; severe criticism of, 146; skepticism about, 103–4, 130; therapeutic value of, 139; trials for, 121–25; use of, in psychiatry, 149

"tuberculin affair," 15; general assessment of, 153–54. *See also* "fraud bacillus"
tuberculin reaction, 99–100; as diagnostic tool, 101, 124, 126; interpretation of, 107; presumed effect of, 100
"tuberculin swindle," 106–9, 137
tuberculosis: etiology of, 14, 71, 81; forms of, 73, 78; human vs. bovine, 70, 111; pathogenesis of, 84; research program for, 71; treatments for, 143; vaccine for, 104. *See also* lupus; phthisis; scrofula
typhoid fever, 191

Uganda, 201, 213
*urocystis cholerae*, 34

vaccination, preventive, 117
vaccine, tuberculosis, trials for, 104
Villemin, Jean Antoine, 14, 49, 74, 78, 81
Virchow Hospital, 163
Virchow, Rudolf, 30, 43, 81, 103; on infectious disease, 86; on inflammation, 48; opposition to tuberculin by, 136, 141; and understanding of microorganisms, 45–46
virulence: attenuation of, 91; Davaine's theory of, 61; Pasteur's view of, 61, 114, 181
vitalism, 24

Waldeyer-Harz, Wilhelm von, 43
Wassermann, August von, 17, 122, 161, 169
weapons, biological, 231
Weigert, Carl, 12, 40, 41, 52, 67
Weigt, August, case of, 123
Wells, H. G., 1
Wöhler, Friedrich, 30
Wollstein, Prussia, 12, 13
wounds, gunshot, 55

zoogloea, 59